Computer Vision-Based Agriculture Engineering

Computer Vision-Based Agriculture Engineering

Han Zhongzhi

CRC Press
Taylor & Francis Group
Boca Raton London New York

CRC Press is an imprint of the
Taylor & Francis Group, an **informa** business

CRC Press
Taylor & Francis Group
52 Vanderbilt Avenue,
New York, NY 10017

First issued in paperback 2021

© 2020 by Taylor & Francis Group, LLC
CRC Press is an imprint of Taylor & Francis Group, an Informa business

No claim to original U.S. Government works

ISBN-13: 978-0-367-25430-8 (hbk)
ISBN-13: 978-1-03-208921-8 (pbk)

Library of Congress Cataloging-in-Publication Data

Names: Zhongzhi, Han, 1981- author.
Title: Computer vision-based agriculture engineering / Han Zhongzhi.
Description: New York, NY: CRC Press, Taylor & Francis Group, [2020] |
Includes bibliographical references and index. |
Summary: "Computer Vision-Based Agriculture Engineering is a summary of the author's work over the past 10 years. Professor Han has presented his most recent research results in the book of all 25 chapters. This unique work provides student, engineers and technologists working in research, development, and operations in the agricultural engineering with critical, comprehensive and readily accessible information. The book applies development of artificial intelligence theory and methods including depth learning and transfer learning to the field of agricultural engineering testing"—Provided by publisher.
Identifiers: LCCN 2019025856 (print) | LCCN 2019025857 (ebook) |
ISBN 9780367254308 (hardback) | ISBN 9780429289460 (ebook)
Subjects: LCSH: Agricultural engineering—Technological innovations. |
Agricultur—Remote sensing. | Quality control—Optical methods.
Classification: LCC S675 .Z49 2020 (print) | LCC S675 (ebook) | DDC 631.5/233—dc23\
LC record available at https://lccn.loc.gov/2019025856
LC ebook record available at https://lccn.loc.gov/2019025857

Visit the Taylor & Francis Web site at
http://www.taylorandfrancis.com

and the CRC Press Web site at
http://www.crcpress.com

Contents

Preface...xv
Author xvii

Chapter 1 Detecting Aflatoxin in Agricultural Products by
Hyperspectral Imaging: A Review...1

1.1 Introduction 1
1.2 Main Detecting Methods...3
    1.2.1 Hyperspectral Imaging (HSI)....................................3
    1.2.2 Near-Infrared Spectroscopy (NIRS) 3
    1.2.3 Flow Chart Showing Typical Steps 5
1.3 Detection of Aflatoxin in Agricultural Products 5
    1.3.1 Corn or Maize 5
        1.3.1.1 Detection of Aflatoxin by
             Hyperspectral Images 5
        1.3.1.2 Detection of Aflatoxin by Other Methods 9
        1.3.1.3 Detect Other Fungi 12
    1.3.2 Cereals, Nuts, and Others..13
        1.3.2.1 Wheat, Barley, and Rice 13
        1.3.2.2 Pistachio Nuts, Hazelnuts, Brazil Nuts,
             and Peanuts...14
        1.3.2.3 Chili Pepper...16
1.4 Limitation and Future Trends..18
    1.4.1 Limitation..18
    1.4.2 Future Trends 19
1.5 Conclusions...20
References 20

Chapter 2 Aflatoxin Detection by Fluorescence Index and
Narrowband Spectra Based on Hyperspectral Imaging...................29

2.1 Introduction 29
2.2 Experiment Materials..30
    2.2.1 Sample Preparation and Image Acquisition.............30
    2.2.2 Illumination Compensation and
        Kernel Segmentation 32
2.3 Data Processing and Result Analysis 33
    2.3.1 Fluorescence Index..33
    2.3.2 Recognition and Regression 34
    2.3.3 Narrowband Spectra...35
2.4 Discussion..37

2.5   Conclusions... 39
References 39

Chapter 3 Application-Driven Key Wavelength Mining Method for
              Aflatoxin Detection Using Hyperspectral Data 41

3.1   Introduction 41
3.2   Materials 42
        3.2.1   Experiment Materials 42
        3.2.2   System Integration 43
3.3   Methods 44
        3.3.1   Data Preprocessing 44
        3.3.2   Recognition Methods 46
3.4   Results 47
        3.4.1   Hyperspectral Wave by ASD 47
        3.4.2   Multispectral Images by Liquid Crystal
                Tunable Filter (HTLF) 48
        3.4.3   Hyperspectral Images by GSM 49
3.5   Discussion 51
        3.5.1   Key Wavelengths Selected by Weighted Voting 51
        3.5.2   Sorter Design 53
3.6   Conclusion 54
References 55

Chapter 4 Deep Learning-Based Aflatoxin Detection of Hyperspectral Data ... 57

4.1   Introduction 57
4.2   Materials and Methods 59
        4.2.1   Peanut Sample Preparation 59
        4.2.2   Hyperspectral Imaging System and
                Image Acquisition 59
        4.2.3   Hyperspectral Imaging Preprocessing 60
        4.2.4   CNN of Deep Learning Method 61
4.3   Results and Discussion 62
        4.3.1   Aflatoxin Detection Using Key Band Images 62
        4.3.2   Aflatoxin Detection Using Spectral and Images 63
4.4   Conclusion 65
References 66

Chapter 5 Pixel-Level Aflatoxin Detection Based on Deep Learning and
              Hyperspectral Imaging 69

5.1   Introduction 69
5.2   Materials and Methods 71
        5.2.1   Peanut Sample Preparation 71
        5.2.2   Hyperspectral Imaging System and
                Image Acquisition 71

        5.2.3   Hyperspectral Imaging Preprocessing 72
        5.2.4   CNN of Deep Learning Method 73
    5.3   Results and Discussion 74
        5.3.1   Deep Learning for Training Kernels........................ 74
        5.3.2   Deep Learning for Testing Kernels......................... 75
        5.3.3   Models Compared for All Kernels 77
    5.4   Discussion... 79
    5.5   Conclusions.. 80
    References 80

Chapter 6   A Method of Detecting Peanut Cultivars and Quality Based on
             the Appearance Characteristic Recognition...................................... 83

    6.1   Introduction 83
    6.2   Materials and Method... 84
        6.2.1   Materials for Test.. 84
        6.2.2   Image Acquisition and Pretreatment..................... 84
        6.2.3   The Appearance Characteristic Index of the Seed..... 86
        6.2.4   The Establishment of the Recognition Model........... 86
    6.3   Results and Analysis.. 87
        6.3.1   The Result of Recognition on Peanut Varieties.......... 87
        6.3.2   The Result of Recognition on Peanut Qualities 88
    6.4   Analysis of the Results of Recognition and Detection........... 88
    6.5   Discussions 89
    References 89

Chapter 7   Quality Grade Testing of Peanut Based on Image Processing........... 91

    7.1   Introduction 91
    7.2   Materials and Method... 92
        7.2.1   Materials for Testing 92
        7.2.2   Maintaining the Integrity of the Specifications 93
        7.2.3   Model of Quality Recognition.............................. 94
        7.2.4   Method of Grading.. 94
    7.3   Results and Analysis.. 95
        7.3.1   Analysis of Recognition Results of Grains' Quality ...95
        7.3.2   Analysis of the Result of Specification and Grading...96
    7.4   Conclusions.. 97
    References 97

Chapter 8   Study on Origin Traceability of Peanut Pods Based on
             Image Recognition 99

    8.1   Introduction 99
    8.2   Materials and Method... 100
        8.2.1   Test Materials 100

8.2.2   Methods 101
8.2.3   The Method Used to Optimize 101
8.3   Results and Analysis 102
8.4   Discussions 103
8.5   Conclusion 104
References 104

Chapter 9   Study on the Pedigree Clustering of Peanut Pod's Variety
Based on Image Processing 107

9.1   Introduction 107
9.2   Materials and Method 108
9.2.1   Experimental Materials 108
9.2.2   Methods 109
9.2.2.1   PCA Algorithm 110
9.2.2.2   Clustering Algorithm 110
9.3   Conclusion and Analysis 110
9.3.1   Statistical Characteristics Clustering 110
9.3.2   PCA Clustering 111
9.4   Discussions 113
9.5   Conclusions 114
References 114

Chapter 10   Image Features and DUS Testing Traits for Identification and
Pedigree Analysis of Peanut Pod Varieties 117

10.1   Introduction 117
10.2   Materials and Method 118
10.1.1   Materials 118
10.1.1.1   Peanut Samples 118
10.1.1.2   Image Acquisition 119
10.1.2   Methods 119
10.1.2.1   Feature Extraction 119
10.1.2.2   Analysis and Identification Model 121
10.3   Results and Analysis 122
10.3.1   Feature Selection by Fisher 122
10.3.2   Variety Identification by SVM 123
10.3.3   Paternity Analysis by K-Means 125
10.4   Discussions 126
10.4.1   Biological Basis for Seed Testing with
Appearance 126
10.4.2   Finding Candidate Features for DUS Testing 127
10.5   Conclusions 128
References 128

Chapter 11 Counting Ear Rows in Maize Using Image Processing Method...... 131

    11.1   Introduction 131
    11.2   Materials.. 132
    11.3   Procedure... 132
            11.3.1 Image Obtaining.. 132
            11.3.2 Characteristic Indicators of Corn Varieties.............. 132
            11.3.3 Pretreatment 132
            11.3.4 Construction of the Counting Model...................... 132
    11.4   Results and Analysis.. 134
    11.5   Conclusions... 135
    References 135

Chapter 12 Single-Seed Precise Sowing of Maize Using
Computer Simulation 137

    12.1   Introduction 137
    12.2   Materials and Methods 139
            12.2.1 Mathematical Depiction of the Problem 139
            12.2.2 Method of Analog Simulation................................. 141
                     12.2.2.1 Computer Simulation of Planting Method... 141
                     12.2.2.2 Seedling Missing Spots and Missing
                                   Seedling Compensation 142
    12.3   Results and Analysis.. 143
            12.3.1 Comparison of the Two Planting Methods' Yield.... 143
            12.3.2 Influence of Field Seedling Emergence
                     Rate on Yield... 145
            12.3.3 Interactions between Sensitivity and Field
                     Seedling Emergence Rate 146
            12.3.4 Seedling Missing Spots and Its Distribution Rule ... 146
    12.4   Conclusion 148
    References 149

Chapter 13 Identifying Maize Surface and Species by Transfer Learning 151

    13.1   Introduction 151
    13.2   Materials and Method.. 153
            13.2.1 Image Characteristics... 153
            13.2.2 Workflow Diagram... 153
            13.2.3 Convolutional Neural Network................................ 153
            13.2.4 Transfer Learning... 155
    13.3   Performance of the Model.. 155
    13.4   Performance of Limited Data Model 157
    13.5   Comparison with Manual Method 157
    13.6   Expanding Application... 158

13.7   Discussion 160
13.8   Conclusions and Future Work 163
References 163

Chapter 14   A Carrot Sorting System Using Machine Vision Technique 167

14.1   Introduction 167
14.2   Materials and Methods 168
        14.2.1   Carrot Samples 168
        14.2.2   Grading System 169
        14.2.3   Image Processing and Detection Algorithms 170
                 14.2.3.1   Image Preprocessing and Segmentation ... 170
                 14.2.3.2   Shape Detection Algorithm 171
                 14.2.3.3   Fibrous Root Detection 172
                 14.2.3.4   Surface Crack Detection 175
        14.2.4   Bayes Classifier 175
14.3   Experimental Results 175
        14.3.1   Detection 175
        14.3.2   Fibrous Root Detection 176
        14.3.3   Crack Detection 177
        14.3.4   Time Efficiency 178
14.4   Discussions 178
        14.4.1   Shape Detection 178
        14.4.2   Fibrous Root Detection 178
        14.4.3   Crack Detection 179
        14.4.4   Future Work 179
14.5   Conclusions 180
References 180

Chapter 15   A New Automatic Carrot Grading System Based on
Computer Vision 183

15.1   Introduction 183
15.2   Materials and Methods 184
        15.2.1   Materials 184
        15.2.2   Design of Carrot Grading System 184
        15.2.3   Image Processing and Grading Algorithms 186
                 15.2.3.1   Image Acquisition 186
                 15.2.3.2   Image Preprocessing 187
                 15.2.3.3   Defect Detection Algorithms 188
                 15.2.3.4   Grading Regular Carrots by Size 190
        15.2.4   Control of Grading System 190
15.3   Results 191
        15.3.1   Defect Detection 191
        15.3.2   Regular Carrot Grading 191
        15.3.3   Time Efficiency 191

          15.3.4   Performance Parameter 192
     15.4   Discussion 193
     15.5   Conclusion 194
     References 194

Chapter 16 Identifying Carrot Appearance Quality by Transfer Learning 197

     16.1   Introduction 197
     16.2   Materials and Method 199
          16.2.1   Image Characteristics 199
          16.2.2   Workflow Diagram 199
          16.2.3   Convolutional Neural Network 199
          16.2.4   Transfer Learning 201
     16.3   Performance of the Model 201
     16.4   Comparison with Manual Work 203
     16.5   An Expanding Application 206
     16.6   Discussion 207
     16.7   Conclusions and Future Work 209
     References 210

Chapter 17 Grading System of Pear's Appearance Quality
          Based on Computer Vision 213

     17.1   Introduction 213
     17.2   System Development 214
          17.2.1   Formation of Grading System's Hardware Device ... 214
          17.2.2   Concrete Implement 215
          17.2.3   Testing of Real Object 215
          17.2.4   Software System 216
     17.3   Implementation of Algorithm 217
          17.3.1   Detection of Defects 217
          17.3.2   Feature Extraction 219
          17.3.3   Grade Judgment 219
     17.4   Results and Discussion 221
     17.5   Ending Words 223
     References 223

Chapter 18 Study on Defect Extraction of Pears with Rich Spots and Neural
          Network Grading Method 225

     18.1   Introduction 225
     18.2   Material and Method 226
          18.2.1   Experimental Materials 226
          18.2.2   Image Preprocessing 227
               18.2.2.1   Background Removing and
                              Outline Extraction 227

18.2.2.2 Removal of Spots on the Surface 228
18.2.2.3 Extraction of Defective Parts 229
18.2.3 Feature Extraction and Recognition 229
18.2.3.1 Scalarization of National Standard 229
18.2.3.2 Extraction of Fruit Type and
Defective Part 230
18.2.3.3 Judgment of ANN Grade 230
18.3 Results and Analysis 230
18.3.1 Effect of Spot Removal and Defect Extraction 230
18.3.2 Grade Judgment of Fruit Type and Defect 231
18.3.3 Comprehensive Grade Judgments 231
18.3.4 Influencing Factors of Grading of
Comprehensive Quality 231
18.4 Discussions 233
References 233

Chapter 19 Food Detection Using Infrared Spectroscopy with
k-ICA and k-SVM: Variety, Brand, Origin, and Adulteration 235

19.1 Introduction 235
19.2 Materials and Methods 237
19.2.1 Materials 237
19.2.2 Algorithm 238
19.2.3 Flowchart 240
19.3 Results and Discussion 240
19.3.1 Different Features Selection Method 241
19.3.2 Different Recognition Models 242
19.3.3 Different Samples or Features 243
19.4 Summary 245
References 245

Chapter 20 Study on Vegetable Seed Electrophoresis Image
Classification Method 247

20.1 Introduction 247
20.2 Material and Method 248
20.2.1 Experimental Materials 248
20.2.2 Method 248
20.3 Results and Analysis 250
20.3.1 Recognition of Corp 250
20.3.2 Cluster Analysis 252
20.4 Discussions 254
20.5 Conclusions 255
References 256

Chapter 21 Identifying the Change Process of a Fresh Pepper by
Transfer Learning 257

    21.1 Introduction 257
    21.2 Materials and Method 258
         21.2.1 Image Characteristics 258
         21.2.2 Workflow Diagram 259
         21.2.3 Convolutional Neural Network 259
         21.2.4 Transfer Learning 260
    21.3 The Performance of the Model 261
    21.4 Computer vs Humans 263
    21.5 Expanding Application 264
    21.6 Discussion 265
    21.7 Conclusions and Future Work 267
    References 267

Chapter 22 Identifying the Change Process of Fresh Banana by
Transfer Learning 271

    22.1 Introduction 271
    22.2 Materials and Method 272
         22.2.1 Image Dataset 272
         22.2.2 Convolutional Neural Network 274
         22.2.3 Transfer Learning 274
         22.2.4 Experimental Setup 275
    22.3 The Performance of Model 276
         22.3.1 Experimental Result 276
         22.3.2 Computer vs Humans 277
         22.3.3 Expanding Application 277
    22.4 Discussion 279
    22.5 Conclusions 280
    References 280

Chapter 23 Pest Recognition Using Transfer Learning 283

    23.1 Introduction 283
    23.2 Materials and Methods 284
         23.2.1 Materials 284
         23.2.2 Introduce of the Model 285
    23.3 Result and Analysis 286
         23.3.1 Pests Recognition Result by Transfer
                Learning Model 286
         23.3.2 Comparison of the Model with Traditional Methods... 286
         23.3.3 Comparison of the Model with Human Expert 286
         23.3.4 Universal of the Transfer Learning Model 288

23.4   Discussion...291
        23.4.1   Image Numbers and Image Capture Environment... 291
        23.4.2   Image Background and Segmentation......................291
        23.4.3   Similar Outline Disturb..291
23.5   Conclusion 293
References 293

Chapter 24 Using Deep Learning for Image-Based Plant Disease Detection 295

24.1   Introduction 295
24.2   Materials and Methods 296
        24.2.1   Dataset...296
        24.2.2   Introduced Model 297
        24.2.3   Experiment Environment 299
        24.2.4   Experiment Result...299
24.3   Model Universal Adaptability 299
        24.3.1   Big Datasets Validation 299
        24.3.2   Small Datasets Validation 300
        24.3.3   Artificial Recognition..301
24.4   Discussion..302
        24.4.1   Effect of Data Size 302
        24.4.2   Image Background 303
        24.4.3   Symptom Variations..303
        24.4.4   Machine Recognition and Human Recognition.......304
References 304

Chapter 25 Research on the Behavior Trajectory of Ornamental Fish
            Based on Computer Vision...307

25.1   Introduction 307
25.2   Experimental Materials and Methods 309
        25.2.1   The Experimental Device 309
        25.2.2   Preprocessing 310
        25.2.3   The Positioning of the Fish 310
        25.2.4   Reduction of Actual 3D Coordinates 311
25.3   Three-Dimensional Trajectory Analysis 313
25.4   Discussion..318
25.5   Conclusion 319
References 319

Index..321

Preface

This is a book on the application of computer vision technology in agricultural engineering. The book covers a wide range of aspects, mainly divided into three major areas: computer vision technology for agricultural product appearance quality detection and related equipment development; key indicators of agricultural product safety (*aflatoxin*) detection technology research; and detection of other factors such as pest, fruit and vegetable change process, and some agronomic traits (*maize*). The research in these fields involves computer vision, image processing, spectroscopy and spectral analysis, artificial intelligence, deep learning, pattern recognition, neural networks, and other disciplines. In order to develop practical equipment, automatic control, mechanical design and manufacture, optical engineering, and so on may also be involved. This book is a summary of the author's work over the past 10 years, and it is rich in content, which can be used as a textbook or reading material for undergraduates and postgraduates, as well as a reference for researchers in related fields. I believe that the publication of this book has a positive significance for the development of precision agriculture and intelligent agriculture.

This book consists of 25 chapters, Chapters 1–5 are about the detection of aflatoxin, which involves the detection of key safety indicators of agricultural products. Hyperspectral imaging is used. The problem is extremely complex, especially for inhomogeneous objects. Chapters 5–10 are about peanut appearance detection, including quality detection and variety detection. Among them, quality detection is the basis of sorter design. Variety detection is one of the basic techniques for identifying true and false seeds and breeding of different varieties. Chapters 11–13 are about maize, which includes image detection of ear rows, simulation of single-grain maize sowing, and identification of maize varieties using deep learning. Chapters 14–16 are about carrots. The design methods of two kinds of carrot sorters and a method for recognizing the appearance quality of carrots through deep learning are introduced. Chapters 17–22 are about fruits and vegetables. It includes pears, bananas, peppers, and etc. There are many methods used such as conventional image processing, near-infrared spectroscopy, electrophoresis, and deep learning. The storage process of pepper and banana is also studied in detail. Chapters 23–24 study the detection of crop diseases and pests based on transfer learning. Chapter 25 studies fish behavior.

It can be said that the author of this book is a scientist with very rich experience in the field of agricultural engineering. Since 2008, he has led and completed the projects of Shandong Natural Science Foundation (Youth Foundation and Surface Project, Grant No. ZR2009DQ019, ZR2017MC041) and National Natural Science (Youth Foundation and Surface Project, Grant No. 31201133, 31872849), and since 2010, he has led and completed the projects of Qingdao Science and Technology Plan four times (Grant No. 11-2-3-20-nsh, 14-2-3-52-nsh, 16-6-2-34-nsh, 19-6-1-66-nsh). Since 2012, he began to engage in the transformation of scientific and technological achievements. He is the leader of the key R&D projects of Shandong Province (No. 2019GNC1 06037) and he had completed the National Innovation Fund for

Small and Medium-sized Enterprises (No. 13C26213713582) relying on cooperative companies. As a technical leader, he has accomplished many special projects such as the transformation of achievements into application in Shandong Province and the pioneering talents of innovation and entrepreneurship in Qingdao. At the same time, he is the instructor of five national innovation projects for college students.

The author thanks the above funds for their support. In addition, he thanks the teachers and students of his research group and the authors of all the references. Finally, this book is dedicated to his two daughters, wife, and mother.

It should be noted that some chapters of this book are written by the author's students under his guidance. Among them are Chapters 14 and 15 written by Deng Limiao; Chapters 13, 16, and 21 by Zhu Hongfei; Chapter 22 by Ni Jiangong; Chapters 23 and 24 by Wang Dawei; and Chapter 25 by Zhu Zhaohu. My hearty thanks to them.

MATLAB® is a registered trademark of The MathWorks, Inc. For product information, please contact:
The MathWorks, Inc.
3 Apple Hill Drive
Natick, MA 01760-2098 USA
Tel: 508-647-7000
Fax: 508-647-7001
E-mail: info@mathworks.com
Web: www.mathworks.com

Author

Han Zhongzhi, Ph.D. (1981–) was born in Junan County, Shandong Province, China. He is a full-time professor at Qingdao Agricultural University, supervisor for master's students, third-level candidate of "1361" talent engineering, head of modern agricultural intelligent equipment innovation team, chief expert of Qingdao agricultural intelligent equipment expert workstation, evaluation expert of National Natural Science Foundation and National IoT Special Fund, Intel-certified visual computing engineer, member of International Computer Association (ACM) Expert Committee, reviewer of many journals such as *Computers and Electronics in Agriculture*, and is part of the editorial committee of *Higher Education Research and Practice*. His main research interest is computer vision intelligent detection in agricultural products.

Professor Han started his career in Qingdao Agricultural University in 2006. During 2012–2016, he completed his doctoral research in Computer technology and resource information engineering in China University of Petroleum (East China). In 2014 and 2018, he was successively promoted to associate professor and full-time professor, respectively. In 2019, he came to Chiba University in Japan as a visiting scholar. In the development of an intelligent sorter for agricultural products, Professor Han led to the development of four series of 12 machine products. As the first accomplisher, he has published more than 50 journal papers (>20 cited by Science Citation Index and Engineering Index [SCI/EI]) and a book titled *Computer Vision Intelligent Detection of Agricultural Products* in Chinese. He has been granted 7 invention patents, 17 utility model patents, 13 software copyrights, and won 5 awards of science and technology above prefecture level.

1 Detecting Aflatoxin in Agricultural Products by Hyperspectral Imaging
A Review

Agriculture products that are contaminated by mycotoxins, especially aflatoxin (AF), are toxic to domestic animals and humans as feed or food because they are a known carcinogen associated with liver and lung cancers. Traditionally, methods for the detection and quantification of AF are based on analytical tests including thin-layer chromatography (TLC) and high-performance liquid chromatography (HPLC), among others. However, these are costly and time-consuming. Hyperspectral imaging (HSI) and near-infrared spectroscopy (NIRS) are considered as rapid, nondestructive techniques for the detection of mycotoxin contamination. The main objective of this study was to review the application of different types of imaging and spectral techniques to detect mycotoxin contamination in a variety of agriculture products, such as cereals (corn or maize), different kinds of nuts, and other crops (such as red chili pepper) in the recent 10 years. AF contamination and HSI, components of the detecting system, the basic method principles, and the limitation and future trends are also discussed.

1.1 INTRODUCTION

AFs, one kind of mycotoxins produced by fungi, are toxic secondary metabolites of *Aspergillus flavus* and *Aspergillus parasiticus*. AF-contaminated grains are toxic to domestic animals when ingested in feed and are a known carcinogen associated with liver and lung cancers in humans. Mycotoxins are small (MW ~700), toxic chemical products formed as secondary metabolites by a few fungal species that readily colonize crops and contaminate them with toxins in the field or after harvest – some 350 mold species produce a large range of secondary metabolites (over 300, of which ~30 are toxic) (Turner et al., 2009). Among molds that can attack these foods, *A. flavus* and *A. parasiticus* are important because they can produce AFs that are considered a potent natural toxin (Eduardo, 2011). Temperature; water availability; plant nutrition; infestation of weeds, birds, and insects; plant density; crop rotation; drought stress; presence of antifungal compounds; fungal load; microbial competition; substrate composition; and mold strain capacity to produce AF are some important factors. As a result of variable conditions that can occur during pre- and postharvest, the AF contamination level among crops and nuts within the same lot can have an extremely uneven distribution. AF contamination has been reported for grains such as corn, wheat, rice, and barley, and nuts such as pistachio nuts, Brazil nuts, hazelnuts, and peanuts (Marroquín, et al., 2014; Selvaraj et al., 2015).

AF contamination has been considered one of the world's most serious food safety problems (Robens, 2008). The United Nations Food and Agriculture Organization estimates annual global losses from mycotoxins at 1 billion tons of foodstuffs. The primary organisms impacted by mycotoxins are plants. Currently, about 25% of agricultural crops worldwide are contaminated by these metabolites (Levasseur, 2012). Consequently, AF levels in food and feed are regulated by the Food and Drug Administration in the United States, allowing 20 ppb (parts per billion) limits in food and 100 ppb in feed for interstate commerce (USDA, 2002, Ministry of Health, PRC, 2011). Aflatoxin B1 (AFB1) has been recognized by the International Agency of Research on Cancer as a group 1 carcinogen for animals and humans, and the *Official Journal of the European Union* has established action levels for AFB1 presence in all feed materials between 5 and 20 ppb (Fernández et al., 2009).

AFs are mycotoxins of major significance, and hence there has been significant research on a broad range of analytical and detection techniques that could be useful and practical. Traditionally, many methods have been used in detecting mycotoxin contamination in fresh and minimally processed agriculture or food products, such as TLC and HPLC (Tong et al., 2011). These analytical tests require the destruction of samples and are costly and time-consuming. Thus, the ability to detect AF in a rapid, nondestructive way is crucial to the grain and nut industry, particularly to the corn industry.

AF has ultraviolet (UV) fluorescence characteristics and the characteristics of surface and superficial distribution. This makes it possible for imaging (especially hyperspectral images, HSIs) and spectroscopy (especially NIRS) to detect AF. It was reported that AF emits fluorescence when excited with UV light (Carnaghan et al., 1963). In addition, A. flavus-infected grains also emit bright greenish-yellow fluorescence (BGYF) under UV excitation. The relationship between A. flavus infection, BGYF, and AF was reported in a study with cotton seeds (Marsh et al., 1969). HSIs and NIRS are considered as two rapid, nondestructive techniques for detection of mycotoxin contamination. Other methods, such as fluorescence spectral imaging, can also detect the presence of a contaminant (such as AF in corn) (Yao et al., 2010d). Certainly, chemically analyzed methods (TLC and HPLC) can provide calibration method for image or spectroscopy analysis. It is important to note that spectral techniques are suitable for homogeneity objects, and image techniques are more suitable for non-homogeneity objects, and certainly it is not absolute (Baiano, 2017).

This chapter provides a review about mycotoxin detection method in cereals and cereal foodstuffs, and analytical methods proposed for their determination especially from 2007 to the present. The main objective of this study was to review the application of different types of imaging techniques to detect AF contamination in a variety of agriculture products. Recent developments and potential applications of HSI or other methods such as NIRS and fluorescence spectral imaging are discussed which replace current labor-intensive, time-consuming methods by providing a quick, accurate, and low-cost alternative for detecting AFs. The basic principles and components of the detecting system are also discussed.

1.2 MAIN DETECTING METHODS

1.2.1 HYPERSPECTRAL IMAGING (HSI)

HSI can not only be used in biological and microbiological applications (Zavattini et al., 2003; Gowen et al., 2015; He and Sun, 2015) but can also be extensively used in quality and safety assessment of food and agricultural products (Chen et al., 2013; Zhang et al., 2012; Liu et al., 2013; Moghaddam et al., 2013; Wu and Sun, 2013a, b). HSI is an emerging process analytical tool for food quality and safety control (Gowen et al., 2007), classifies and monitors quality of agricultural materials (Mahesh et al., 2015), and is an effective tool for quality analysis of fish and other seafood (Cheng et al., 2014). HSI can also be used for online agricultural and food product inspection (Yoon et al., 2010) and food processing automation (Park et al., 2002). HSI can be used in agricultural inspection and detection (Ononye et al., 2010), especially for nuts. Nakariyakul and Casasent (2011) identified internally damaged almond nuts using hyperspectral imagery. Zhu et al. (2007) inspected walnut shell and meat differentiation using fluorescence hyperspectral imagery with ICA-kNN optimal wavelength selection.

HSI technology offers a noninvasive approach toward screening for food safety inspection and quality control based on its spectral signature (Yao et al., 2010d). However, the high dimensionality of these data requires feature extraction or selection for good classifier generalization. For fast and inexpensive data collection, several features (λ responses) can be used (Casasent and Chen, 2004). The performance of support vector machine (SVM) classification with and without feature selection is assessed. Confusion matrices of different configurations are used for comparison, demonstrating that the multi-classifier system with nonuniform feature selection performs well, achieving an overall accuracy of 84% (Samiappan et al., 2013).

A typical general overview of the HSI system is used in this study is illustrated in Figure 1.1. The instruments in our experiment include a field spectrometer or Fourier-transform near-infrared (FT-NIR) spectrometer, a liquid crystal tunable filter, a grating spectrometer module, an SCMOS CCD, a 365 nm UV LED or a 355 nm laser, and an electric displacement platform. Besides these, some optical lens, a mercury lamp, and a Teflon panel were also used in our experiments.

1.2.2 NEAR-INFRARED SPECTROSCOPY (NIRS)

NIRS is also considered as a rapid, nondestructive, reliable, sensitive, and cost-effective technique, which could be used for characterizing the chemical composition (identifying functional groups) of various microorganisms (Bhat, 2013). Fourier-transform infrared (FTIR) spectroscopy was used to differentiate (based on functional groups; spectral range between 3,500 and 500 cm^{-1}) between different genera of fungi (mainly *Aspergillus* sp. and *Mucor* sp.), capable of producing mycotoxins. Irrespective of the overall similarities between the spectra of different fungi, there is a unique spectrum for each one with precise differences in the functional groups. Sorting systems based on optical methods have the potential to rapidly detect and physically remove seeds severely contaminated by fungi. Surface characteristics of seeds, like discoloration caused by fungi, are generally

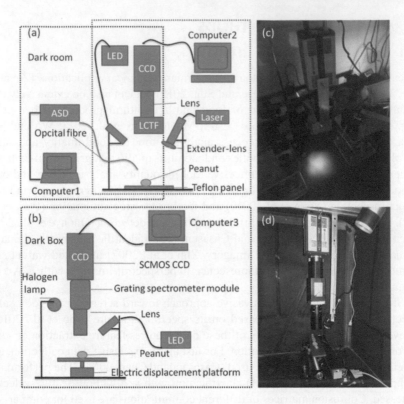

FIGURE 1.1 The general overview of data collection systems. (a) General overview of ASD spectral system and multispectral images system, (b) general overview of HSI system, (c) physical picture of ASD spectral system and multispectral images system, and (d) physical picture of HSI system (Zhongzhi and Limiao, 2018).

detectable in the visible range of the electromagnetic spectrum, whereas internal attributes are detectable in the near-infrared (NIR) range (Pasikatan and Dowell, 2001).

Besides this, Liang et al. (2015) detected fungal infection in almond kernels using NIR reflectance spectroscopy, Moscetti et al. (2014) detected mold-damaged chestnuts by NIRS, Kaya et al. (2014) discriminated moldy peanuts with reference to AF using Fourier transform infrared spectroscopy, coupled with attenuated total reflectance unit (FTIR-ATR) system, Ambrose et al. (2016) compared the nondestructive measurement of corn seed viability using FT-NIR and Raman spectroscopy, and Everard et al. (2014) compared hyperspectral reflectance and fluorescence imaging (FI) techniques for detection of contaminants on spinach leaves. Lee et al. (2015) gave an empirical evaluation of three vibrational spectroscopic methods including Raman, FT-NIR reflectance, and FTIR for detection of AFs in maize. As Agelet and Hurburgh (2014) indicated that there are some limitations in the applications of NIRS for single seed analysis, spectral techniques are suitable for homogeneity materials such as liquid, semi-liquid (Baiano, 2017), or powdered food products.

1.2.3 Flow Chart Showing Typical Steps

A typical general flowchart of the method we used is illustrated in Figure 1.2. For the hyperspectral wave data of system (Figure 1.1), first we preprocess this data using spectral smoothing method by the Savitzky–Golay algorithm. Then, we select the key wavelengths using four methods (Fisher, SAP, BestFirst, Ranker) from the whole spectral wavelengths. Finally, we use four classifiers (Random Forest, SVM, KNN, BP-ANN) as the recognition models.

1.3 DETECTION OF AFLATOXIN IN AGRICULTURAL PRODUCTS

Method and detection system studies for detection of AF in agricultural products with mainly HSI are most active in recent years. Figure 1.3a illustrates the number of papers on different kinds of agricultural products which are retrieved from the database of Elsevier® using subject words (image or spectral and AF) and removed the apparently unrelated literature from 2007. It was found that maize is the first crop to be studied, and Yao's group is one of the famous teams in maize research (Figure 1.3b).

1.3.1 Corn or Maize

1.3.1.1 Detection of Aflatoxin by Hyperspectral Images

Yao's research group (Mississippi State University, USA) is one of the best teams in this research area, and DiCrispino, Hruska and Tao are members of Yao's group. Using a fluorescence HSI system, Yao et al. (2010b) examined the relationship

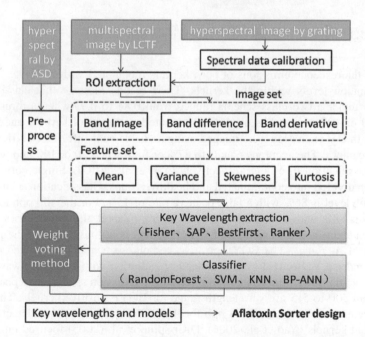

FIGURE 1.2 A typical general flowchart of our method (Zhongzhi and Limiao, 2018).

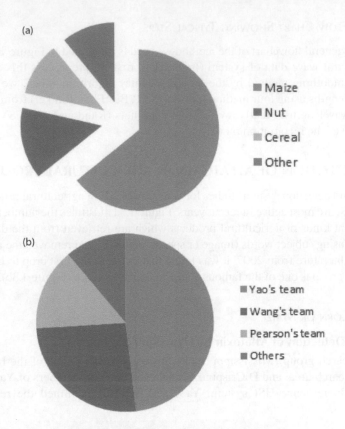

FIGURE 1.3   (a) Main research groups and (b) main study teams.

between fluorescence emissions of corn kernels inoculated with *A. flavus* and AF contamination levels within the kernels. Their results indicate that classification accuracy under a two-class schema ranged from 0.84 to 0.91 when a threshold of either 20 or 100 ng g^{-1} was used. Contaminated and control kernels under long-wavelength UV excitation were imaged using a visible near-infrared (VNIR) hyper-spectral camera. The accuracy equals to 87% or 88% when 20 or 100 ppb was used as the classification threshold, respectively (Yao et al., 2010d). Single corn kernels have been divided into contaminated and healthy groups. Classification accuracy for the 20 ppb level is 86% with a false positive rate of 15%. For the 100 ppb level, the accuracy is 88% with a false positive rate of 16% (Yao et al., 2010c). Corn kernels have been examined for evidence of BGYF, which is an indication of the possible presence of *A. flavus*. The BGYF positive kernels were manually picked out and imaged under a VNIR HSI system under UV radiation with excitation wavelength centered at 365 nm. Initial results exhibited strong emission spectra with peaks centered from 500 to 515 nm wavelength range for BGYF positive kernels. The mean AF concentration level was 5,114 ppb for the BGYF positive and undetectable for the normal kernels (Yao et al., 2006). DiCrispino et al. (2005) focused on observing the changes in the *A. flavus* spectral signature over an 8-day growth period.

The study used a VNIR HSI system for data acquisition. The results showed that while *A. flavus* gradually progressed along the experiment timeline, the day-to-day surface reflectance of *A. flavus* displayed significant difference in discrete regions of the wavelength spectrum. Tao utilized the visible/near-infrared (Vis/NIR) spectroscopy over the spectral range of 400–2,500 nm to detect AF13-inoculated corn kernels and the corresponding AF contamination (Tao et al., 2018). Based on 180 corn kernels, both chemometric methods of partial least squares discriminant analysis (PLS-DA) and principal component analysis combined with quadratic discriminant analysis (PCA-QDA) were used to develop the classification models. The best accuracies in identifying the AF-contaminated corn kernels achieved were 90.9% and 88.9%, with the classification thresholds of 20 and 100 ppb, respectively. They developed a high-speed dual-camera system for batch screening of AF contamination of corns using multispectral FI (Han et al., 2018). Each camera simultaneously captures a single band fluorescence image (436 and 532 nm) from corn samples. The processing time for each screening was about 0.7 s, and an optimal result of 98.65% was achieved for sensitivity and 96.6% for specificity. The device is shown in Figure 1.4.

In the case of a whole corn cob, Yao et al. (2013a) discuss the development of a HSI system for whole-corn ear imaging and use a rotational stage to turn the corn ear. Spectral analysis based on contaminated "hot" pixel classification showed a distinct spectral shift/separation between contaminated and clean ears with fluorescence peaks at 501 and 478 nm, respectively. All inoculated and naturally infected control ears had fluorescence peaks at 501 nm that differed from uninfected corn ears (Hruska et al., 2013).

Naturally occurring *A. flavus* strains can be either toxigenic or atoxigenic, indicating their ability to produce AF or not. Maize ears were inoculated with AF13, a toxigenic strain of *A. flavus*, and AF38, an atoxigenic strain of *A. flavus*, at dough stage of development and harvested 8 weeks after inoculation. After harvest, the kernels with 20 and 100 ppb thresholds were grouped into 'contaminated' and 'healthy', respectively. Yao et al. (2012, 2013b) found that the contaminated kernels all had longer peak wavelength than did the healthy ones. Results from the discriminant analysis classification indicated an overall higher classification accuracy for the 100 ppb threshold on the germ side (94.4%). The germ side was also more useful at discriminating the healthy kernels from the contaminated kernels for the 20 ppb threshold. Zhu et al. (2015) used fluorescence and reflectance VNIR HSIs to classify AF-contaminated corn kernels. In this study, UV excitation and reflectance hyperspectral imagery with halogen illumination were acquired. By PCA and K-nearest neighbors algorithm, the best overall accuracy achieved was 92.67%.

In the last 2 years, using HSI, Wang's research group in China Agricultural University conducted a series of research results on AF detection on maize. Using PCA and factorial discriminant analysis (FDA), a minimum classification accuracy of 88% was achieved for the validation set and verification set. It can be used to detect AFB1 at concentrations as low as 10 ppb when applied directly on the maize surface (Wang et al., 2014). Based on region of interest (ROI) of spectral image, the first two PCs were found to indicate the spectral characteristics of healthy and infected maize kernels, respectively. The wavelengths of 1,729 and 2,344 nm were also identified to indicate AFB1 exclusively (Wang et al., 2015a). By using five principal components

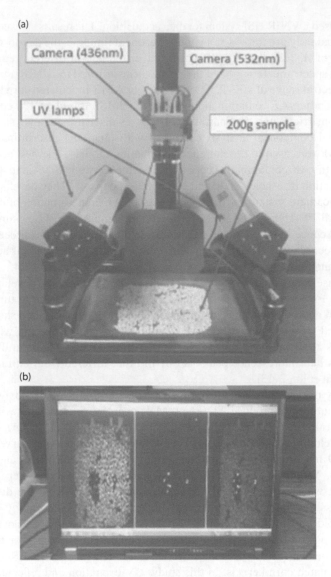

FIGURE 1.4 Image acquisition device. (a) Schematic representation of dual-camera system and (b) AlfaView™ dual-camera system (Han et al., 2018).

(PCs) from PC4 to PC8 as the input of the spectral angle mapper (SAM) classifier, accuracies of the three varieties of kernels reached 96.15%, 80%, and 82.61%, respectively, if kernels contained either high (>100 ppb) or low (<10 ppb) levels of AF (Wang et al., 2015b). Also, they studied the feasibility of detecting AFB1 on inoculated maize kernels using Vis/NIR HSI. By linear discriminant analysis (LDA) and PCA, an overall classification accuracy of 98% was achieved (Wang et al., 2015c). Chu X is a member of Wang's group and his graduate student found four optimum wavelengths (812.42, 873.00, 900.36, and 965.00 nm) for the discrimination model

(Chu et al., 2014). In addition to this, Chu et al. (2017) detected AFB1 in individual maize kernels using short-wave infrared (SWIR) (1,000–2,500 nm) HSI. PCA was applied on average spectra of each kernel and SVM classification methods to qualitatively classify the AFB1 contamination levels (<20, 20–100, ≥100 ppb) in single kernels. The classification accuracies were 83.75% and 82.50% for calibration and validation set, respectively, and five wavelengths (1,317, 1,459, 1,865, 1,934, and 2,274 nm) were selected as the characteristic wavelengths.

Kimuli is a member of Wang's team. Kimuli et al. (2018b) used Vis/NIR HSIs (400–1,000 nm) to classify AFB1-contaminated maize kernels. Based on 600 kernels and four maize varieties of four AFB1 solutions (10, 20, 100, and 500 ppb), FDA showed the ability to predict AFB1 contamination of each variety with more than 96% validation accuracy while prediction for AFB1 contamination group membership of pooled samples reached 98% accuracy in validation. Meanwhile, Kimuli et al. (2018a) used another SWIR HSI system (1,000–2,500 nm) to detect AFB1 on the above-mentioned materials and produced the best PLS-DA classification model with an accuracy of 100% and 96% in calibration and validation, respectively.

Besides this, to detect maize inoculated with toxigenic and atoxigenic strains with hyperspectral imagery, a genetic algorithm (GA) and an SVM were successfully applied for the classification. Under halogen light sources, on average, 83% of the toxigenic fungus pixels and 74% of the atoxigenic fungus pixels were correctly classified; while under UV light sources, 67% of the toxigenic fungus pixels and 85% of the atoxigenic fungus pixels were correctly classified (Jin et al., 2009). Using SWIR (1,100–1,700 nm) HSI technique and a PLS-DA model, Kandpal et al. (2015) studied the examination of AFB1 on corn kernels. The overall classification accuracy yielded from the developed model was 96.9%. Cheng et al. (2019) used a custombuilt UV-Vis-NIRS system (from 304 to 1,086 nm by an increment of 0.5 nm) to classify single corn kernels by AF level. The random forest model had a sensitivity of 87.1% and a specificity of 97.7% in the training set and a sensitivity of 85.7% and a specificity of 97.3% in the test set. Spectral regions around 390, 540, and 1,050 nm are found to be important for classification.

Table 1.1 illustrates the potential of some imaging technique and detection methods for AF in corn.

1.3.1.2 Detection of Aflatoxin by Other Methods

Fluorescence spectral imaging is an alternative method to detect the presence of a contaminant (such as AF in corn) of target material. Yao et al. (2010d) provided a US patent application. The fluorescence from the UV-excited target passes through a filter to a spectral imaging camera. Analyzed by a computer, AF detection in contaminated corn kernels is based on peak fluorescence and peak fluorescence shift in the spectral range from 451 to 500 nm. In another paper (Yao et al., 2011), they described how narrow-band fluorescence indices were developed for AF contamination detection based on single corn kernel samples. The indices were based on two bands extracted from full-wavelength fluorescence hyperspectral imagery. The two band results were later applied to two large-sample experiments with 25 g and 1 kg of corn per sample. The detection accuracies were 85% and 95% when 100 ppb threshold was used. Yao et al. (2011) proposed a technique that automatically

TABLE 1.1

Summary of Detecting AF by HSIs

Acquisition Mode	Wavelength Range	Main Method	Main Result	Reference
HSI	400–600 nm 900–1,000 nm	Multiple linear regression; discriminant analysis	Accuracy 0.84~0.91	Yao et al. (2010b)
Hyperspectral camera	VNIR	Maximum likelihood; binary encoding	Accuracy 87%–88%	Yao et al. (2010a)
Hyperspectral camera	VNIR	SAM	Accuracy 86%–88%	Yao et al. (2010c)
Hyperspectral camera	VNIR	365 nm UV radiation Observational method	Spectra peaks 500–515 nm	Yao et al. (2006)
Hyperspectral camera	VNIR	Observational method	Spectrum changed day to day	DiCrispino et al. (2005)
Hyperspectral camera	VNIR	Design method	Design a system for whole corn ear	Yao et al. (2013a)
Hyperspectral camera	VNIR	"Hot" pixel classification	Spectral peaks at 501 and 478 nm	Hruska et al. (2013)
HSI	Fluorescence	Corn ears were inoculated	Fungal inoculums yielded better results	Yao et al. (2012)
HSI	Fluorescence	Discriminant analysis	Accuracy 94.4%	Yao et al. (2013b)
HSI	VNIR	PCA; K-nearest neighbors	Accuracy 92.67%	Zhu et al. (2015)
HSI	1,000 and 2,500 nm	PCA; FDA	Accuracy 88%	Wang et al. (2014)
HSI	NIR	PCA; SAM	Accuracy 92.3%	Wang et al. (2015a)
Vis/NIR hyperspectra	400–1,000 nm	FDA; PCA	Accuracy 98%	Wang et al. (2015c)
HSI (SWIR)	1,000–2,500 nm	PCA; SAM	80%–96.15%	Wang et al. (2015b)
HSI	400–1,000 nm	SNV; Fisher	813, 873, 900, 965 nm 87.4% and 80.9%	Chu et al. (2014)
HSI	1,000–2,500 nm	PCA; SVM	1,317, 1,459, 1,865, 1,934, 2,274 nm 83.75% and 82.50%	Chu et al. (2017)
HSIs	UV light	PCA; GA; B-distance; SVM	67%–99%	Jin et al. (2009)

(Continued)

TABLE 1.1 (*Continued*)
Summary of Detecting AF by HSIs

Acquisition Mode	Wavelength Range	Main Method	Main Result	Reference
Hyperspectral system	1,100–1,700 nm	PLS; FDA	96.9%	Kandpal et al. (2015)
Vis/NIR HSIs	400–1,000 nm	FDA	98%	Kimuli et al. (2018b)
SWIR HSI system	1,000–2,500 nm	PLS-DA	96%	Kimuli et al. (2018a)
UV-Vis-NIR	304 –1,086 nm	Random forest	390, 540, and 1,050 nm 97.7% and 97.3%	Cheng et al. (2019)

detects AF-contaminated corn kernels by using dual-band imagery. The method exploits the fluorescence emission spectra from corn kernels captured under 365 nm UV light excitation. The preliminary results shown there demonstrate the potential of our technique for AF detection. Contaminated and control kernels under long-wavelength UV excitation were imaged using a VNIR hyperspectral camera. Three narrow-band fluorescence indices were developed and tested in this study. It was found that the highest correlation was −0.81 with the normalized difference fluorescence index (NDFI). The two bands used for the NDFI were 437 and 537 nm (Yao et al., 2013c). The use of key wavelengths for contamination detection would be helpful for developing rapid and noninvasive inspection systems.

Multispectral is another alternative method. Stasiewicz et al. (2017) used multispectral kernel sorting to reduce AFs and fumonisins in Kenyan maize. They built a mathematical model relating reflectance at nine distinct wavelengths (470–1,550 nm) and achieved 77% sensitivity and 83% specificity to identify kernels with AF >10 ng g^{-1} and fumonisin >1,000 ng g^{-1}.

Transmittance spectra (500–950 nm) and reflectance spectra (550–1,700 nm) were analyzed to determine if they could be used to distinguish AF contamination in single whole corn kernels. Spectra were obtained on whole corn kernels exhibiting various levels of BGYF. Spectra were analyzed using discriminant analysis and PLS regression. More than 95% of the kernels were correctly classified as containing either high (>100 ppb) or low (<10 ppb) levels of AF (Pearson et al., 2001). However, for automated high-speed detection and sorting, instrumentation that uses single-feature reflectance spectra may be more practically implemented. To achieve accurate sorting, single kernel reflectance spectra (500 1,700 nm) were analyzed to select the optimal pair of optical filters to detect mycotoxin-contaminated corn during high-speed sorting. In a laboratory setting, and with the kernels stationary, absorbances at 750 and 1,200 nm could correctly identify >99% of the kernels as AF-contaminated (>100 ppb) or uncontaminated (Pearson et al., 2004). For the Kansas corn, the sorter was able to reduce AF levels by 81% from an initial average of 53 ppb.

NIRS is an excellent candidate for a rapid and low-cost method for the detection of AFs in cereals. A total of 152 samples were involved and analyzed for AF content. The best predictive model to detect AFB1 in maize was obtained using standard normal variate and detrending (SNVD) as scatter correction [$r^2 = 0.80$ and 0.82; standard error of cross-validation (SECV) = 0.211 and 0.200 for grating and FT-NIRS instruments, respectively]. The results of spectroscopic models developed have demonstrated that NIRS technology is an excellent alternative for fast AFB1 detection in cereals (Fernández et al., 2009). In a blind study (Gordon et al., 1997), ten corn kernels showing BGYF in the germ or endosperm and ten BGYF-negative kernels were correctly classified as infected or not infected by Fourier transform infrared photoacoustic spectroscopy (FTIR–PAS). Ten major spectral features were interpreted and assigned by theoretical comparisons of the relative chemical compositions of fungi and corn. Using NIR imaging, people can investigate other fungi in maize as follows.

1.3.1.3 Detect Other Fungi

A. flavus and other pathogenic fungi display typical infrared spectra which differ significantly from spectra of substrate materials such as corn (Gordon et al., 1997). Various types of spectroscopic techniques such as infrared (IR), Raman, NIR, nuclear magnetic resonance (NMR), and UV can be used for fungi and mycotoxins detection (Singh et al., 2011). NIRS is an excellent candidate for a rapid and low-cost method for the detection of other fungi in cereals.

It was found that two NIR reflectance spectral bands centered at 715 and 965 nm could correctly identify 98.1% of asymptomatic kernels and 96.6% of kernels, respectively, showing extensive discoloration and infected with *A. flavus, Aspergillus niger, Diplodia maydis, Fusarium graminearum, Fusarium verticillioides,* or *Trichoderma viride* (Pearson and Wicklow, 2006). These two spectral bands can easily be implemented on high-speed sorting machines for removal of fungal-damaged grain.

A desktop spectral scanner equipped with an imaging-based spectrometer ImSpector–Specim V10, working in the VNIR spectral range (400–1,000 nm), was used. The results show that the HSI is able to rapidly discriminate commercial maize kernels infected with toxigenic fungi from uninfected controls (Del Fiore et al., 2010). Another experiment indicated that early detection of fungal contamination and activity is possible. In this test, HSIs of clean and infected kernels were acquired using a SisuChema hyperspectral push-broom imaging system with a spectral range of 1,000–2,498 nm (Williams et al., 2012a and b).

FTIR–PAS is a highly sensitive probe for fungi growing on the surfaces of individual corn kernels. However, the photoacoustic technique has limited potential for screening bulk corn because currently available photoacoustic detectors can accommodate only a single intact kernel at a time. Transient-infrared spectroscopy (TIRS), on the other hand, is a promising new technique that can acquire analytically useful infrared spectra from a moving mass of solid materials (Gordon et al., 1999).

Fumonisins and *F. verticillioides* are other kinds of important mycotoxins. NIR could accurately predict the incidence of kernels infected by fungi, and by *F. verticillioides* in particular, as well as the quantity of fumonisin B1 in the meal (Berardo et al., 2005). The best predictive ability for the percentage of global fungal infection and *F. verticillioides* was obtained using a calibration model utilizing maize kernels ($r^2 = 0.75$ and SECV = 7.43) and maize meals ($r^2 = 0.79$ and SECV = 10.95), respectively. FT-NIR is a suitable method to detect fumonisin FB1 + FB2 in corn meal and to discriminate safe meals from those contaminated equipped with an integration sphere (Gaspardo et al., 2012). Also, multispectral image analysis method can predict milled maize fumonisin contamination. Maize samples were grounded and imaged under ten different LED lights with emission centered at wavelengths ranging from 720 to 940 nm. The digital images were converted into matrices of data to compute comparative indexes. A three-layer feed-forward neural network was trained to predict mycotoxin content from the calculated indexes. The results showed a significant correlation between predictions from image analysis and the concentration of the mycotoxin fumonisin as determined by chemical analysis (Firrao et al., 2010).

NIR HSI combined with multivariate image analysis (MIA) is a powerful tool for the evaluation of growth characteristics of fungi. In Williams et al.'s (2012) study, they evaluated the growth characteristics of three Fusarium species by NIR and MIA. Reflectance and transmittance visible and NIRS were used to detect fumonisin in single corn kernels infected with *F. verticillioides*. Kernels with >100 and <10 ppm could be classed accurately as fumonisin positive or negative, respectively (Dowell et al., 2002).

1.3.2 CEREALS, NUTS, AND OTHERS

1.3.2.1 Wheat, Barley, and Rice

AFs produced by *A. flavus* are commonly found in human and animal foods including wheat, barley, and rice. Nasir and Jolley (2002) reported a simple, portable, and rapid fluorescence polarization (FP) assay for AF determination in grains. Spiked popcorn samples were analyzed by FP. FP results of naturally contaminated samples correlated well with HPLC ($r^2 = 0.97$). FP analysis of spiked popcorn samples (with a mixture of B1/B2/G1/G2, 7/1/3/1, w/w) gave a good correlation with spiked values ($r^2 = 0.99$). However, FP consistently underestimated the AF contents. NIRS is an excellent candidate for a rapid and low-cost method for the detection of AFs in cereals.

Healthy and fungal-damaged wheat kernels infected by the species of storage fungi, namely *Penicillium* spp., *Aspergillus glaucus*, and *A. niger*, were scanned using a short-wave NIR HSI system in the 700–1,100 nm wavelength range and an area scan color camera. From the color images, a total of 179 features (123 color and 56 textural) were extracted, and the top features selected were used as input to the statistical classifiers. The LDA classifier correctly classified 97.3%–100.0% healthy and fungal infected wheat kernels, using the combined HSI features and the top ten features selected from 179 color and textural features of the color images as input (Singh et al., 2012). Senthilkumar et al. (2016a) detected fungal infection and Ochratoxin A contamination in stored wheat using NIR HSI. PCA was applied to the two-dimensional data, and based on the highest factor loadings, 1,280, 1,300, and 1,350 nm were identified as significant wavelengths. All the three classifiers differentiated healthy kernels from fungal-infected kernels with a classification accuracy of more than 90%. The peak at 1,480 nm was identified only in the Ochratoxin A-contaminated samples. The Ochratoxin A-contaminated samples can be detected with 100% classification accuracy.

In the case of barley, the best predictive model was developed using SNVD on the dispersive NIRS instrument ($r^2 = 0.85$ and SECV = 0.176) and using spectral data as log $1/R$ for FT-NIRS ($r^2 = 0.84$ and SECV = 0.183) (Fernández et al., 2009). Senthilkumar et al. (2016b) detected fungal infection and Ochratoxin A contamination in stored barley using NIR HSI. The two-dimensional data corresponding to each fungal-infected sample and Ochratoxin A-contaminated sample were subjected to PCA for data reduction and to identify significant wavelengths. The significant wavelengths 1,260, 1,310, and 1,360 nm corresponding to *A. glaucus*, *Penicillium* spp., and non-Ochratoxin A producing *Penicillium verrucosum*-infected kernels and wavelengths 1,310, 1,360, and 1,480 nm corresponding to Ochratoxin A-contaminated kernels were obtained based on the highest PC factor loadings. The three classifiers differentiated sterile kernels with a classification accuracy of more than 94%, fungal-infected kernels with more than 80% at initial periods of fungal infection and attained 100% classification accuracy after 4 weeks of fungal infection.

Based on NIRS, with a wavelength range between 950 and 1,650 nm, spectra were obtained on 106 rice samples, by reflection mode, including 90 naturally contaminated samples and 16 artificially contaminated samples. For yellow-green *Aspergillus* infection, the most accurate predictive statistical model was developed using a pretreated (maximum normalization) NIR spectra, with the following statistical characteristics [$r = 0.437$, standard error of prediction (SEP) = 18.723%, and bias = 4.613%] (Sirisomboon et al., 2013). The results showed that the NIRS could be used to detect aflatoxigenic fungal contamination in rice with caution and the technique should be improved to get a better prediction model.

1.3.2.2 Pistachio Nuts, Hazelnuts, Brazil Nuts, and Peanuts

Traditionally, immunoaffinity column clean-up with HPLC is a technique for the detection of AFs in nuts such as pistachio and cashew (Pearson et al., 1999).

The presence of greenish-yellow fluorescence on nuts indicates the presence of AF. A dual-wavelength fiber-optic fluorescence photometer was tested by measuring the sorting index k log10 (I490/I420) of nuts exhibiting different kinds of fluorescence under long-wavelength UV excitation, where I490 and I420 are the fluorescence at 490 and 420 nm, respectively (McClure and Farsaie, 1980). In 1981, an automatic electro-optical sorter was designed and developed for removing BGY (bright greenish yellow) in fluorescent pistachio nuts (Farsaie et al., 1981). In 1988, a high-speed fiber-optic sorter was used to separate the three categories of nuts (BGY, purple, others) based on the fluorescent properties of the nuts (Haghighi, 1988). The feasibility of using BGYF in pistachio nuts as a discriminating factor for the identification of *A. flavus*-infested nuts, at harvest and in postharvest, is investigated. Results show a strong relationship between BGYF and AF content at harvest (Hadavi, 2005). The relationship between inside-brown kernels and AF presence is confirmed. The early-split nuts are the most contaminated nuts, growth split nuts are less contaminated, and pistachios with sound hulls are almost clean. Besides this, the relationship between the appearances of brown color in the intercotyledonary plane of pistachio nuts is confirmed earlier as a good indicator of high-AF contaminations. All individual pistachios were classified based on BGYF presence on shell before the examination of the intercotyledonary area for brown color. The results indicate that the early-split pistachios had the highest BGYF incidence among all cultivars (Amani et al., 2011).

Besides BGYF presence, a machine vision system based on visible color image was developed to separate stained pistachio nuts (Pearson, 1995). This system may be used to reduce the labor involved with manual grading or to remove AF-contaminated product from low-grade process. And the system had a minimum overall error rate of 14% for the bichromatic sorter reject stream and 15% for the small shelling stock stream. Haff and Pearson (2006) reported a technique using NIRS, which was developed for selecting the optimal spectral bands for use in dual-wavelength sorting machines commonly found in food processing plants. It is believed that it could be applied to any commodity sorted using commercially available, dual-wavelength, NIR sorting devices. Thermal images have beneficial information which can be used for diagnostic purposes. A new algorithm (Kheiralipour et al., 2013), threshold-based classification (TBC), was developed to analyze thermal images of healthy and fungal-infected pistachio kernels in MATLAB® 2010a environment. Results showed that TBC algorithm can classify successfully healthy and infected pistachio kernels without considering the infection stages.

Soil is a reservoir for *A. flavus* and *A. parasiticus*, fungi that commonly colonize peanut seeds and produce carcinogenic AFs. Densities of these fungi in soil vary greatly among fields and may influence the severity of peanut infection (Horn, 2006). Transmittance NIR spectra (500–1,500 nm) of individual peanut was measured to detect the internally moldy nuts. The moldy nuts, the appearance of which had little difference from the sound nuts by visual observations, could be distinguished from each other by comparing the transmittance ratio of 700–1,100 nm by NIR spectrometry. Because of the higher incidence of AF contamination on these moldy nuts, taking out the internally moldy nuts detected by NIR could drastically reduce the AF content of the overall lot (Hirano et al., 1998).

Multispectral imaging system can be used to detect aflatoxin-contaminated hazelnut kernels. Classification accuracies of 92.3% were achieved for aflatoxin-contaminated and -uncontaminated hazelnuts. The AF concentrations were decreased from 608 to 0.84 ppb for tested hazelnuts by removing the nuts that were classified as AF-contaminated (KalKan et al., 2011). Their algorithm was also used to classify fungal-contaminated and -uncontaminated hazelnut kernels, and an accuracy of 95.6% was achieved for this broader classification. They also indicate that the reflectance images at 460–500 nm are the most discriminative bands for AF contamination (Kalkan and Yardimci, 2008).

As the criteria for sorting, characteristics of in-shell Brazil nuts and their relationship to AF contamination have been proposed (De Mello and Scussel, 2007). Physical methods for mechanically in-shell Brazil nuts sorting by color, size, density, and inner deterioration were developed to assess the nut quality and reduce AF contamination. Utilizing NIR spectrophotometry, the nuts' inner deterioration was detected, with no need of de-shelling them at the wavelength range of

2,200–2,500 nm. Any nut measurement detected, lower or higher than those sorting settings, was considered off-standard and rejected (De Mello and Scussel, 2009).

Jiang et al. (2016) used NIR HSIs (970 and 2,570 nm) to identify moldy peanuts. Thus, the content of carcinogenic substance is detected indirectly. PCA was mainly used in the spectral dimension to select sensitive bands. The results illustrated that the accuracy was 87.14% in learning image and 98.73% in validation image for identifying the moldy kernels. Qiao et al. (2017) utilized spectral-spatial characteristics in SWIR HSIs to classify and identify fungi-contaminated peanuts. They used analysis of variance (ANOVA), non- parametric weighted feature extraction (NWFE), SVM, and image-segmentation methods and achieved 94.2%–99.73% accuracies in pixel-wise classification.

Wu et al. (2019) used laser-induced fluorescence spectroscopy (FS) (400–610 nm) for discrimination of AFB1-contaminated pistachio kernels. Based on 250 kernels and four low-concentration AFB1 solutions (5, 10, 20, and 50 ppb), good discriminant ability (accuracy ≥ 98.4%) was achieved for AFB1 contamination using PCA and SVM. They also used imaging and spectroscopic techniques (including NIRS, midinfrared spectroscopy (MIRS) and conventional imaging techniques [color imaging (CI) and HSI, and FS/FI] for determination of toxigenic fungi and AFs in nuts and dried fruits (Wu et al., 2018). An interesting classification and determination results can be found in both static and on/in-line real-time detection for contaminated nuts and dried fruits.

1.3.2.3 Chili Pepper

Chili pepper may also be contaminated by AFs during harvesting, production, and storage. In order to differentiate AF-contaminated chili peppers from uncontaminated ones, a compact machine vision system based on HSI and machine learning is proposed. Both UV and halogen excitations are used. It was observed that, with the proposed features and selection methods, robust and higher classification performance was achieved with fewer numbers of spectral bands enabling the design of simpler machine vision systems (Ataş et al., 2012). They also proposed a compact machine vision system based on HSI and machine learning for the detection of AF-contaminated chili peppers. We used the difference images of consecutive spectral bands along with individual band energies to classify chili peppers into AF-contaminated and -uncontaminated classes. Both UV and halogen illumination sources were used in the experiments (Atas et al., 2011).

A multispectral imaging system was applied to detect AF-contaminated red chili peppers. Classification accuracies of 80% were achieved for AF-contaminated and -uncontaminated classes. The AF concentrations were decreased from 38.26 to 22.85 ppb for red chili peppers by removing the peppers that were classified as AF-contaminated (KalKan et al., 2011).

FT-NIR spectroscopy can be used for rapid, nondestructive quantification of AFB1 in red chili powder. The feasibility of measuring AFB1 in red chili powder was investigated by using FT-NIR spectroscopy in diffuse reflectance mode combined with appropriate chemometric techniques. AF-free chili powder samples were spiked with a known amount of AFB1 ranging from 15 to 500 µg kg^{-1} and used for calibration model building based on PLS regression algorithm (Tripathi and Mishra,

2009). In another paper, the potential of the NIRS technique for the analysis of red paprika for AFB1, Ochratoxin A, and total AFs is explored (Hernández et al., 2008). NIRS technique using a fiber-optic probe offers an alternative for the determination of these three parameters in paprika, with an advantageously lower cost and higher speed as compared with the chemical method.

In the work proposed by Rajalakshmi and Subashini (2012), the chili pepper samples were collected, and the X-ray, multispectral images of the samples were processed using image processing methods. The computational intelligence method was used in addition to image processing to provide the best, high-performance, and accurate results for detecting the mycotoxin level in the samples collected (Table 1.2).

TABLE 1.2

Main Summary of Detecting Fungal Contaminated by Images or Spectroscopy

Product	Type of Toxins	Mode (Wavelength)	Analysis Methods	Main Result	Reference
Wheat	*Aspergillus*	HSI (950–1,650 nm)	Partial least squares regression (PLSR)	$r = 0.437$, SEP = 18.723%, and bias = 4.613%	Sirisomboon et al. (2013)
Wheat	Fungal; Ochratoxin A	NIR HSI	PCA quadratic discriminant	Key bands: 1,280, 1,300, 1,350; 90%–100%	Senthilkumar et al. (2016a)
Wheat	Fungal	Imaging system in the 700–1,100 nm	LDA	97.3%–100.0%	Singh et al. (2012)
Barley	Fungal; Ochratoxin A	NIR HSI	PCA three classifiers	1,260, 1,310, 1,360 nm 80%–100%	Senthilkumar et al. (2016b)
Barley	AF	Dispersive NIRS instrument	SNVD	$r^2 = 0.85$ and SECV = 0.176	Fernández et al. (2009)
Rice	Aflatoxigenic fungal	Infrared spectroscopy (950–1,650 nm)	Maximum normalization	$r = 0.437$, SEP = 18.723%	Sirisomboon et al. (2013)
Hazelnut	AF	Multispectral imaging system	Two-dimensional local discriminant	92.3%–95.6%	KalKan et al. (2011)
Hazelnut	AF	Multispectral imaging system	Statistical observation	Key bands: 460–500 nm	Kalkan and Yardimci (2008)
Brazil nuts	AF	Infrared spectroscopy (2,200–2,500 nm)	Statistical observation	Building a sorting machine	De et al. (2000)
Peanuts	Moldy	HSIs (970–2,570 nm)	PCA; segmentation	87.14%–98.73%	Jiang et al. (2016)

(Continued)

TABLE 1.2 (*Continued*)

Main Summary of Detecting Fungal Contaminated by Images or Spectroscopy

Product	Type of Toxins	Mode (Wavelength)	Analysis Methods	Main Result	Reference
Peanuts	Fungi	SWIR HSIs	ANOVA, NWFE, SVM	94.2%–99.73%	Qiao et al. (2017)
Chili peppers	AF	HSI (400–720 nm)	HBBE; PCA; SVM–RFE	Key band: 400; 420; 90%	Ataş et al. (2012)
Chili peppers	AF	HSI (400–720 nm)	Neural network connection weights	87.5%	Atas et al. (2011)
Red chili peppers	AF	Multispectral imaging system	Two-dimensional local discriminant	80%	KalKan et al. (2011)
Red chili powder	AFB1	FT-NIR spectroscopy	PLS	RMSECV = 0.654% $r^2 = 96.7$	Tripathi and Mishra (2009)
Pepper	Toxin	Multispectral image etc.	Neural networks Fuzzy logic	Provide the best performance	Rajalakshmi and Subashini (2012)
Red paprika	AFB1	NIRS	Modified PLS	Regression squared, RSQ = 0.955 SEP = 0.2 µg kg^{-1}	Hernández et al. (2008)
Nuts and dried fruits	Toxigenic fungi and AFs	Imaging and spectroscopic techniques	Static and on/in-line real-time detection	Interesting classification and determination results	Wu et al. (2018)
Pistachio kernels	AFB1	Laser-induced FS (400–610 nm)	PCA and SVM	Accuracy ≥ 98.4%	Wu et al. (2019)

HBBE, hierarchical bottleneck backward elimination; RFE, recursive feature elimination.

1.4 LIMITATION AND FUTURE TRENDS

1.4.1 LIMITATION

Spectral analysis (hyperspectral FI, NIRS, and other spectral methods) has numerous potential applications in food and feed industry both to prevent the grains' and nuts' AF contamination, and to preserve the grains and nuts safely. Nevertheless, application will become even more widespread in grains and nuts industry if cheaper spectral detectors (HFI or NIR) are developed. In addition to this, the system has to overcome a number of shortcomings such as HSI performance should be improved for its usage in online classification of grains and nuts. The high dimensionality of HSI data requires feature extraction or selection for good classifier generalization. Band selections may be another good method to reduce the dimensionality. As the present system's speed is slow, when dealing with the large-scale classification of grains or nuts, it is necessary to conduct a study to enhance the speed of the operation.

Spectral imaging is a noncontact and nondestructive method of AF contamination detection which has a number of applications in the grain industry. Even though the spectral equipment is expensive, it lacks the software for specific detection. Always under long-wavelength UV excitation, the BGYF can be obtained from grains' and nuts' AF contamination. However, the use of UV light is a threat for the operator. Hence, further studies need to be conducted to improve the spectral imaging system, which could offer an alternate solution for these drawbacks. In addition to this, spectral imaging is still being used in laboratory or experimental stages; hence, further research should be focused on producing low-cost spectral camera to explore the potential capabilities of this tool for facilitating to adopt in large-scale industrial applications.

The performance of spectral system is affected by many environmental conditions such as temperature, wind, solar radiation, cold weather, and moisture content. Therefore, a spectral system that can tolerate all these environmental effects needs to be developed to improve the accuracy of measurements.

1.4.2 FUTURE TRENDS

Besides detecting AF or any other fungal toxins, HSI and spectroscopy were also widely used in other fields of agriculture: Jaillais et al. (2015) detected *Fusarium* head blight contamination in wheat kernels by multivariate imaging; Barbedo et al. (2015) detected *Fusarium* head blight in wheat kernels using HSI; Ravikanth et al. (2015) differentiated contaminants from wheat using NIR HSI; Serranti et al. (2013) used HSI method to detect *Fusarium*-damaged, yellow berry, and vitreous Italian durum wheat kernels; Ambrose et al. (2016) proposed a high-speed measurement of corn seed viability using HSI; Wu et al. (2016) studied the optimal algorithm prediction of corn leaf component information based on HSI; Teena et al. (2014) used NIR HSI to identify fungal-infected date fruits; and Kandpal et al. (2016) used NIR HSI system coupled with multivariate methods to predict viability and vigor in muskmelon seeds. Dale et al. (2013) discriminated grassland species and their classification in botanical families by laboratory-scale NIR HSI, and Lohumi et al. (2016) applied HSI for characterization of intramuscular fat distribution in beef.

In recent years, some other techniques have also been used for AF detection, such as Lee et al. (2014) applied Raman spectroscopy for qualitative and quantitative analyses of AFs in ground maize samples, and they have achieved good results. Shen et al. (2018) detected *Aspergillus* spp. contamination levels in peanuts by spectroscopy and electronic nose. However, the results demonstrate that E-nose methods are a little short of NIRS for the detection of fungal contamination levels in peanuts.

Figure 1.5 illustrates the number of papers that were retrieved from the database of Google Scholar® using different key words. NIRS was mainly used in homogeneity materials, and seeds of agricultural products are often nonhomogeneous substances. So in this chapter, HSI technology is mainly discussed.

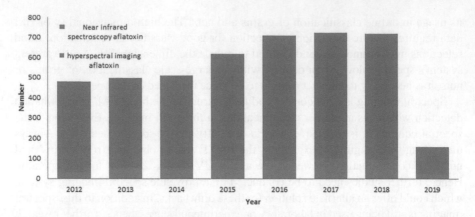

FIGURE 1.5   The number of papers retrieved from the database of Google Scholar.

The classic pattern-recognition problem usually includes two stages: feature extraction and classifier. With the development of deep learning technology, it may be a fashionable way to apply deep learning to the identification of mycotoxin. The authors are trying this out, and preliminary and effective results have been obtained.

1.5 CONCLUSIONS

Spectral imaging technology is currently used in the food industry for detecting AF and other mycotoxins produced by fungi such as *A. flavus*, *A. parasiticus*, *A. niger*, and *Fusarium* spp. However, with increasing concerns related to food security demanding lower food losses and greater awareness of food safety issues among consumers, uses of more sophisticated methods are demanded. This has spurred considerable interest in the research into more advanced uses of spectral imaging technology as discussed and evaluated in this chapter. The success of some investigations in distinguishing microorganisms even at the species level offers great potential for early noninvasive detection that would enable intervention strategies that would improve quality and safety control of foods. Noninvasive methods that could analyze large quantities of crops could also significantly reduce the risk from mycotoxins, and the studies cited in this chapter have demonstrated positive data. Further methodology development including feature extraction and band selection algorithms and recognition algorithm such as artificial neural network (ANN) and SVM to enhance the sensitivity and accuracy of the results would increase the potential for widespread application.

REFERENCES

Agelet L E, Hurburgh C R. Limitations and current applications of near infrared spectroscopy for single seed analysis. *Talanta*, 2014, 121: 288–299.
Amani M, Hadavi E, Tajabadipour A, et al. Distribution of brown kernel as an index for aflatoxin presence among three cultivars of pistachio. *International Symposium on Mycotoxins in Nuts and Dried Fruits 963*, 2011, pp. 143–146.

Ambrose A, Kandpal L M, Kim M S, et al. High speed measurement of corn seed viability using hyperspectral imaging. *Infrared Physics & Technology*, 2016, 75: 173–179.

Ambrose A, Lohumi S, Lee W H, et al. Comparative nondestructive measurement of corn seed viability using Fourier transform near-infrared (FT-NIR) and Raman spectroscopy. *Sensors and Actuators B: Chemical*, 2016, 224: 500–506.

Ataş M, Yardimci Y, Temizel A. A new approach to aflatoxin detection in chili pepper by machine vision. *Computers and Electronics in Agriculture*, 2012, 87: 129–141.

Atas M, Yardimci Y, Temizel A. Aflatoxin contaminated chili pepper detection by hyperspectral imaging and machine learning. *SPIE Defense, Security, and Sensing. International Society for Optics and Photonics*, Orlando, FL, 2011: 80270F.

Baiano A. Applications of hyperspectral imaging for quality assessment of liquid based and semi-liquid food products: A review. *Journal of Food Engineering*, 2017, 214: 10–15.

Barbedo J G A, Tibola C S, Fernandes J M C. Detecting Fusarium head blight in wheat kernels using hyperspectral imaging. *Biosystems Engineering*, 2015, 131: 65–76.

Berardo N, Pisacane V, Battilani P, et al. Rapid detection of kernel rots and mycotoxins in maize by near-infrared reflectance spectroscopy. *Journal of Agricultural and Food Chemistry*, 2005, 53(21): 8128–8134.

Bhat R. Potential use of fourier transform infrared spectroscopy for identification of molds capable of producing mycotoxins. *International Journal of Food Properties*, 2013 (just-accepted).

Carnaghan R B, Hartley R D, O'Kelly J. Toxicity and fluorescence properties of the flatoxins (Aspergillusflavus). *Nature*, 1963, 14(200): 1101–1101.

Casasent D, Chen X W. Aflatoxin detection in whole corn kernels using hyperspectral methods. *Optical Technologies for Industrial, Environmental, and Biological Sensing. International Society for Optics and Photonics*, 2004: 275–284.

Chen Q, Zhang C, Zhao J, et al. Recent advances in emerging imaging techniques for non-destructive detection of food quality and safety. *Trends in Analytical Chemistry*, 2013, 52: 261–274.

Cheng X, Vella A, and Stasiewicz M J. Classification of aflatoxin contaminated single corn kernels by ultraviolet to near infrared spectroscopy. *Food Control*, 2019, 98: 253–261.

Chu X, Wang W, Yoon S C, et al. Detection of aflatoxin B1 (AFB1) in individual maize kernels using short wave infrared (SWIR) hyperspectral imaging. *Biosystems Engineering*, 2017, 157: 13–23.

Chu X, Wang W, Zhang L D, et al. Hyperspectral optimum wavelengths and Fisher discrimination analysis to distinguish different concentrations of aflatoxin on corn kernel surface. *Guang Pu Xue Yu Guang Pu Fen Xi= Guang Pu*, 2014, 34(7): 1811–1815.

Dale L M, Thewis A, Boudry C, et al. Discrimination of grassland species and their classification in botanical families by laboratory scale NIR hyperspectral imaging: Preliminary results. *Talanta*, 2013, 116: 149–154.

De Mello F R, Scussel, V M. Characteristics of in-shell Brazil nuts and their relationship to aflatoxin contamination: criteria for sorting. *Journal of Agricultural and Food Chemistry*, 2007, 55(22): 9305–9310.

De Mello F R, Scussel, V M. Development of physical and optical methods for in-shell Brazil nuts sorting and aflatoxin reduction. *Journal of Agricultural Science*, 2009, 1(2): 1–14.

Del Fiore A, Reverberi M, Ricelli A, et al. Early detection of toxigenic fungi on maize by hyperspectral imaging analysis. *International Journal of Food Microbiology*, 2010, 144(1): 64–71.

DiCrispino K, Yao H, Hruska Z, et al. Hyperspectral imagery for observing spectral signature change in Aspergillus flavus. *Optics East 2005. International Society for Optics and Photonics*, 2005: 599606-1–599606-9.

Dowell F E, Pearson T C, Maghirang E B, et al. Reflectance and transmittance spectroscopy applied to detecting fumonisin in single corn kernels infected with Fusarium verticil-lioides. *Cereal Chemistry*, 2002, 79(2): 222–226.

Everard C D, Kim M S, Lee H. A comparison of hyperspectral reflectance and fluorescence imaging techniques for detection of contaminants on spinach leaves. *Journal of Food Engineering*, 2014, 143: 139–145.

Farsaie A, McClure W F, Monroe R J. Design and development of an automatic electro-optical sorter for removing BGY [bright-greening yellow] fluorescent pistachio nuts [Aflatoxin contamination control]. *Transactions of the ASAE [American Society of Agricultural Engineers]*, 1981, 24: 1372–1375.

Fernández-Ibañez V, Soldado A, Martínez-Fernández A, et al. Application of near infrared spectroscopy for rapid detection of aflatoxin B1 in maize and barley as analytical quality assessment. *Food Chemistry*, 2009, 113(2): 629–634.

Firrao G, Torelli E, Gobbi E, et al. Prediction of milled maize fumonisin contamination by multispectral image analysis. *Journal of Cereal Science*, 2010, 52(2): 327–330.

Gaspardo B, Del Zotto S, Torelli E, Cividino S R, Firrao G, Della Riccia G, Stefanon B. A rapid method for detection of fumonisins B1 and B2 in corn meal using Fourier transform near infrared (FT-NIR) spectroscopy implemented with integrating sphere. *Food Chemistry*, 2012, 135(3): 1608–1612.

Gordon S H, Jones R W, McClelland J F, et al. Transient infrared spectroscopy for detection of toxigenic fungi in corn: Potential for on-line evaluation. *Journal of Agricultural and Food Chemistry*, 1999, 47(12): 5267–5272.

Gordon S H, Schudy R B, Wheeler B C, et al. Identification of Fourier transform infrared photoacoustic spectral features for detection of *Aspergillus flavus* infection in corn. *International Journal of Food Microbiology*, 1997, 35(2): 179–186.

Gowen A A, Feng Y, Gaston E, et al. Recent applications of hyperspectral imaging in microbiology. *Talanta*, 2015, 137: 43–54.

Gowen A A, O'Donnell C P, Cullen P J, et al. Hyperspectral imaging—An emerging process analytical tool for food quality and safety control. *Trends in Food Science & Technology*, 2007, 18(12): 590–598.

Hadavi E. Several physical properties of aflatoxin-contaminated pistachio nuts: Application of BGY fluorescence for separation of aflatoxin-contaminated nuts. *Food Additives and Contaminants*, 2005, 22(11): 1144–1153.

Haff R P, Pearson T C. Spectral band selection for optical sorting of pistachio nut defects. Transactions of the *ASABE*, 2006, 49(4): 1105–1113.

Haghighi B. Aflatoxin detection in single pistachio nuts and in different fluorescent categories. Research Bulletin of Isfahan University, 1988, p. 2.

Han D, Yao H, Hruska Z, Kincaid R, Ramezanpour C, Rajasekaran K, Bhatnagar D. Development of high speed dual-camera system for batch screening of aflatoxin contamination of corns using multispectral fluorescence imaging. *Sensing for Agriculture and Food Quality and Safety X. International Society for Optics and Photonics*, 2018, 10665: 106650J.

He H J, Sun D W. Hyperspectral imaging technology for rapid detection of various microbial contaminants in agricultural and food products. *Trends in Food Science & Technology*, 2015, 46(1): 99–109.

Hernández-Hierro J M, García-Villanova R J, González-Martín I. Potential of near infra-red spectroscopy for the analysis of mycotoxins applied to naturally contaminated red paprika found in the Spanish market. *Analytica chimica acta*, 2008, 622(1): 189–194.

Hirano S, Okawara N, Narazaki S, Near infra red detection of internally moldy nuts. *Bioscience, Biotechnology, and Biochemistry*, 1998, 62: 102–107.

Horn B W. Relationship between soil densities of Aspergillus species and colonization of wounded peanut seeds. *Canadian Journal of Microbiology*, 2006, 52(10): 951–960.

Hruska Z, Yao H, Kincaid R, et al. Fluorescence imaging spectroscopy (FIS) for comparing spectra from corn ears naturally and artificially infected with aflatoxin producing fungus. *Journal of Food Science*, 2013, 78(8): T1313–T1320.

Jaillais B, Roumet P, Pinson-Gadais L, et al. Detection of Fusarium head blight contamination in wheat kernels by multivariate imaging. *Food Control*, 2015, 54: 250–258.

Jiang J, Qiao X, He R. Use of near-infrared hyperspectral images to identify moldy peanuts. *Journal of Food Engineering*, 2016, 169: 284–290.

Jin J, Tang L, Hruska Z, et al. Classification of toxigenic and atoxigenic strains of Aspergillus flavus with hyperspectral imaging. *Computers and Electronics in Agriculture*, 2009, 69(2): 158–164.

Kalkan H, Beriat P, Yardimci Y, et al. Detection of contaminated hazelnuts and ground red chili pepper flakes by multispectral imaging. *Computers and Electronics in Agriculture*, 2011, 77(1): 28–34.

Kalkan H, Yardimci Y. Detection of contaminated hazelnuts by multispectral imaging. *IEEE 16th Signal Processing, Communication and Applications Conference, 2008. SIU 2008*. IEEE, 2008, pp. 1–4.

Kandpal L M, Lee S, Kim M S, et al. Short wave infrared (SWIR) hyperspectral imaging technique for examination of aflatoxin B 1 (AFB 1) on corn kernels. *Food Control*, 2015, 51: 171–176.

Kandpal L M, Lohumi S, Kim M S, et al. Near-infrared hyperspectral imaging system coupled with multivariate methods to predict viability and vigor in muskmelon seeds. *Sensors and Actuators B: Chemical*, 2016, 229: 534–544.

Kaya-Celiker H, Mallikarjunan P K, Schmale D, et al. Discrimination of moldy peanuts with reference to aflatoxin using FTIR-ATR system. *Food Control*, 2014, 44: 64–71.

Kheiralipour K, Ahmadi H, Rajabipour A, et al. Development of a new threshold based classification model for analyzing thermal imaging data to detect fungal infection of pistachio kernel. *Agricultural Research*, 2013, 2(2): 127–131.

Kimuli D, Wang W, Jiang H, Zhao X, Chu, X. Application of SWIR hyperspectral imaging and chemometrics for identification of aflatoxin B1 contaminated maize kernels. *Infrared Physics & Technology*, 2018a, 89: 351–362.

Kimuli D, Wang W, Lawrence K C, Yoon S C, Ni X, Heitschmidt, G W. Utilisation of visible/near-infrared hyperspectral images to classify aflatoxin B1 contaminated maize kernels. *Biosystems Engineering*, 2018b, 166: 150–160.

Lee K M, Davis J, Herrman T J, et al. An empirical evaluation of three vibrational spectroscopic methods for detection of aflatoxins in maize. *Food Chemistry*, 2015, 173: 629–639.

Lee K M, Herrman T J, Yun U. Application of Raman spectroscopy for qualitative and quantitative analysis of aflatoxins in ground maize samples. *Journal of Cereal Science*, 2014, 59(1): 70–78.

Levasseur-Garcia C. Infrared spectroscopy applied to identification and detection of microorganisms and their metabolites on cereals (corn, wheat, and barley). In Aflakpui G, editor, *Agricultural Science* (pp. 9–196). New York: Intech, 2012.

Liang P S, Slaughter D C, Ortega-Beltran A, et al. Detection of fungal infection in almond kernels using near-infrared reflectance spectroscopy. *Biosystems Engineering*, 2015, 137: 64–72.

Liu D, Zeng X A, Sun D W. Recent developments and applications of hyperspectral imaging for quality evaluation of agricultural products: A review. *Critical Reviews in Food Science and Nutrition*, 2013 (just-accepted).

Lohumi S, Lee S, Lee H, et al. Application of hyperspectral imaging for characterization of intramuscular fat distribution in beef. *Infrared Physics & Technology*, 2016, 74: 1–10.

Mahesh S, Jayas D S, Paliwal J, et al. Hyperspectral imaging to classify and monitor quality of agricultural materials. *Journal of Stored Products Research*, 2015, 61: 17–26.

Marroquín-Cardona A G, Johnson N M, Phillips T D, et al. Mycotoxins in a changing global environment–a review. *Food and Chemical Toxicology*, 2014, 69: 220–230.

Marsh P B, Simpson M E, Ferretti R J, Merola G V, Donoso J, Craig G O, Trucksess M V, Work P S. Mechanism of formation of a fluorescence in cotton fiber associated with aflatoxin in the seeds at harvest. *Journal of Agricultural and Food Chemistry*, 1969, 17: 468–472.

McClure W F, Farsaie A. Dual-wavelength fiber optic photometer measures fluorescence of aflatoxin contaminated pistachio nuts. *Transactions of the ASAE*, 1980, 23(1): 204–207.

Micotti da Gloria E (2011). Aflatoxin contamination distribution among grains and nuts. In Torres-Pacheco I, editor, *Aflatoxins-Detection, Measurement and Control* (pp. 75–90). London: InTech.

Ministry of Health, PRC. *GB 2761-2011 The National Food Safety Standards of mycotoxins in food limited* (In Chinese). Beijing: China Standards Press, 2011.

Moghaddam T M, Razavi S M A, Taghizadeh M. Applications of hyperspectral imaging in grains and nuts quality and safety assessment: A review. *Journal of Food Measurement and Characterization*, 2013, 7(3): 129–140.

Moscetti R, Monarca D, Cecchini M, et al. Detection of mold-damaged chestnuts by near-infrared spectroscopy. *Postharvest Biology and Technology*, 2014, 93: 83–90.

Nakariyakul S, Casasent D P. Classification of internally damaged almond nuts using hyperspectral imagery. *Journal of Food Engineering*, 2011, 103(1): 62–67.

Nasir M S, Jolley M E. Development of a fluorescence polarization assay for the determination of aflatoxins in grains. *Journal of Agricultural and Food Chemistry*, 2002, 50(11): 3116–3121.

Ononye A E, Yao H, Hruska Z, et al. Calibration of a fluorescence hyperspectral imaging system for agricultural inspection and detection. *SPIE Defense, Security, and Sensing. International Society for Optics and Photonics*, 2010: 76760E-1–76760E-11.

Park B, Lawrence K C, Windham W R, et al. Hyperspectral imaging for food processing automation. *International Symposium on Optical Science and Technology. International Society for Optics and Photonics*, 2002, pp. 308–316.

Pasikatan M C, Dowell F E. Sorting systems based on optical methods for detecting and removing seeds infested internally by insects or fungi: A review. *Applied Spectroscopy Reviews*, 2001, 36(4): 399–416.

Pearson S M, Candlish A A G, Aidoo K E, et al. Determination of aflatoxin levels in pistachio and cashew nuts using immunoaffinity column clean-up with HPLC and fluorescence detection. *Biotechnology Techniques*, 1999, 13(2): 97–99.

Pearson T C. Machine vision system for automated detection of stained pistachio nuts. *Photonics for Industrial Applications. International Society for Optics and Photonics*, 1995: 95–103.

Pearson T C, Wicklow D T. Detection of corn kernels infected by fungi. *Transactions of the ASABE*, 2006, 49(4): 1235–1245.

Pearson T C, Wicklow D T, Maghirang E B, et al. Detecting aflatoxin in single corn kernels by transmittance and reflectance spectroscopy. *Transactions of the American Society of Agricultural Engineers*, 2001, 44(5): 1247–1254.

Pearson T C, Wicklow D T, Pasikatan M C. Reduction of aflatoxin and fumonisin contamination in yellow corn by high-speed dual-wavelength sorting. *Cereal Chemistry*, 2004, 81(4): 490–498.

Qiao X, Jiang J, Qi X, et al. Utilization of spectral-spatial characteristics in shortwave infrared hyperspectral images to classify and identify fungi-contaminated peanuts. *Food Chemistry*, 2017, 220: 393–399.

Rajalakshmi M, Subashini P. A study on non-destructive method for detecting toxin in pepper using neural networks. arXiv preprint arXiv:1208.2092, 2012.

Ravikanth L, Singh C B, Jayas D S, et al. Classification of contaminants from wheat using near-infrared hyperspectral imaging. *Biosystems Engineering*, 2015, 135: 73–86.

Robens J. Aflatoxin-recognition, understanding, and control with particular emphasis on the role of the agricultural research service. *Toxin Reviews*, 2008, 27(3–4): 143–169.

Samiappan S, Bruce L M, Yao H, et al. Support vector machines classification of fluorescence hyperspectral image for detection of aflatoxin in corn kernels, 2013. www.researchgate.net/

Selvaraj J N, Lu Z, Yan W, et al. Mycotoxin detection—Recent trends at global level. *Journal of Integrative Agriculture*, 2015, 14(11): 2265–2281.

Senthilkumar T, Jayas D S, White N D G, et al. Detection of fungal infection and Ochratoxin A contamination in stored barley using near-infrared hyperspectral imaging. *Biosystems Engineering*, 2016a, 147: 162–173.

Senthilkumar T, Jayas D S, White N D G, et al. Detection of fungal infection and Ochratoxin A contamination in stored wheat using near-infrared hyperspectral imaging. *Journal of Stored Products Research*, 2016b, 65: 30–39.

Serranti S, Cesare D, Bonifazi G. The development of a hyperspectral imaging method for the detection of Fusarium-damaged, yellow berry and vitreous Italian durum wheat kernels. *Biosystems Engineering*, 2013, 115(1): 20–30.

Shen F, Wu Q, Liu P, et al. Detection of Aspergillus spp. contamination levels in peanuts by near infrared spectroscopy and electronic nose. *Food Control*, 2018, 93: 1–8.

Singh C B, Jayas D S, Paliwal J, et al. Fungal damage detection in wheat using short-wave near-infrared hyperspectral and digital colour imaging. *International Journal of Food Properties*, 2012, 15(1): 11–24.

Singh C B, Jayas D S, Saeger S. Spectroscopic techniques for fungi and mycotoxins detection. In De Saeger S, editor, *Determining Mycotoxins and Mycotoxigenic Fungi in Food and Feed* (pp. 401–414). Cambridge: Woodhead Publishing, 2011.

Sirisomboon C D, Putthang R, Sirisomboon P. Application of near infrared spectroscopy to detect aflatoxigenic fungal contamination in rice. *Food Control*, 2013, 33(1): 207–214.

Stasiewicz M J, Falade T D O, Mutuma M, et al. Multi-spectral kernel sorting to reduce aflatoxins and fumonisins in Kenyan maize. *Food Control*, 2017, 78: 203–214.

Tao F, Yao H, Zhu F, Hruska Z, Liu Y, Rajasekaran K, Bhatnagar D. Feasibility of using visible/near-infrared (Vis/NIR) spectroscopy to detect aflatoxigenic fungus and aflatoxin contamination on corn kernels. In *2018 ASABE Annual International Meeting. American Society of Agricultural and Biological Engineers*, 2018, p. 1.

Teena M A, Manickavasagan A, Ravikanth L, et al. Near infrared (NIR) hyperspectral imaging to classify fungal infected date fruits. *Journal of Stored Products Research*, 2014, 59: 306–313.

Tong, P, Zhang, L, Xu, J J, Chen, H Y. Simply amplified electrochemical aptasensor of ochratoxin A based on exonuclease-catalyzed target recycling. *Biosensors and Bioelectronics*, 2011. 29(1), 97–101.

Tripathi S, Mishra H N. A rapid FT-NIR method for estimation of aflatoxin B1 in red chili powder. *Food Control*, 2009, 20(9): 840–846.

Turner N W, Subrahmanyam S, Piletsky S A. Analytical methods for determination of mycotoxins: A review. *Analytica Chimica Acta*, 2009, 632(2): 168–180.

USDA. 2002. *Aflatoxin Handbook*. Washington, DC: USDA Grain Inspection, Packers, and Stockyards Administration. www.gipsa.usda.gov/publications/fgis/handbooks/ afl_ins-phb.html.

Wang W, Heitschmidt G W, Ni X, et al. Identification of aflatoxin B1 on maize kernel surfaces using hyperspectral imaging. *Food Control*, 2014, 42: 78–86.

Wang W, Heitschmidt G W, Windham W R, et al. Feasibility of detecting Aflatoxin B1 on inoculated maize kernels surface using Vis/NIR hyperspectral imaging. *Journal of Food Science*, 2015a, 80(1): M116–M122.

Wang W, Lawrence K C, Ni X, et al. Near-infrared hyperspectral imaging for detecting Aflatoxin B 1 of maize kernels. *Food Control*, 2015b, 51: 347–355.

Wang W, Ni X, Lawrence K C, et al. Feasibility of detecting Aflatoxin B1 in single maize kernels using hyperspectral imaging. *Journal of Food Engineering*, 2015c, 166: 182–192.

Williams P J, Geladi P, Britz T J, et al. Growth characteristics of three Fusarium species evaluated by near-infrared hyperspectral imaging and multivariate image analysis. *Applied Microbiology and Biotechnology*, 2012a, 96(3): 803–813.

Williams P J, Geladi P, Britz T J, et al. Investigation of fungal development in maize kernels using NIR hyperspectral imaging and multivariate data analysis. *Journal of Cereal Science*, 2012b, 55(3): 272–278.

Wu D, Sun D W. Advanced applications of hyperspectral imaging technology for food quality and safety analysis and assessment: A review—Part I: Fundamentals. *Innovative Food Science & Emerging Technologies*, 2013a, 19: 1–14.

Wu D, Sun D W. Advanced applications of hyperspectral imaging technology for food quality and safety analysis and assessment: A review—Part II: Applications. *Innovative Food Science & Emerging Technologies*, 2013b, 15: 15–28.

Wu Q, Wang J, Wang C, et al. Study on the optimal algorithm prediction of corn leaf component information based on hyperspectral imaging. *Infrared Physics & Technology*, 2016, 78: 66–71.

Wu Q, Xie L, Xu H. Determination of toxigenic fungi and aflatoxins in nuts and dried fruits using imaging and spectroscopic techniques. *Food Chemistry*, 2018, 252: 228–242.

Wu Q, Xu J, Xu H. Discrimination of aflatoxin B1 contaminated pistachio kernels using laser induced fluorescence spectroscopy. *Biosystems Engineering*, 2019, 179: 22–34.

Yao H, Hruska Z, Brown RL, Cleveland TE. Hyperspectral bright greenish yellow fluorescence (BGYF) imaging of aflatoxin contaminated corn kernels. *Proceedings of SPIE, Optics for Natural Resources, Agriculture, and Foods*, 2006, 6381: 63810.

Yao H, Hruska Z, Kincaid R, et al. Correlation and classification of single kernel fluorescence hyperspectral data with aflatoxin concentration in corn kernels inoculated with Aspergillus flavus spores. *Food Additives and Contaminants*, 2010a, 27(5): 701–709.

Yao H, Hruska Z, Kincaid R D, et al. Method and detection system for detection of aflatoxin in corn with fluorescence spectra. U.S. Patent Application, 12/807,673d, 2010b.

Yao H, Hruska Z, Kincaid R, et al. Spectral angle mapper classification of fluorescence hyperspectral image for aflatoxin contaminated corn. *2nd Workshop on Hyperspectral Image and Signal Processing: Evolution in Remote Sensing (WHISPERS), 2010*. IEEE, 2010c, pp. 1–4.

Yao H, Hruska Z, Kincaid R, et al. Single aflatoxin contaminated corn kernel analysis with fluorescence hyperspectral image. *SPIE 7676, Sensing for Agriculture and Food Quality and Safety II*, 2010d, 7676: 76760D.

Yao H, Hruska Z, Kincaid R, et al. Development of narrow-band fluorescence index for the detection of aflatoxin contaminated corn. *Proceedings of SPIE, Sensing for Agriculture and Food Quality and Safety III*, 2011, 8027: 80270D.

Yao H, Hruska Z, Kincaid R, et al. Utilizing fluorescence hyperspectral imaging to differentiate corn inoculated with toxigenic and atoxigenic fungal strains. *SPIE Defense, Security*, and Sensing, 2012, 8369: 83690B.

Yao H, Hruska Z, Kincaid R, et al. Detecting maize inoculated with toxigenic and atoxigenic fungal strains with fluorescence hyperspectral imagery. *Biosystems Engineering*, 2013a, 115(2): 125–135.

Yao H, Hruska Z, Kincaid R, et al. Hyperspectral image classification and development of fluorescence index for single corn kernels infected with Aspergillus flavus. *Transactions of the ASABE*, 2013b, 56(5): 1977–1988.

Yao H, Kincaid R, Hruska Z, et al. Hyperspectral imaging system for whole corn ear surface inspection. *SPIE Defense, Security, and Sensing. International Society for Optics and Photonics*, 2013c: 87210H-1–87210H-8.

Yoon S C, Park B, Lawrence K C, et al. Development of real-time line-scan hyperspectral imaging system for online agricultural and food product inspection. *SPIE Defense, Security, and Sensing. International Society for Optics and Photonics*, 2010: 76760J-1–76760J-11.

Zavattini G, Vecchi S, Leahy R M, et al. A hyperspectral fluorescence imaging system for biological applications. *Nuclear Science Symposium Conference Record, 2003 IEEE*, 2003, 2: 942–946.

Zhang R, Ying Y, Rao X, et al. Quality and safety assessment of food and agricultural products by hyperspectral fluorescence imaging. *Journal of the Science of Food and Agriculture*, 2012, 92(12): 2397–2408.

Zhongzhi H, Limiao D. Application driven key wavelengths mining method for aflatoxin detection using hyperspectral data. *Computers and Electronics in Agriculture*, 2018, 153: 248–255.

Zhu B, Jiang L, Jin F, Qin L, Vogel A, Tao Y, Walnut shell and meat differentiation using fluorescence hyperspectral imagery with ICA-kNN optimal wavelength optimal wavelength selection. *Sensing and Instrumentation for Food Quality and Safety*, 2007, 1: 123–131.

Zhu F, Yao H, Hruska Z, et al. Classification of corn kernels contaminated with aflatoxins using fluorescence and reflectance hyperspectral images analysis. *SPIE Sensing Technology+ Applications. International Society for Optics and Photonics*, 2015: 94880M-1–94880M-6.

2 Aflatoxin Detection by Fluorescence Index and Narrowband Spectra Based on Hyperspectral Imaging

Aflatoxin is a kind of highly toxic and carcinogenic substance with the characteristic of ultraviolet (UV) fluorescence. To explore the application of hyperspectral imaging technology in aflatoxin detection, a hyperspectral imaging system is built under 365 nm UV light. The hyperspectral images of 250 peanut kernel samples in 33 bands (400–720 nm) with five kinds of concentrations are collected. An object-oriented point-source illumination compensation method is proposed for compensating hyperspectral illumination and four fluorescence indexes such as Radiation Index (RI), Difference Radiation Indexes (DRIs), Ratio Radiation Index (RRI), and Normalized Difference Radiation Index (NDRI) are constructed. Radial basis function–support vector machine (RBF–SVM) model is constructed based on grid search to recognize and make regression analysis on the degree of aflatoxin contamination. The DRIs have the optimal performance, the accuracy rate of fivefold cross validation of SVM is 95.5%, and the mean square error (MSE) and R are 0.0223 and 0.9785, respectively, for testing data. Based on DRIs, narrowband spectra are searched by Fisher, Plus-l-remove-r, and band correlation coefficient. The narrowband spectrum obtained by Fisher's optimization research is optimal, and the spectrum band is 410–430 nm. On this narrowband, the accuracy rate, MSE, and R are 87.2%, 0.27418, and 0.86732, respectively, for testing data using the same model of SVM. Perhaps this narrowband spectrum can be easily transplanted to the online aflatoxin detection production line. The results of the current research are of positive significance for the research of agricultural products such as aflatoxin grain sorting and online fast detection device.

2.1 INTRODUCTION

Aflatoxin is a highly toxic carcinogen, the toxicity of which is 68 times as that of arsenic. It is the strongest chemical carcinogen found so far and the main cause of liver cancer. It widely exists in peanuts, corn, and their products. The National Standards of China and the United States have set strict limits for the level of aflatoxin in foods, and the limited standards are 20 and 100 ppb in food grade and feed

29

grade, respectively (Ministry of Health, PRC, 2011, USDA, 2002). Aflatoxin is a metabolic product of some of the parasitic fungi in *Aspergillus* spp. It has the characteristic of fluorescence emission: aflatoxin B1 and B2 emit blue fluorescence, while aflatoxin G1 and G2 emit green fluorescence under UV light. At present, aflatoxin is detected mainly by biochemical methods, including thin-layer chromatography, high-performance liquid chromatography, micro-column method, and enzyme-linked immunosorbent assay (Teena et al., 2013). Although the detection accuracy is high, the detection means and instruments are complex and the detection speed is slow. So, these methods cannot be used in online detection.

In recent years, spectrum detection has become a new method for aflatoxin detection and has received widespread attention. Ataş et al. studied the hyperspectral imaging analysis technique of aflatoxin contamination in the cayenne pepper bought from the market under two light sources: halogen lamp and UV lamp. They proposed a feature selection method based on gray-level histogram quantization and neural network weights optimization. They found that the feature wavelength played a key role in detection and the recognition rate of leave-one-out method under 420 nm reached 85% (Ataş et al., 2012). The team of H. Yao mainly focused on the high-spectrum study of aflatoxin contamination in the maize. Based on the image of single grain through artificial aflatoxin cultivation, they studied the maximum likelihood estimation and binary encoding method. They found that the fluorescence intensity decreased with the increase in the aflatoxin content and there was a phenomenon of peak shift (Yao et al., 2013). Based on these two points, they applied for a US patent (Yao et al., 2010). However, the detection was still made manually instead of automatically. In China, until 2014, the research group of Wang Wei studied the problems in aflatoxin detection using the hyperspectral data of the United States Agricultural Research Department (USDA, ARS) and pointed out that the aflatoxin detection rate of naturally contaminated grains was 87.5% through common CCD imaging. They optimized five hyperspectral wavelengths from 700 to 1,100 nm and the prediction accuracy rate of artificial aflatoxin contamination concentration reached 88.3% (Zhang et al., 2014; Chu et al., 2014).

In this chapter, the detection of artificial aflatoxin contamination is studied using a hyperspectral imaging system based on the aflatoxin's fluorescence characteristic under UV light. An online fast aflatoxin detection method is found through the optimization of fluorescence index and narrowband spectra.

2.2   EXPERIMENT MATERIALS

2.2.1   Sample Preparation and Image Acquisition

Usually, halogen lamp is mainly used in reflectance spectroscopy and UV lamp to study the fluorescence phenomenon of samples. Considering the fluorescence characteristic of aflatoxin, a UV lamp is used in this study to examine the contribution of different light sources.

The peanut samples used in the experiments are the type of Silihong and bought from the local market. According to the National Standards of China and America (Ministry of Health, PRC, 2011, USDA, 2002), the contamination limits for food grade and feed grade are 20 and 100 ppb, respectively. Here, five kinds of acetonitrile solution, 10, 20, 50, 100, and 10,000 ppb, have been prepared, and aflatoxin B1 is bought from Hairun Detection Center of Qingdao. Then, a certain amount of solution is dropped onto the surface of the 50 peanut samples. Thus, the number of total peanut samples is 250. To avoid the effect of background fluorescence, a kind of kraft paper is used as the background and the peanut samples are pasted on the kraft paper using a double-sided adhesive tape. All the samples are prepared in the laboratory of the College of Food Science and Engineering, Qingdao Agricultural University.

Images are collected in a dark room in the spectrum laboratory of Qingdao Photoelectric Institute of Chinese Academy of Sciences. The collection device used in the experiment is a portable imaging instrument as shown in Figure 2.1. The imaging instrument is composed of liquid crystal tunable filter (LCTF) and a regular CCD camera. The spectral range is 400 ~ 720 nm, and the spectral bandwidth is 10 nm. A 365 nm high-power LED UV lamp (LUYOR-3404 desktop UV lamp of American LUYOR, and 7000 lumens sample illumination) is used as the illumination. The size of the image cubes it can collect is $1,392 \times 1,040 \times 33$ pixels.

Due to the small visual collection angle of the hyperspectral imager, samples need to be moved when photographed under the lamp. To avoid the fluorescence effect of the double-sided adhesive tape when UV fluorescence is photographed, a background cardboard with nine grooves are specially made. When UV images are photographed, the grain samples are taken down from the double-sided adhesive tape board and put on the background board with the face contaminated by aflatoxin outside. Under UV lamp, the images of the 250 samples are taken successively. The total data size is 3.35GB.

The analysis software of hyperspectral images is ENVI4.7 (ITT Visual Information Solutions, Boulder, CO) and MATLAB® 2012b (Math Works, USA).

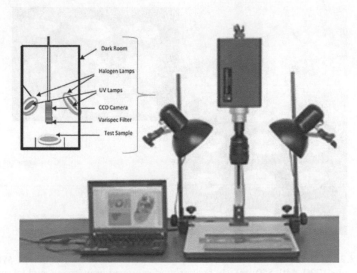

FIGURE 2.1 Image collection device.

To avoid oversaturation and undersaturation, the matching acquisition software can automatically adjust the exposure time of each wavelength according to the image's overall brightness.

Images in Figure 2.2a are pseudo color images synthesized with false color (720, 440, 550 nm) based on our experiences. In this figure, the fluorescence phenomenon of aflatoxin is obvious.

2.2.2 ILLUMINATION COMPENSATION AND KERNEL SEGMENTATION

From the hyperspectral images, it can be seen that the kernel region accounts only for a small part of the whole image and most of the region is background. The background region doesn't make any contribution to the data analysis and result interpretation. On account of only the kernel region, data are useful, so the ROI (region of interest) is defined as the kernel region. The ROI can be selected manually or automatically through programming. Here, we use the pseudo color composite image to segment the kernel region, and the subsequent processing is all made based on the ROI.

In Figure 2.2a, it can be found that the fluorescence phenomenon is obvious on the pseudo color images synthesized by 720, 440, and 550 nm. However, the illumination of the images is uneven. The reason is that the illumination source is point light source, so it is necessary to do the illumination compensation first. Traditional illumination compensation algorithm is not for the detection objects. Although its compensation accuracy is high, it is very complex (Gao et al., 2002). For the average gray value of one kernel to be similar to the other kernel on the same backboard, it is feasible to adjust the average gray level of the kernel region to the same level.

Based on the above, we propose a method for ROI selection and illumination compensation.

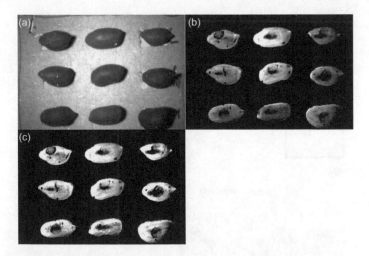

FIGURE 2.2 Process of illumination compensation: (a) pseudo color image synthesized with false color (720, 440, 550 nm), (b) background-removed gray image, and (c) image after illumination compensation.

1. Preprocessing: A series of preprocessing operations are made on the pseudo color images which are de-noising, contrast enhancement, and filling the highlighted region with adjoining pixels, then the images' background is removed by R-1.2G calculation, and the kernel's unicom region is calculated. The background-removed images are shown in Figure 2.2b.
2. Illumination compensation: The average gray value of each grain's unicom region, $a(k)$, is calculated for each band. Here, k is the number of the grain's unicom regions. Then, $I'(x,y,k) = I(x,y,k)*(\max(a)/a(k))$ is calculated for each pixel in the unicom region. Here, I and I' are the gray values of the original pixel and new pixel, respectively. The enhanced image is shown in Figure 2.2c.
3. Image segmentation: The kernels are sorted from row to column and the image data blocks of 33 bands in the kernel region (the upper left corner's coordinates, the length and width of grain region) are segmented and stored in order. So, one by one, the hyperspectral imaging blocks of 250 kernels have been obtained.

2.3 DATA PROCESSING AND RESULT ANALYSIS

2.3.1 FLUORESCENCE INDEX

In order to find the really useful spectrum, we proposed four appropriate fluorescence indexes, such as RI, DRI, RRI, and NDRI.

Under the lamp, each pixel value is the reflected radiation brightness, and the UV image is mainly reflected by the fluorescence brightness. After quantizing, each pixel can be referred to the gray value of each pixel. For the k-th band, the single band's average gray value (or named Radiation Index, RI) is defined in equation 2.1. Here, $I_k(x,y)$ is defined as the radiation intensity of the pixel at coordinate (x, y).

$$RI_k = \sum_x \sum_y I_k(x,y) / x \cdot y \quad k = 1,2,...,33 \tag{2.1}$$

On this basis, we extract the following fluorescence index:

Difference Radiation Indexes (DRI)

$$DRI_k = \sum_x \sum_y [I_{k+1}(x,y) - I_k(x,y)] / (x \cdot y) \quad k = 1,2,...,32 \tag{2.2}$$

Ratio Radiation Index (RRI)

$$RRI_k = \sum_x \sum_y [I_{k+1}(x,y) / I_k(x,y)] / (x \cdot y) \quad k = 1,2,...,32 \tag{2.3}$$

Normalized Difference Radiation Index (NDRI)

$$NDFI_k = \sum_x \sum_y \frac{I_{k+1}(x,y) - I_k(x,y)}{I_{k+1}(x,y) + I_k(x,y)} / (x \cdot y) \quad k = 1,2,...,32 \tag{2.4}$$

Here, RI_k corresponds to the average gray level of the pixels $I_k(x, y)$ in the k-th band. And the spectral indexes defined by equations 2.1–2.4 represent the average values of each band. Thus, compared with all of each pixel's gray value, the information is reduced effectively.

2.3.2 RECOGNITION AND REGRESSION

SVM was first proposed by Corinna Cortes and Vapnik in 1995. It has shown many special advantages in solving small-sample cases and nonlinear high-dimensional pattern recognition problems. It can be applied to many machine learning problems such as function fitting. However, the vector sets are usually difficult to be divided in the low-dimensional space, and the solution is to map them to high-dimensional space by a kernel function. However, at the same time, it will lead to the increase of calculation complexity, and the kernel function can be used neatly to solve this problem. In this chapter, the RBF-SVM (Chang and Lin, 2011) kernel function based on grid optimization has been used to solve the aflatoxin's detection and regression problems.

First, according to the five concentrations, the samples have been divided into five classes, and SVM is used in the recognition process. The recognition results are shown in Table 2.1. The data on the upper row denote the performance of the training set and the data on the bottom row the performance of the testing set. It can be found that three kinds of band operations are better than pure RI, and DRI was the best index of the four fluorescence indexes, which reached up to 95.5%. At the same time, SVM is used to investigate the regression ability of four band indexes. In terms of MSE and R for concentration regression, DRI is also the best.

Then, based on DRI, SVM regression (SVR) and classical least square regression (LSR) method are used to investigate the regression ability of different concentrations. Figure 2.3 illustrates the regression effect of the two methods. Overall, the regression effect of SVM is much superior than partial least squares (PLS). Similar conclusions can be drawn from the training sets (Figure 2.3a), test sets (Figure 2.3b), or regression errors (Figure 2.3c).

Through the above two analyses, we can get a better model of aflatoxin detection, that is, DRI and SVM regression.

TABLE 2.1
The SVM Classification Capability of Three Fluorescence Indexes

Fluorescence Index	Dataset	CCR (%)	MSE	R
RI	Training	25.7	1.8011	0.9990
	Testing	22. 0	1.9488	0.7974
DRI	Training	98.0	0.0446	0.9888
	Testing	95.5	0.0223	0.9785
RRI	Training	81.0	0.1774	0.9148
	Testing	54.0	0.6890	0.6708
NDRI	Training	44.0	1.0921	0.4700
	Testing	38.0	1.6800	0.3051

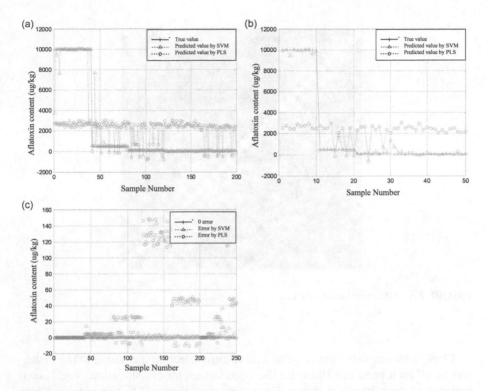

FIGURE 2.3 Regression effect of SVM and PLS: (a) training set, (b) testing set, and (c) error.

2.3.3 Narrowband Spectra

This study is based on the whole bands we collected, but it is very difficult for the image system used on the production line to collect so many band signals within a short time. To collect narrowband spectra may be a very practical method for it can be imaged flashily. If the narrowband contains much information and has little relevance between each other, then it will have greater difference of object spectral.

Band correlation is one of the most important methods to measure the information between two different bands. As there is no relationship between the band relevance and the arrangement of the matrix, we first get the images of each single kernel in the same band and normalize them to the same size. And then, the images of the 250 kernels are stitched to form an image of the same band. This operation is carried out for 33 bands. Second, the DRI is calculated, and relevant calculations about the two band difference indexes are carried out. Finally, the correlation coefficient matrix is obtained (Figure 2.4). The largest coefficient is 0.7657 corresponding to the bands of 27 and 28. The smaller the correlation between the bands, the larger the information of band combination. That is, the band combination information quantity is inversely proportional to the correlation coefficient.

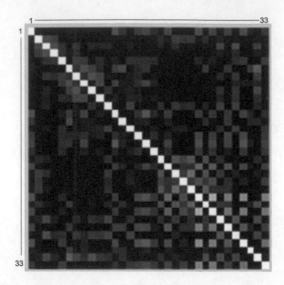

FIGURE 2.4   DFI correlation matrix.

Fisher's feature selection is a classical feature selection method, and the Fisher weight of each band can illustrate the contribution for classification. The Fisher weights of each band's difference are shown in Figure 2.5. Obviously, 2 and 3 are the highest bands based on the difference between their weights (about 410–430 nm). The higher the Fisher weight, the higher the identification ability of this band index.

Greedy mountaineering methods include the sequential forward method, sequential backward method, and plus-r-remove-q method. In sequential forward method, a feature is selected from the unselected features and added to the selected feature set. This newly selected feature set can make the objective function get the largest value (for example, the classifier gets the largest correction rate). This process is repeated until the number of selected features reaches a certain value. Similarly, sequential backward method starts with the set of all features, deleting one of the least important features at a time. At last, the remaining feature set can also make the objective function get the largest value. Both the sequential forward and sequential backward methods make simple serial search when many feature combinations are omitted. Once a feature has been added to (or removed from) the feature set, it cannot be removed (or added) again. To make up for the disadvantage of non-backtracking, local backtracking can be added in the searching process, and this process is called plus-r-remove-q method. The r features are added to the feature set in the sequential forward method, and then the q features are removed in the sequential backward method. Using this method, we selected two key bands which are 19 and 30.

Table 2.2 illustrates the discrimination capability of spectral bands selected by the above three methods. It can be seen that the recognition rate and regression

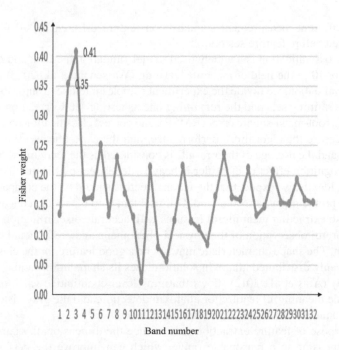

FIGURE 2.5 Fisher's weight coefficients.

TABLE 2.2

The Discrimination Capability of Spectral Bands Selected by the Three Methods

	Band 1	Band 2	Dataset	CCR (%)	MSE	R
Fisher	2	3	Training	93.1	0.13136	0.93486
			Testing	87.2	0.27418	0.86732
Plus-r-remove-q	19	30	Training	89.2	0.24472	0.88444
			Testing	74.4	0.7679	0.66644
Correlation coefficient	27	28	Training	82.9	0.3297	0.83666
			Testing	56.8	0.95018	0.57224

effect obtained by Fisher's method are the best. The two best key bands are 2 and 3 representing the wavelength of 410–430 nm. This is consistent with objective laws, because aflatoxin can emit blue–green fluorescence when stimulated by UV light, and the fluorescence wavelength is around 420 nm.

2.4 DISCUSSION

We can use formulas 2–4 to calculate three fluorescence indexes, namely, DRI, RRI, and NDRI. Each fluorescence index is based on 32 feature vectors. To avoid the effect of different measurement standards on feature differences, all features are standardized before feature selection.

K-fold cross validation is a widely used model validation method, and k is always set with 5 or 10 in the field of machine learning (Wassenaar et al., 2003). With five-fold cross validation, we divide the experiment samples into five groups: four of them are used as training sets and the remaining one as testing set. Correlation coefficient method (R), root-mean-square error (RMSE), and correct classification rate (CCR) are used to measure the algorithm's performance, and the experiments were conducted five times, and the average of their results is considered as the recognized final result.

The recognition effect of classifier is heavily dependent on the quality of feature extraction. Ideally, we expect that the feature vectors should be the compact description of the problem that is recognized. Our problem is to detect the concentration of aflatoxin, so extracting meaningful features with accurate discrimination is needed. It needs careful observation of the physical phenomena and professional knowledge of aflatoxin. The shape characteristic may not be a good feature for the classification, because evenly distributed aflatoxin solution loses its shape information. A correlational study (Ataş et al., 2012) shows that gray-level quantitative value of aflatoxin may provide meaningful feature for aflatoxin detection, and the gray histogram distribution changes with different aflatoxin contents.

The purpose of feature extraction is to reduce the dimensionality curse and the feature dimension to a reasonable range, which can improve the performance of the classifier and increase the recognition speed. Moreover, it is also helpful for understanding the mechanism behind the problem. There are many methods for the selection of feature band, such as segmented principal component analysis, leapfrog algorithm, and genetic algorithm. The feature selection method used in this chapter is based on Fisher's method, and it is proved to be a simple and practical method. And optimum wavelength has explicit physical meaning. Compared with the principal component analysis, its recognition effect depends on the feature contribution rate of the first few key wavelengths.

In order to enhance the detection performance (speed and accuracy), many aspects need to be studied further. Brighter light can be obtained with shorter integral time, so a larger power UV lamp may be needed. Meanwhile, to improve signal's quality, a photomultiplier tube CCD camera can get higher quantum efficiency, and the quantum efficiency of EMCCD 0 is more than 90% while an ordinary PCO1600 camera's efficiency is only 40% (www.photomet.com.cn/). All these will further improve the detection accuracy. In addition, the data acquisition of this chapter is based on full wavelength (400–700 nm), and it takes about 3 min to collect one image. However, only a small part of the spectrum shows the value of detection. It is feasible to study an algorithm using only a few wavelengths. A fluorescence index method using only two wavelengths is proposed in this chapter. Thus, the data acquisition time can be reduced dramatically. The detection time can be reduced by installing switching selector devices of wavelength, such as LCTFs, EMCCD wavelength addressing capabilities, or using a rotating wheel with band-pass recorders for fixed band.

Ataş et al. (2012) reported that the selected key band is 420 nm for the pepper powder bought from the market, and Yao et al. (2013) reported that the selected key band is 437 nm for corn kernels. In this chapter, the optimal narrowband index is

410–430 for peanuts, which is consistent with that of Ataş et al. (2012) and Yao et al. (2013). Of course, different contamination objects may have different detection key wavelengths (Samiappan et al., 2013). But, it may be in a rough range.

There are some highlighted areas in the images which are not aflatoxin. We analyzed the highlighted areas on the background and found that the double-sided adhesive in the sample preparation process and the dust on the kernel's on-sample surface can produce yellow–green fluorescence. So, the ROI of this chapter is based on the region of the kernel which includes peanut region and aflatoxin region. The objective of this is to eliminate the influence of background. And of course, we can also use the aflatoxin concentration area as ROI (Chu et al., 2014). But, in the early stage, some common software such as ENVI is needed to select the kernel's ROI region manually, and the selection accuracy is dependent on the human eye and proficiency, and that will lack objectivity. Of course, the choice of ROI can also be made automatically by the false color synthesis area. However, the judgment result of this method is not accurate because of the effect of environmental light and dust.

2.5 CONCLUSIONS

In this chapter, we study the rapid detection method of the peanut aflatoxin hyperspectral image. First, to solve the uneven illumination problem of point light source, we proposed an illumination compensation method to compensate for the illumination of 33 bands. Then, we selected kernel region as ROI, and the hyperspectral images are segmented for each single kernel. On the basis of single band's average gray value (or named Radiation Index, RI), three kinds of fluorescence indexes were put forward, and the aflatoxin concentration recognition accuracy of each fluorescence index under five kinds of concentration is compared. It can be concluded that the discrimination capability of DRI fluorescence index is the strongest compared with that of the other three indexes. Moreover, comparing the selected key wavelength by three band selection methods, the wavelength feature selected by Fisher's method is the best and the optimal wavelength is 410–430 nm. The optimum fivefold cross-validation results obtained from the training set and testing set are 93.1 and 87.2, respectively, through two narrowband ratios. This narrowband spectrum can be transplanted easily to the online aflatoxin detection production line. The results of this study have positive significance for online sorting and rapid detection of aflatoxin in agricultural products.

REFERENCES

Ataş M, Yardimci Y, Temizel A. A new approach to aflatoxin detection in chili pepper by machine vision. *Computers and Electronics in Agriculture*, 2012, 87: 129–141.

Chang C C, Lin C J. LIBSVM: A library for support vector machines, 2011. http://www.csie.ntu.edu.tw/-cjlin/libsvm.

Chu X, Wang W, Zhang L D, et al. Hyperspectral optimum wavelengths and Fisher discrimination analysis to distinguish different concentrations of aflatoxin on corn kernel surface. *Spectroscopy and Spectral Analysis*, 2014, 34(7): 1811–1815.

Gao J-Z, Ren M-W, Yang J-Y. A practical and fast method for non-uniform illumination correction. *Journal of Image and Graphics*, 2002, 7(6): 548–552.

Ministry of Health, PRC. *GB 2761-2011 The National Food Safety Standards of mycotoxins in food limited* (In Chinese). Beijing: China Standards Press, 2011.

Samiappan S, Bruce L M, Yao H, et al. Support vector machines classification of fluorescence hyperspectral image for detection of aflatoxin in corn kernels. *5th IEEE Workshop on Hyperspectral Image & Signal Processing: Evolution in Remote Sensing*, 2013, pp. 1–4.

Teena M, Manickavasagan A, Mothershaw A, et al. Potential of machine vision techniques for detecting fecal and microbial contamination of food products: A review. *Food and Bioprocess Technology*, 2013, 6(7): 1621–1634.

USDA. 2002. *Aflatoxin Handbook*. Washington, DC: USDA Grain Inspection, Packers, and Stockyards Administration. www.gipsa.usda.gov/publications/fgis/handbooks/afl_insphb.html.

Wassenaar H J, Chen W, Cheng J, Sudjianto A. Enhancing discrete choice demand modeling for decision-based design. *ASME 2003 Design Engineering Technical Conferences and Computers and Information in Engineering Conference*, Chicago, IL. DETC'03, 2003.

Yao H, Hruska Z, Kincaid R, et al. Hyperspectral image classification and development of fluorescence index for single corn kernels infected with *Aspergillus flavus*. *Transactions of the ASABE*, 2013, 56(5): 1977–1988.

Yao H, Hruska Z, Kincaid R D, et al. Method and detection system for detection of aflatoxin in corn with fluorescence spectra. U.S. Patent Application 12/807,673, 2010.

Zhang N, Liu W, Wang W, et al. Image processing method of corn kernels mildew infection and aflatoxin levels. *Journal of the Chinese Cereals and Oils Association*, 2014, 29(2): 82–88.

3 Application-Driven Key Wavelength Mining Method for Aflatoxin Detection Using Hyperspectral Data

The first step of developing aflatoxin intelligent sorter is to determine the key wavelengths for aflatoxin detection. In order to find more accurate wavelengths, in this chapter, three kinds of sensor system are built separately: the first sensor is hyperspectrometer by ASD spectrometer, the second is the multispectral camera system based on liquid crystal tunable filter (LCTF), and the third one is the hyperspectral camera based on grating spectrometer module (GSM). Under 365 nm UV LED illumination, using these three systems, three hyperspectral datasets of peanut samples, 45, 41, and 73, before and after aflatoxin contamination have been collected separately. In order to select the best key wavelengths, four feature selection methods (Fisher, SPA, BestFist, and Ranker) and four classifier models (KNN, SVM, BP-ANN, Random Forest) were analyzed and compared. Using all selected wavelengths based on different datasets, a weighted voting method was proposed and ten key wavelengths (440, 380, 410, 460, 420, 370, 450, 490, 700, and 600 nm) were selected. Based on the best model (Random Forest), the best integrated average recognition rate is 94.5%. And then, using these key wavelengths and the best classification model, a new design system for aflatoxin sorter base on a polygon mirror was proposed. Although the structure of this system is simple, its detection accuracy is high, which can be applied to online sorting of aflatoxin detection.

3.1 INTRODUCTION

Aflatoxin is a highly toxic carcinogen, and its toxicity is about 68 times as that of arsenic. It is the strongest chemical carcinogen found so far and is the biggest cause of liver cancer. It exists widely in peanut, corn, and their products. It has been limited strictly in the National Standards of China and the United States, and the standard limits are, respectively, 20 and 100 ppb in food grade and feed grade (Ministry of Health, PRC, 2011; USDA, 2002). Aflatoxin is a metabolic product of some of the parasitic fungi in *Aspergillus* spp. It has the characteristic of fluorescence emission, and aflatoxin B1 and B2 emit blue fluorescence and aflatoxin G1 and G2 green fluorescence under ultraviolet light. At present, aflatoxin is detected mainly by biochemical methods, including thin-layer chromatography, high-performance liquid chromatography, micro-column method, and enzyme-linked immunosorbent assay (Bao et al., 2001). Although the detection accuracy is high, the detection means and instruments are complex and the detection speed is slow. So, these methods cannot be used in online detection.

In recent years, spectrum detection has become a new method for aflatoxin detection, which has drawn widespread attention (Teena et al., 2013). Hyperspectral wave, multispectral images, and hyperspectral images are three main tools for nondestructive exploration of agricultural products and food. The research team of Yao mainly focused on aflatoxin contamination in maize by hyperspectral data. They studied the spectra of single grain through artificial aflatoxin cultivation using the maximum likelihood estimation and binary encoding method based on hyperspectral image. And they found that the best fluorescence reflection peak of aflatoxin is between 437 and 537 nm, the recognition rate is 87%–88%, and the fluorescence intensity decreased with the increase in the content of aflatoxin, and there was a phenomenon of peak shift (Yao et al., 2013). And based on these two points, they applied for a US patent (Yao et al., 2010). But the device was still made manually instead of automatically. Ataş et al. (2012) used the multispectral imaging analysis technique to study aflatoxin contamination in the cayenne pepper bought from the market under two light sources (halogen lamp and ultraviolet lamp) and proposed a feature selection method based on gray-level histogram quantization and neural network weight optimization. The wavelength feature that plays a key role in detection was found and the recognition rate of leave-one-out method under the 420 nm reached 85%. In China, the research group of Wang Wei studied the aflatoxin detection problem using the hyperspectral images of the United State Agricultural Research Department (USDA, ARS) and pointed out that the aflatoxin detection rate of naturally contaminated grains was 87.5% through common CCD imaging. They optimized five high spectral wavelengths from 700 to 1,100 nm, and the prediction accuracy rate of artificial aflatoxin contamination concentration in five highest spectral wavelengths reached 88.3% (Zhang et al., 2014).

The above literature shows that different instruments and algorithms will affect the detection effect. Which one is the best? In this chapter, in order to find the accurate key wavelengths for aflatoxin detection, three kinds of datasets are collected from hyperspectrometer, multispectral camera, and hyperspectral camera separately. By analyzing the performance of four feature selection methods (Chu et al., 2014) and four classifier models (Samiappan et al., 2013), we attempt to propose a new weighted voting feature selection method to find the key feature wavelengths and combine with the best classification model to design a new type of peanut aflatoxin sorter.

3.2 MATERIALS

3.2.1 Experiment Materials

The peanuts were bought from a local market of Shandong, China, in 2016. The variety of peanut is four red. From these peanuts, kernels that have a good appearance and almost the same in size had been selected as samples (about 1 g). We divided these samples into three groups (45, 41, and 73 samples) randomly for different experiments.

We prepared aflatoxin solution using aflatoxin B1 (from China national strain center) and acetonitrile (from market) in the ratio of 1:50. And then, we dropped the aflatoxin solution onto the peanut surface using a pipette (one drop of the solution is about 1 μL) and injected one to five drops randomly into different samples. Different numbers of drops on one kernel will result in different contents of aflatoxin. If one drop, the kernel contains about 20 ppb; if two drops, the kernel contains about 40 pph; and so on. This means that aflatoxin content in one kernel is between 20 and 100 ppb ($\mu g\ kg^{-1}$), and sample kernels of each various levels are about 1/5.

Here, one drop is 1 μL, the aflatoxin in the solution is 1/50, so pure aflatoxin in one drop is $1\mu L*(1/50) = 1*10^{-6}\ L*(1/50) = 20*10^{-9}L$. According to the specific gravity of solution, it is 1; the weight of aflatoxin is $20*10^{-9}kg$; one peanut kernel is about $1g = 1*10^{-3}kg$; and one drop on one kernel means there is $(20*10^{-9}kg)/(1*10^{-3}kg) = (20*10^{-6}kg)/(1*10^{-3}kg) = 20\ mg\ kg^{-1}$ (ppb).

Different numbers of drops represent different levels of aflatoxin contamination. The purpose of dropping different drops is to investigate the generalization ability of the models used in this chapter. We just divided the test results into two categories: contaminated and uncontaminated.

3.2.2 System Integration

The instruments in our experiment include a field spectrometer (FieldSpec 3, ASD, USA), an LCTF (VariSpec Cri, USA), a GSM (V8E, Dualix Spectal Imaging, China), a SCMOS CCD (ORCA-Flash2.8 C11440-10C, Hamamatsu, Japan), a 365 nm ultraviolet LED (HRC-UV-4A, Shenzhen Weihailixin Technology Development Co., Ltd., China), and a 355 nm laser (Triple frequency Nd:YAG laser, the eleventh Institute of China Electronics Technology Group) and an electric displacement platform (HG-32-TA, Beijing UH Guano Dakota Co., Ltd., China). Besides these, some optical lens, a mercury lamp, and a Teflon panel were also used in our experiments. Based on these instruments and materials, three experimental systems have been established.

The first system including the ASD spectrometer and the 365 nm LED illumination is illustrated in the black dotted box of Figure 3.1a. The spectral range of the spectrometer is 350–2,500 nm and 1 nm spectral bandwidth. The spectrum of each sample of 45 peanuts has been collected ten times before and after aflatoxin contamination; thus, there are 900 spectra totally. It should be noted that the instrument needs whiteboard calibration before obtaining the data.

The second hardware of the image acquisition system is composed of a SCMOS CCD camera and a VariSpec liquid LCTF, as illustrated in the dotted box of Figure 3.1a. Using this system, multispectral images of 41 samples ranging from 400 to 720 nm were collected before and after aflatoxin contamination. The images of the first 20 samples have been collected under 27 bands (400, 405, 410, 415, 420, 425, 430, 435, 440, 445, 450, 460, 470, 480, 490, 500, 520, 540, 560, 580, 600, 620, 640, 660, 680, 700, and 720 nm) under UV 365 nm LED illumination, and the images of the last 21 sample were collected under 29 bands (above 27 band plus 530, 535 nm) and two illumination sources (UV 365 nm LED illumination and UV 355 nm laser sources).

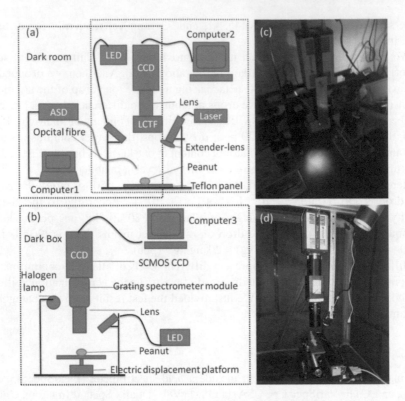

FIGURE 3.1   The general overview of three data collection systems: (a) general overview of ASD spectral system and multispectral images system, (b) general overview of hyperspectral images system, (c) physical picture of ASD spectral system and multispectral images system, and (d) physical picture of hyperspectral images system.

For the convenience of the experiment, the first and second systems have been put together and the physical picture is illustrated in Figure 3.1c.

The third system contains a GSM (292–1,200 nm, about 1 nm spectral bandwidth), a SCMOS CCD, and a UV 365 nm LED illumination, as illustrated in Figure 3.1b and d. The electric displacement platform was used for moving the sample step by step to photo the picture one by one. An additional halogen lamp was used for illumination when we replace samples. In total, there were 73 peanut samples imaged before and after contamination.

3.3   METHODS

3.3.1   Data Preprocessing

Hyperspectral wave data selected by system 1 is illustrated in Figure 3.2a, in which each line represents the mean of one peanut of ten times. The light line indicates contaminated peanuts, and the dark line indicates peanuts that are not contaminated.

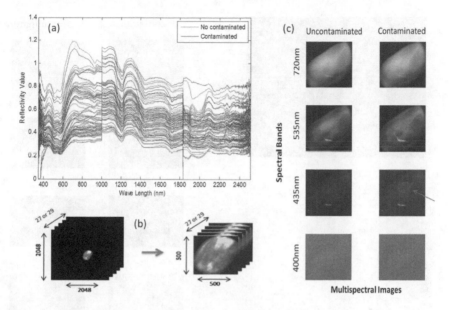

FIGURE 3.2 Hyperspectral data and preprocessing of multispectral images: (a) hyperspectral wave data selected by system 1, (b) extract the ROI of system 2, and (c) four different band images of one sample before and after contamination of system 3.

The original multispectral images collected by system 2 include not only the image of peanut but also the background. So, we should extract the region of interest (ROI) of peanut from background image first (Figure 3.2b). Through this step, the original images (2,048*2,048*27 or 29 bands) were converted to the ROI images (500*500*27 or 29 bands). Figure 3.2b illustrates four different band images of one sample before and after contamination. In the contaminated image of 435 nm, the area of aflatoxin can be seen faintly (position of arrow).

For hyperspectral images collected by system 3, it should be noted that the first step is to determine the wavelength of different band images using a mercury lamp. The process of calibration is shown in Figure 3.3. First, we illuminated the lens (under the GSM) using a mercury lamp directly, and then, we rotated the GSM to make the light line thinner and brighter, while maintaining them as horizontal as possible (Figure 3.3a). Third, we took the image by SCMOS CCD and calculated brightness value of each row (Figure 3.3b). According to the special emission wavelengths of the mercury lamp (404, 435, 546, and 578 nm), it is easy to determine the position of each wavelength in the image which is the 237, 304, 534, and 600 rows. Then, we can calculate other wavelengths (ranged from 292 to 850 nm) of each rows in the image based on the linear relationship of wavelengths and rows.

In system 3, for each movement (about 0.02 cm) of the electric displacement platform, the camera takes an image (1,200*1,600 pixels). For one peanut sample (about 1.5 cm), it should move 750 steps and take 750 images. For one sample, the size of the hyperspectral cube is 1,200 rows * 1,600 columns * 750 pages. In order to reduce the data size, we first extracted the peanut area and compressed the image

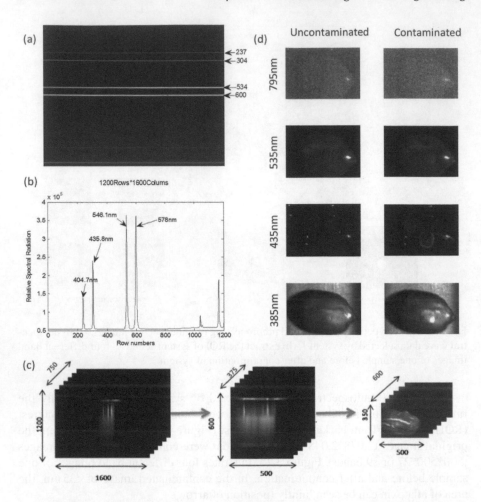

FIGURE 3.3 Wavelength calibration and preprocessing of the hyperspectral images: (a) emission images of mercury lamp, (b) special emission wavelengths of mercury lamp, (c) preprocessing of hyperspectral images, and (d) four different wavelengths.

into 600 rows * 500 columns * 375 pages, and then rotated the image cube to 375 rows * 500 columns * 600 pages. Now, each page of images represents specific wavelength image of peanuts. The above image processing procedure is shown in Figure 3.3c. Figure 3.3d illustrates four different band images of one sample before and after contamination. From the image of 435 nm, the area of aflatoxin contamination can be seen more obviously.

3.3.2 RECOGNITION METHODS

For the hyperspectral wave data of system 1, first, we preprocess this data using spectral smoothing method based on the Savitzky–Golay algorithm (Schafer, 2011). And then, we select the key wavelengths using four methods (Fisher, SAP,

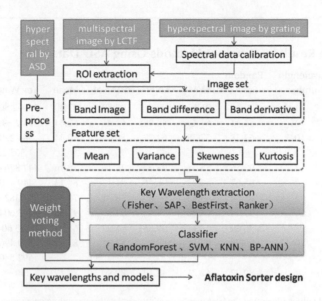

FIGURE 3.4 Flowchart of our method.

BestFirst, Ranker) from the whole spectral wavelengths. Finally, we use four classifiers (Random Forest, SVM, KNN, BP-ANN) as the recognition models. In order to improve the performance of the examination of aflatoxin, fivefold cross validation (Wiens et al., 2008) was used for classification (Figure 3.4).

For multispectral and hyperspectral images acquired by systems 2 and 3, after extracting the ROI, we transform each image of individual band to other two kinds of images. One is the band difference image of consecutive spectral band image. The other is the derivative image of consecutive spectral band image. These two kinds of images are described in expressions 3.1 and 3.2:

$$DI_k(x,y) = I_{k+1}(x,y) - I_k(x,y) \tag{3.1}$$

$$DE_k(x,y) = \left[I_{k+1}(x,y) - I_k(x,y) \right]./ I_k(x,y) \tag{3.2}$$

Here, $I_k(x,y)$ corresponds to pixel gray level at image point (x,y) for the kth spectral band. And then, for each band of three kinds of images set (I_k, DI_k, DE_k), we calculated the mean, variance, skewness, and kurtosis, respectively, and the four statistics were used as feature set for key wavelength extraction and classifier which was the same as the steps illustrated in hyperspectral wave data of system 1.

3.4 RESULTS

3.4.1 Hyperspectral Wave by ASD

Table 3.1 illustrates the key wavelength selection and classification result using four kinds of wavelength selection method and four classifiers based on 900 spectra

TABLE 3.1

Recognition Result by Different Methods Using ASD Data

Methods	Wavelength Number	Random Forest	SVM	KNN	BP-ANN	Mean	Key Wavelength (nm)
Null	2,451	100.0	91.11	100.0	98.89	97.50	350–2,500
SPA	3	54.44	54.44	48.89	62.44	62.44	393 2,396 1,721 1,205
	10	67.78	57.78	62.22	80.44	80.44	384 2,480 1,570 1,363 2,481 2,094
Fisher	3	93.56	55.78	93.67	78.16	78.16	447 376 461 449 455
	10	95.67	54.22	96.56	84.13	84.13	449 460 463 380 464
BestFirst	3	95.22	63.33	92.89	78.31	**78.31**	377 382 496 511 589
	10	99.44	64.00	99.44	97.58	**97.58**	593 604 642 748 750
Ranker	3	98.89	55.11	**99.56**	66.27	66.27	1,173 604 1,174 605
	10	99.78	64.11	**100.0**	81.91	81.91	589 496 590 599 592 540
Mean	–	**89.42**	62.21	88.14	80.90	80.17	

Note: The bold values illustrated the best recognition rate in the corresponding categories.

acquired by system 1. In order to investigate the identification performance of different models for less key wavelengths, we use two wavelengths set, one includes ten key wavelengths, and the other includes only top three of the ten wavelengths.

It is found that using the original feature set, all recognition rates of four models are high (above 90%). When the number of features reduces, the recognition performance deteriorates, and different feature selection methods affect the performance of classification. Among the four feature selection methods, BestFirst method has the best average recognition rate (78.31%, 97.58%), and, among the four classifier models, Random Forest has the best recognition rate (89.42%). Among them, the combination of Ranker and KNN is the most prominent. Using ten key wavelengths (1,115, 546, 1,116, 547, 531, 438, 532, 441, 534, and 482 nm) can achieve the recognition rate of 100%. By only using three top wavelengths (1,115, 546, and 1,116 nm), the recognition rate can reach 99.56%. This result almost approximates to the result of using all wavelengths (2,451 wavelengths). However, the recognition rate of KNN on one dataset does not show if the model has the best generalization ability, so different datasets should be evaluated.

3.4.2 Multispectral Images by Liquid Crystal Tunable Filter (HTLF)

The image dataset collected by system 2 includes three sub-datasets. One is the first 20 samples under 365 nm LED illumination. The second is the last 21 samples under 365 nm LED illumination, and the third is the last 21 samples under 355 nm laser illuminations. The first sub-dataset has 27 band images, and the latter two have 29 band images. After we calculated the four statistics (mean, variance, skewness, and kurtosis) for each three kinds of band images (including I_k, DI_k, and DE_k), the feature size of the first sub-dataset is 316 (27*4+26*4+26*4) and the last two is 340 (29*4+28*4+28*4).

TABLE 3.2
Recognition Result by Different Methods Using Multispectral Image Data

Methods	Number	Random Forest	SVM	KNN	BP-ANN	Mean	Key Wavelength (nm)
Null	316	98.75	82.50	97.50	91.22	92.49	400–720
SPA	3	80.00	72.50	81.25	70.00	75.94	410 430 500 540 425
	10	96.25	75.00	72.50	93.17	84.23	450 490 700 620 580
Fisher	3	91.25	82.50	83.75	92.44	87.49	400 720 700 680 660
	10	97.50	82.50	96.25	98.54	93.70	640 620 600 580 560
BestFirst	3	95.00	66.25	92.5	92.68	86.61	410 415 420 425 445
	10	100.0	60	91.25	98.78	87.51	490 500 520 420 425
Ranker	3	98.75	87.5	100.0	94.88	95.28	440 445 450 460 440
	10	98.75	91.25	100.0	95.12	96.28	410 405 400 700 680
Mean	–	95.14	77.78	90.56	91.87	88.84	

Table 3.2 illustrates the mean classification results and the key wavelength selection using four kinds of wavelength selection method and four classifiers on the three sub-datasets. It is found that using the original feature set, the recognition rates of the four models are high (mean 92.49%). When the number of features reduces, the recognition performance deteriorates. Among the four feature selection methods, the Ranker method has the best recognition rate (average 95.28%, 96.28%) on different number of features, and among the four classifier models, Random Forest has the best recognition rate (mean 95.14%). Among them, the Ranker-KNN is most prominent (100%). Overall, the average recognition rate of multispectral data is 88.84% which is higher than that of ASD spectrometer (80.17%). The reason may be that the multispectral data consider the spatial distribution characteristics of aflatoxin, while the ASD considers only the spectral characteristics.

3.4.3 Hyperspectral Images by GSM

When the spectral resolution and image resolution were further increased, we used the hyperspectral imaging data to further investigate the detection performance of aflatoxin (system 3). Following the same way of Section 3.2, for each kind of band image (including I_k, DI_k, and DE_k), we collected four features (mean, variance, skewness, kurtosis). So for each band, we collected 12 features totally. There are 600 wavelengths and 146 samples (before and after contamination of 73 peanut samples), so the size of feature set is 12*600*146 = 1,051,200. Here, the features of 1–44 samples are depicted in Figure 3.5a, in which the column denotes 600 bands and the row represents the sequence number of features, and each kernel has 24 columns, the 1–12 column represents 12 features before contamination, and the 13–24 indicates the characteristics after. There are 1,056 features of the 44 kernels before and after contamination. Given that different kind of features or band images have different contribution to recognition rate, in order to find the identification ability of each feature and band, we calculated the recognition rate of each band image on four

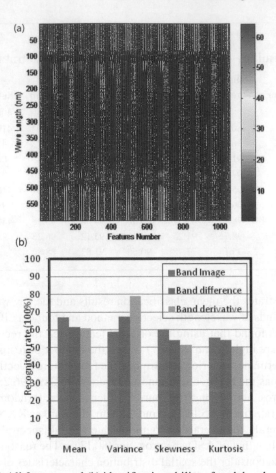

FIGURE 3.5   (a) All features and (b) identification ability of each band.

statistics and find that the average rate of variance of band derivative image is the highest, which is about 78% (Figure 3.5b). So, the following results are all based on the variance of band derivative images.

Then, using four feature selection methods and four classifier models as mentioned above, we selected ten key wavelengths. And based on the tenth and top three wavelengths, we calculated the recognition rate for 146 samples. It can be seen that the SPA-ANN model gets the best identification rate of 92.89% on ten key wavelengths, and the Fisher-ANN model achieves the best identification rate of 92.89% on three key wavelengths. Meanwhile, the ANN shows the best performance (mean 90.85%). But overall, the recognition rate of hyperspectral images (87.39%) is slightly lower than that of multispectral images (88.84%), as shown in Table 3.3. The reason may be that, when the characteristic data volume is too large, the data redundancy is aggravated, and the model recognition rate decreases. But in general, the rate of recognition is relatively stable, the majority of which are more than 85%.

TABLE 3.3

Recognition Rates by Different Methods Using Hyperspectral Image Data

Methods	Number	Random Forest	SVM	KNN	BP-ANN	Mean	Key Wavelength (nm)
Null	316	84.82	87.95	82.14	89.47	86.01	**292–850**
SPA	3	84.82	87.95	82.14	89.47	86.09	702 726 684 676 664
	10	85.27	87.95	89.29	**92.89**	88.85	560 563 529 490 487
Fisher	3	83.48	87.50	85.71	**90.75**	86.86	370 454 426 741 445
	10	84.38	87.95	87.05	92.29	87.92	447 428 443 456 455
BestFirst	3	85.27	87.95	83.93	89.81	86.74	372 384 417 419 421
	10	88.84	87.95	85.71	90.67	88.29	525 548 723 736 745
Ranker	3	85.71	87.95	84.82	90.07	87.14	387 388 384 389 386
	10	83.48	87.95	85.27	92.21	87.23	419 757 416 760 759
Mean		85.12	87.90	85.12	**90.85**	87.39	

Note: The bold values illustrated the best recognition rate in the corresponding categories.

3.5 DISCUSSION

3.5.1 KEY WAVELENGTHS SELECTED BY WEIGHTED VOTING

In Section 3.3, different datasets are used to get different results. Even for the same dataset, the key wavelengths selected by different features selection methods are different. Furthermore, even based on the same key wavelengths, the best identification model is different too. So here come some questions: Which dataset is authoritative? Which key wavelength is more important? Which model is the best to be used in sorter design? Different documents draw different conclusions (Teena et al., 2013; Qiao et al., 2017). There are good questions.

In order to find more precise key wavelengths, it should be considered comprehensively of different features selection methods on different datasets. Here, we propose a weighted voting method that can be described as follows:

Weighted voting value (each band) = sum of vote * weight coefficient

= sum of vote * (average recognition rate * score of wavelength order).

Here, only the UV–visible–near-infrared region, from 350 to 1,200 nm, covers most of the spectral regions of the three systems. We set one voting band region for every 10 nm, and there are 84 band regions. If the key wavelengths selected are in this region, the band region gains one vote. To make it more reasonable, we multiply the sum of votes by a weight coefficient. Here, the weight is defined as the multiplication of the average recognition rate of four models and score of wavelength order. For each wavelength selection method, we have selected ten key wavelengths. If the wavelength is number 1, the score of wavelength is 1. The second is 0.9, the third is 0.8, ..., and the tenth is 0.1. For all three datasets in Section 3.3, using four kind of selection methods, we already have selected 120 key wavelengths. Using this method, we calculated the weighted voting value of each band region. The result is illustrated in Figure 3.6.

The selected top ten key bands are 440, 380, 410, 460, 420, 370, 450, 490, 700, and 600 nm (Figure 3.6). Extraordinarily, the first voted band is 440 nm which has been marked in light color, and it has gained 10 votes, weight voting value of which is 598.1. This wavelength is fit to the excitation wavelength of aflatoxin (which is 435.8 nm) under 365 nm illumination which has been proved by the theory (Bao et al., 2001).

In order to find the best model using these top ten wavelengths, we investigate four classification models based on the above three datasets (Table 3.4). It is found that Random Forest is the best, and the average recognition rate is 94.25%. Meanwhile, the generalization ability of Random Forest is preferable. In addition, BP-ANN is also good, with an average recognition rate of 92%, slightly worse than the Random Forest model. Besides using ten key wavelengths, we also use the top three wavelengths for recognition and it also obtained more high recognition rate. Here, it must be noted that the multispectral image only has eight key wavelengths because the band of multispectral system does not include 370 and 380 nm.

Table 3.4 lists the overall performance of the selection of ten key wavelengths. The recognition rate is generally higher and more stable for most models, and the best model recognition rate is up to 95%. This indicates that the proposed method is insensitive to the acquisition system and reflects the essential properties of aflatoxin detection.

The measuring devices are three kinds of devices used in spectral analysis. Previous studies usually use only one of them, so the results are not the same. Calibration of different instruments is a big problem. In this chapter, the results of three kinds of instruments are compared, and the weighted voting method is proposed for comprehensive analysis. The selected key wavelengths may have better credibility. Although the measuring range of different instruments is different, the number of nonoverlapping bands is low in the voting process, which demonstrates that it has little effect on final results.

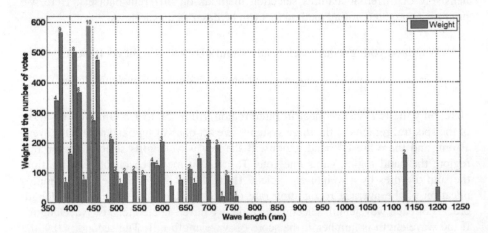

FIGURE 3.6   Key wavelength selected by weight voting.

TABLE 3.4

Recognition Result by Different Methods Using 3 Dataset above

Methods	Number	Random Forest	SVM	KNN	BP-ANN	Key Wavelength (nm)
Hyperspectral	3	92.33	57.33	91.78	86.49	440 380 410 460 420
wave	10	99.44	61.56	99.00	96.58	370 450 490 700 600
Multispectral	3	98.75	50.00	48.75	95.12	440 410 460 420 450
image	8	100.0	50.00	50.00	92.93	490 600 700
Hyperspectral	3	87.05	87.95	85.71	89.73	440 380 410 460 420
image	10	87.95	87.95	84.38	91.44	370 450 490 700 600
Mean	–	**94.25**	65.80	76.60	92.05	–

Note: The bold values illustrated the best recognition rate in the corresponding categories.

3.5.2 SORTER DESIGN

Using the key wavelengths and the best classification model obtained in Section 4.1, we can design peanut aflatoxin intelligent sorter. Here, we give a design framework, which is illustrated in Figure 3.7.

In order to detect two sides of each peanut, the sorter included two optics detection systems which is composed of 365 nm UV LED, polygon mirror (Li et al., 2017), silicon photomultiplier (SiPM) (Grodzicka-Kobylka et al., 2017), and line-scan sensor. Each polygon mirror has three or ten reflection panels, and each panel installs a key wavelength reflection mirror. When peanut kernels fly into detection system, it will be illuminated by UV LED. The reflex light will expose to the polygon mirror. With the rotation of the mirror, the light of different wavelengths will be reflected to SiMP, and the SiPM system will transform the light signal to an electric signal and transmit it to the computer. The classification model we have established above will be installed on computer and will identify which peanut is healthy or contaminated.

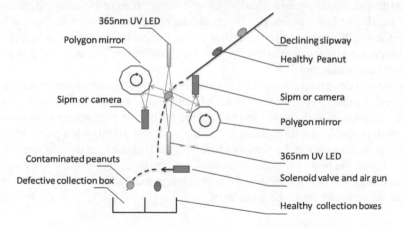

FIGURE 3.7 Design framework of peanut aflatoxin intelligent sorter.

The sorter process is as follows: peanut kernels drop from the slipway, and then they pass through optical detection systems. If a peanut is estimated to be contaminated by the computer, the detection system will control the solenoid valve and an air gun will blow it away to the collection box for defective peanuts. If it is a healthy peanut, detection system will do nothing, and the peanut will be dropped into the collection box for healthy peanuts.

Here, the number of reflecting panels is set to 10. Of course, it can be more or less. When the number increases, it may be more accurate. But it will increase the calculation amount of the computer. Conversely, if the number decreases, it may be more inaccurate. But because of the small amount of data, the computer will have high processing speed and the efficiency of the machine will be high. Particularly, if the number is three, three key wavelengths can be used in design sorter using common color line-scan camera and three key wavelengths optics filters. However, it is important to note that different cameras have different light-sensitive properties. You need to think carefully when choosing a camera. To further improve the ability to detect weak fluorescence, an electron-multiplying CCD (EMCCD) camera is a good choice (Yao et al., 2013).

Using ten key wavelengths, it can get higher accuracy. Of course, only using eight is also acceptable. In this chapter, we only showed the special situation of three and ten. The fewer the bands are selected, the lower the recognition rate will be. At the same time, when the key wavelengths increase, the amount of data to be processed will also increase. It is necessary to find a balance between accuracy and speed.

3.6 CONCLUSION

Different datasets collected by different sensors may have different characters. Even if using the same method, key wavelengths extracted may be different. Theoretically, the same materials will certainly have the same spectra. Spectral is the real reaction of materials. However, different collecting equipment and different environment may make the spectral change. In addition, using different feature extracting method may collect different key wavebands. The modeling complexity causes humans difficult to decide which wavelength or classifier is more important. Especially in engineering design, we always want to use the optimal classification model in order to get the highest recognition rate.

In order to design an aflatoxin intelligent sorter, three datasets have been collected by different sensor systems. Four kinds of feature selection methods and four kinds of classification models were used to test and verify the performance for aflatoxin detection. Based on the three datasets and their results, we proposed a new weighted voting method to decide the final key wavelengths and then we determined the best model. Using the best optimization method, we design a practicable aflatoxin sorter. The key band, the best model, and the sorter design method may have positive meaning for future sorter developing industry.

REFERENCES

Ataş M, Yardimci Y, Temizel A. A new approach to aflatoxin detection in chili pepper by machine vision. *Computers and Electronics in Agriculture*, 2012, 87: 129–141.

Bao L, Zhang P, Lei Z, et al. Comparison between the post-column derivatization with bromine by HPLC and the fluorometric analysis for determination of aflatoxins in peanut. *Science of Inspection and Quarantine*, 2001, 9(11): 18–20.

Chu X, Wang W, Zhang L D, et al. Hyperspectral optimum wavelengths and fisher discrimination analysis to distinguish different concentrations of aflatoxin on corn kernel surface. *Spectroscopy and Spectral Analysis*, 2014, 34(7): 1811–1815.

Grodzicka-Kobylka M, Szczesniak T, Moszyński M. Comparison of SensL and Hamamatsu 4 × 4 channel SiPM arrays in gamma spectrometry with scintillators. *Nuclear Instruments and Methods in Physics Research Section A: Accelerators, Spectrometers, Detectors and Associated Equipment*, 2017, 856: 53–64.

Li Y X, Gautam V, Brüstle A, et al. Flexible polygon-mirror based laser scanning microscope platform for multiphoton in-vivo imaging. *Journal of Biophotonics*, 2017, 10(11): 1526–1537.

Ministry of Health, PRC. *GB 2761-2011 The National Food Safety Standards of mycotoxins in food limited* (In Chinese). Beijing: China Standards Press, 2011.

Qiao X, Jiang J, Qi X, et al. Utilization of spectral-spatial characteristics in shortwave infrared hyperspectral images to classify and identify fungi-contaminated peanuts. *Food Chemistry*, 2017, 220: 393–399.

Samiappan S, Bruce L M, Yao H, et al. Support vector machines classification of fluorescence hyperspectral image for detection of aflatoxin in corn kernels. *Meeting Proceedings*, 2013: 1–4.

Schafer R W. What is a Savitzky-Golay filter? *IEEE Signal Processing Magazine*, 2011, 28(4): 111–117.

Teena M, Manickavasagan A, Mothershaw A, et al. Potential of machine vision techniques for detecting fecal and microbial contamination of food products: A review. *Food and Bioprocess Technology*, 2013, 6(7): 1621–1634.

USDA. *Aflatoxin Handbook*. Washington, DC: USDA Grain Inspection, Packers, and Stockyards Administration, 2002. www.gipsa.usda.gov/publications/fgis/handbooks/afl_insphb.html.

Wiens T S, Dale B C, Boyce M S, et al. Three way k-fold cross-validation of resource selection functions. *Ecological Modelling*, 2008, 212(3): 244–255.

Yao H, Hruska Z, Kincaid R, et al. Hyperspectral image classification and development of fluorescence index for single corn kernels infected with Aspergillus flavus. *Transactions of the ASABE*, 2013, 56(5): 1977–1988.

Yao H, Hruska Z, Kincaid R D, et al. Method and detection system for detection of aflatoxin in corn with fluorescence spectra. U.S. Patent Application 12/807,673, 2010.

Zhang N, Liu W, Wang W, et al. Image processing method of corn kernels mildew infection and aflatoxin levels. *Journal of the Chinese Cereals and Oils Association*, 2014, 29(2): 82–88.

4 Deep Learning-Based Aflatoxin Detection of Hyperspectral Data

Aflatoxin is a kind of virulent and strong carcinogenic substance, and it is found widely in peanuts, maize, and other agricultural products. In order to detect aflatoxin in peanut, we proposed a spectral–spatial (SS) combinative deep learning (DL) method based on hyperspectral images. First, we construct a push-broom hyperspectral imaging system using a grating spectrometer module (GSM), a SCMOS CCD, and an electric displacement platform. Using this system, we collect the hyperspectral images of 73 peanut samples before and after contamination by aflatoxin under 365 nm ultraviolet (UV) illumination. Then, based on Convolutional Neural Network (CNN), we propose the SS combinative DL method. In this method, we select principal components analysis (PCA) images or key band images as the spatial data and calculate the mean of each spectral band (including the derivative of consecutive spectral bands) as the spectral data. It is found that, because the combination of spectral and spatial data takes into account both spectral and spatial information, it obviously improves the performance of recognition rate. The total fivefold cross-validation recognition rate is about 95%, which is higher than k-nearest neighbor (KNN), support vector machine (SVM), and back propagation artificial neural network (BP-ANN). The combination of PCA image and spectral image is better than the combination of key band image and spectral image. Our results suggest that the method we proposed is a rapid, precise, and nondestructive technique for the detection of toxic metabolites in peanuts, and it could be an alternative to manual techniques.

4.1 INTRODUCTION

Aflatoxin – a natural toxic substance produced by fungi and molds found in certain food – is an ongoing threat to the global population. The toxin can be found in a number of food types, but it is most commonly found in nuts – particularly peanuts. It has been limited strictly in the National Standards of P.R. China (Ministry of Health, PRC, 2011) and the United States (USDA, 2002), and the limits of foodstuff and feedstuff were 20 and 100 ppb ($\mu g\ kg^{-1}$), respectively. Research shows that the chronic intake of food infected with aflatoxin can increase the risk of dying from liver cancer to 66%. It is classified as a group 1 carcinogenic agent and is estimated to be 68 times more deadly than arsenic. Aflatoxin generally is produced in damp environments such as storehouses that are not kept below a certain humidity standard, and the level of this toxic substance increases fast when the growth conditions are comfortable for the fungi that produces it, infecting other food and products. As it is colorless and tasteless, it can be extremely difficult to identify; what is more, the substance can withstand temperature up to 280°C, meaning that it cannot be destroyed or removed by cooking in a high temperature. As a result, many traditional methods such as high-performance liquid chromatography (HPLC) have proven ineffective at detecting or removing aflatoxin (Bao et al., 2001).

On account of aflatoxin could be recognized by superficial distribution, in recent years, hyperspectral data has become a new method for aflatoxin detection, which has received widespread attention (Teena et al., 2013), particularly in *maize* (Yao et al., 2013; Hruska et al., 2013; Zhang et al., 2014; Chu et al., 2014; Yao et al., 2010; Samiappan et al., 2013). Among them, Hruska et al. (2013) studied the fluorescence imaging spectroscopy for comparing spectra from corn ears naturally and artificially infected with aflatoxin-producing fungus. Two image bands (437 and 537 nm) were used to calculate a ratio of the fluorescence, and a threshold of 0.5 was used to segment the contaminated ("hot" pixel) and good pixels. They found that the spectral analysis based on contaminated "hot" pixel classification showed a distinct spectral shift/separation between contaminated and clean ears with fluorescence peaks at 501 and 478 nm. All of inoculated and naturally infected ears had fluorescence peaks at 501 nm that differed from uninfected corn ears. Besides *maize*, Ataş et al. (2012) studied the hyperspectral imaging analysis technique of aflatoxin contamination in the cayenne pepper under two light sources (halogen lamp and UV lamp). They found that characteristic wavelength plays a key role in detection, and the recognition rate reached 85% under the 420 nm using leave-one-out and PCA–SVM method. Qiao et al. (2017) utilized the short-wave infrared hyperspectral images to classify and identify aflatoxin-contaminated peanuts. Using nonparametric weighted feature extraction (NWFE) and SVM as the classifier, the result shows that the classification accuracy of peanut pixels is above 90% in validate images. There is few literature in the study of peanut aflatoxin using hyperspectral data. The above researches were based on key wavelengths or key characters, and then they used a model like SVM (Chang and Lin, 2011) to classify which kernel has been contaminated. Whether the number of key wavelength or characters are enough, there have no criterion. We can only use the recognition rate to estimate.

In recent years, DL becomes a superexcellent classifier in machine learning field, which was widely used in image classification including CNN, deep belief net (DBN), stacked autoencoder (SAE), and many models. As a classical algorithm, Lecun et al. proposed CNN in 1998, but the effect is not good enough. Hinton et al. (2006) proposed a DBN, and then DL (including CNN) made breakthrough progress. Recently, DL method was successfully used in hyperspectral data. Chen et al. (2014) used SAE and logistic regression for hyperspectral data and found it is better than linear SVM and radial basis function–support vector machine (RBF–SVM). The correct recognition rate is about 94% using three datasets. Until 2016, Go program AlphaGo of DeepMind company won the South Korea world go champion Li Shishi. As an explosive news, DL began to attract worldwide attention. In their work, Silver et al. (2016) achieved success with the aggregate score of 4:1 using value networks and policy networks based on CNN and strategy of Monte Carlo tree search (MCTS). However, to the best of our knowledge, no reports of using DL to rapidly detect quality and safety of food have been found.

In this chapter, we propose an SS–DL method based on PCA images, key band images, and spectral features as the import data, and we use CNN as the DL model for aflatoxin detection based on hyperspectral images. The objectives of this study are to find: (1) whether spectral information or spatial information is more important in aflatoxin-detected problems? In other words, whether using a few specific key bands selected by feature extraction can be good enough? (2) How DL method worked when using in aflatoxin-detected problems, and whether it is good enough or better than traditional recognition models?

4.2 MATERIALS AND METHODS

4.2.1 PEANUT SAMPLE PREPARATION

One breed of healthy peanut (Luhua 11#) was bought from a local market, purchased in 2016, origin from Shandong, China. Of them, 73 peanut kernels had been selected as samples which have good appearance and same size roughly. We configured aflatoxin solution with aflatoxin B1 (from China national strain center) and acetonitrile (from the market) in the ratio of 1:50. And then, we dropped the aflatoxin solution onto the peanut surface using a pipette (one drop of solution is about 1 μL). We divided these samples into three groups randomly and inoculated one, three, or five drops of solution, respectively. This means that aflatoxin content in one peanut (about 1 g) is about 20, 60, or 100 ppb (μg kg^{-1}). Although we divide the samples into three groups according the degree of contamination, we do not distinguish them in the following test, because these three groups are all contaminated kernels, which are to investigate the generalization ability of our models.

4.2.2 HYPERSPECTRAL IMAGING SYSTEM AND IMAGE ACQUISITION

The system contains a GSM (V8E, Dualix Spectral Imaging, China, 292–865 nm about 1 nm spectral bandwidth), an SCMOS CCD (ORCA-Flash2.8 C11440-10C, Hamamatsu, Japan), an electric displacement platform (HG-32-TA, Beijing UH Guano Dakota Co., Ltd., China), and a UV 365 nm LED illumination source (HRC-UV-4A, Shenzhen Weihai lixin Technology Development Co., Ltd., China). Using these instruments, we establish a push-broom hyperspectral imaging system, and Figure 4.1a and b are the real photo of experimental device and its structural representation, respectively.

The electric displacement platform was used for moving the sample step by step (0.02 cm), in order to take photos one by one. By each moving, it outputs a 15 V electrical level which drives the CCD to take pictures. The exposure time of CCD was set to 0.1 s, and the image resolution was set to 1,200*1,600 pixels (about 2 MB). This picture is only one line image of the sample with a wavelength range of 292–850 nm. One peanut kernel is about 1.5 cm, and the electric displacement platform had to move 750 times. So, for one sample, we can acquire 750 images.

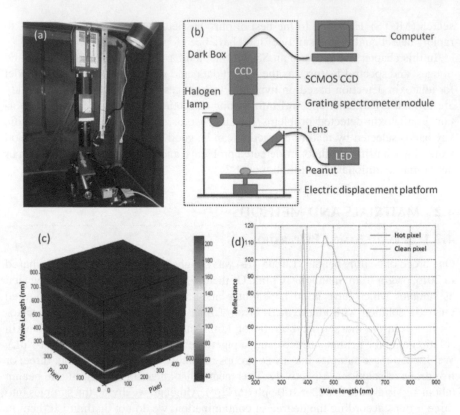

FIGURE 4.1 System integration and selected data: (a) experimental equipment, (b) the structure of the experimental equipment, (c) hyperspectral image cube, and (d) mean spectral of hot and clean pixels.

In the experiment, we put the uncontaminated peanuts in line on the Teflon plate, and then shot their images one by one. Subsequently, we dropped aflatoxin solution on the peanut surface using pipettes and captured the image once again. That is to say, we collect the hyperspectral images twice (before and after contamination). There were 73 samples, so we take photos 146 times totally. The raw data size (146 hyperspectral images cube) is about 220 GB.

4.2.3 HYPERSPECTRAL IMAGING PREPROCESSING

For each image (1,200*1,600 pixels), the 1,200 lines correspond to the wavelengths (292–865 nm) and the 1,600 columns refer to the pixels of one line about 0.02 cm on peanut surface. Using 750 images of one peanut, we synthesized a hyperspectral image cube whose size is 1,200 pixels*1,600 pixels*750 pages. We rotated the cube to 1,600*750*1,200, and now the 1,200 pages (each image is 1,600*750) represented wavelength band image.

In order to compress the size of the cube, we only select peanut area about 500*375 pixels and the wave band have been compressed to quartern (1/4). So, for each peanut, the cube is turned to 500*375*300. Figure 4.1c illustrates the cube matrix of one of the samples, and all the hyperspectral cubes are the raw datasets used in this study.

Using the two hyperspectral images before and after contamination of the peanut kernel, we can calculate the area of aflatoxin contamination. First, we obtain the region of kernel through image binarization based on the visible wave bands (such as 600 nm band image). Then, for each pixel on the peanut area, we calculate the spectral angle using a method proposed by Du et al. (2004) (Spectral Information Divergence Spectral Angle, SID_SA). If the spectral angular difference is greater than 0.03, we label these pixels as "hot pixel" or contaminated pixels on contaminated peanut; otherwise, they are labeled "clean," "good," or uncontaminated pixels. The number of all the hot and clean pixels is recorded, and then the mean wave line is plotted as shown in Figure 4.1d, in which the mean spectrum curve of hot pixels is light line and clean pixels of one peanut are dark line. It was obvious that there is a peak at 450 nm approximately, and the reflectance of contaminated pixels is higher than uncontaminated 450–550 nm.

4.2.4 CNN OF DEEP LEARNING METHOD

The algorithm flow for aflatoxin detection of our method is illustrated in Figure 4.2a. In the first step, we collect the hyperspectral images by the instrument in Figure 4.1a and b, and then calibrate and extract the region of interest (ROI) of the peanut or contaminated pixels. In the second step, we use three kinds of characters that are calculated as import dataset. One is the mean value of each band and the derivative of consecutive spectral bands. The second is the key band image (like 432 nm) by Fisher, and the third is the PCA images. In the third step, we use traditional recognition models (such as KNN, SVM, and BP-ANN) and modern recognition model

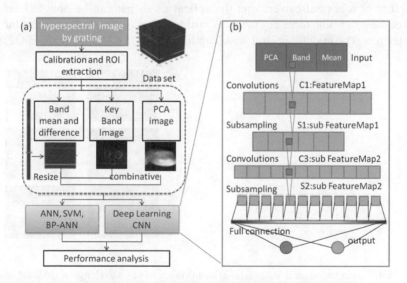

FIGURE 4.2 Flowcharts of (a) all methods and (b) deep learning methods.

(CNN DL) as the classifier to detect aflatoxin, and then we analyze the performance of different datasets (including our SS combinative features) and different models.

Here, we use CNN model as the representative of DL method. The network framework of the CNN model is illustrated in Figure 4.2b. There are five hidden layers: the first layer is input layer, the second is convolution layer, the third is subsampling layer, the fourth is convolution layer, and the fifth is subsampling layer. The output layer is full connection layer. In our problem, the input is the dataset calculated from hyperspectral images and the output is 0 or 1, which presents whether the peanut is contaminated or not.

4.3 RESULTS AND DISCUSSION

4.3.1 AFLATOXIN DETECTION USING KEY BAND IMAGES

In traditional recognition model, the first step is feature selection [using Fisher, successive projections algorithm (SPA), PCA, etc.], and the second step is using classifier (KNN, BP-ANN, SVM, etc.) for recognition. Ordinarily, in agricultural products' detection, a camera or a spectrograph must be used. The camera can provide spatial information, but it only has one (gray) or three (RGB) bands; hence, spectral information is insufficient. Otherwise, spectrograph can be considered the spectral information, but it cannot distinguish spatial distribution, so it is just suitable for homogeneous objects. In this problem of the detection of peanut aflatoxin, not only the spatial information but also the spectral information should be considered. In addition, in order to detect some specific ingredients, such as aflatoxin, some narrowband should be needed (Yao et al., 2011). As the spectrum becomes wider, the recognition rate goes down.

Here, using the Fisher method (Malina, 1987), we select some key bands that are illustrated in Figure 4.3a for aflatoxin classification. Considering the imaging capability of a normal camera and the optical filter that can be bought from the market, we determine three key bands in visible region (in the dotted box) as the key wavelengths. The middle spectral wavelengths are 432(15), 514(23), and 600(22) nm,

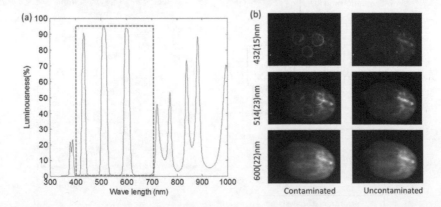

FIGURE 4.3 Key bands and images. (a) Transmittance of key wavelengths; (b) Band images of three key bands.

and here, the numbers within parentheses is bandwidths. The three band images of one peanut which have been contaminated or not are illustrated in Figure 4.3b, and the contaminated image is in the right column and healthy image is in the left. It is obvious that, in 432 nm image, the region contaminated by aflatoxin can be seen by naked eye; here, the three white circles mean the sample is inoculated by three drops of aflatoxin solution.

Using these three key bands, we examine the recognition performance under different resolution ratios from 1*1 to 224*224, and the results are provided in Table 4.1 using a fivefold cross-validation method (Wiens et al., 2008) by different models. They are KNN algorithm, SVM algorithm, error BP-ANN algorithm, and CNN DL algorithm. It is observed that the recognition capability from top to bottom is DL, SVM, ANN, and KNN, and the recognition performance of all models has been improved with the increase in image resolution. Nevertheless, we can see that, when the image resolution is highest, the recognition rate of DL model is only about 80%. It isn't good enough. Most of the recognition models do not achieve good recognition results when the image resolution is low. The reason may be that the spatial resolution provided is not sufficient. But it could not be too large – too large resolution means longer processing time. Another reason may be that three key wavelengths are not enough. To improve the recognition efficiency, it needs to add more spectral information to input data and maintain a certain spatial resolution.

4.3.2 AFLATOXIN DETECTION USING SPECTRAL AND IMAGES

First, we establish three datasets as input. The first dataset is the mean reflectance of each individual band image (Figure 4.3a) and the derivative of consecutive spectral band (Figure 4.3b), the second dataset is the key band images, and the third is the PCA images. The first three components (the cumulative variance is above 99%) can representative vast majority of information of the hyperspectral images (Figure 4.3c). It is obvious that the first PCA image implicate the largest information values of the samples. In Figure 4.3d, we illustrate the first three PCA images.

In order to use the spectral information and the spatial information of each wavelength sufficiently, we establish two compound datasets which combined spectra

TABLE 4.1
Recognition Result by Three Different Bands (432, 514, 600 nm)

Image Resolution	KNN	BP-ANN	SVM	DL
1*1*3	55.00	55.83	59.17	73.66
7*7*3	60.83	58.33	64.17	74.16
14*14*3	62.50	53.33	64.50	75.83
28*28*3	66.67	59.17	65.83	76.67
56*56*3	63.33	60.00	73.50	77.50
78*78*3	64.17	66.67	73.66	78.00
112*112*3	69.16	70.00	78.00	81.00
224*224*3	74.16	75.83	81.00	82.50

with image. We used the image (no matter be the key image or the PCA image) and spectral (the mean/derivative spectral band), these are called SS-combined features. Since the performance of the identified model is independent of the order of the identified data, we put the spectral data directly behind the image data. To unify the input data size, we adjusted the image and the spectrum to 28*28 pixels.

Using these five kinds of datasets as input data, we test the performance of DL. In order to verify the superiority of the algorithm, we also compare other models like KNN, SVM, and ANN. To identify it, we use the fivefold cross-validation method, that is, randomly select 4/5 samples as the training set, and the remaining fifth as a test set, and then loop five times. The output of these models is two classes which are healthy and contaminated (0 or 1). The results are illustrated in Table 4.1 (Figure 4.4).

Overall, among the first three traditional models, KNN, RBF–SVM, and BP-ANN, the recognition rate of SVM is the best and KNN is the worst. Furthermore, the DL is apparently higher than the three. It is obvious that the method of DL can be used to improve the detection rate of aflatoxin kernels.

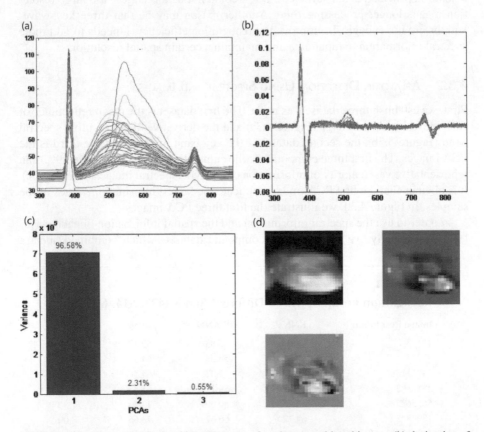

FIGURE 4.4 Datasets for classifier: (a) mean of each spectral band image, (b) derivative of consecutive spectral band image, (c) cumulative variance of the first three PCA, and (d) first three PCA images.

TABLE 4.2
Recognition Result by Different Methods

Dataset	Spectral and Image	KNN	BP-ANN	SVM	DL
Spectral	400–720 nm	62.14	86.43	88.63	88.83
Key band	432 nm	60.83	65.00	67.17	76.67
PCA image	PCA1	87.50	88.33	88.83	92.50
Spectral and key band	SS	71.66	76.67	77.50	87.50
Spectral and PCA	SS	92.50	95.00	95.67	95.83

Among the first three individual datasets, PCA image feature is better than spectral features, and the key band image performs worst. The reason may be that the PCA image considers not only the image spatial resolution but also the spectral resolution of the image. Meanwhile, the SS-combined features are better than individual features. Therein, the combination of PCA image and the spectral feature is the best.

When we use the SS dataset (especially the combined of spectral and PCA) and DL model constructed in this chapter, the overall recognition rate is about 90%, which is much higher than the detection method of traditional cameras (only use three band images) in agricultural products detection (the results are provided in Table 4.2, and it is about 70%–80%).

This conclusion indicates that, for aflatoxin detection of peanut, the spectral information can play a more important role than spatial image information; meanwhile, the spatial distribution of the image should also be considered in the detection. Taking advantage of the current DL as a new method for detection of agricultural products, it is more effective than traditional methods.

However, it should be pointed out that the dataset constructed in this chapter is not big enough, and it does not take the full advantages of the algorithm of DL, and further we will expand the dataset, which will be the next step of our work.

4.4 CONCLUSION

Aflatoxin is a highly toxic, carcinogenic substance, which widely exists in peanut and its products. In this chapter, we first built our own set of hyperspectral imaging systems using grating module, SCOMS camera, and electric displacement platform. And then, we acquired 146 hyperspectral image cubes of 73 peanut samples before and after contaminated with aflatoxin.

In order to test the aflatoxin contamination, we proposed a SS combined with the CNN method (SSCNN). By comparing different feature datasets and different identification models, we found that the combination of spectral and spatial feature set and the DL method is the best. Here, we used the PCA and spectral feature. The recognition rate reached 95.83% by CNN DL method which was significantly higher than other method. This conclusion indicated that the method of this chapter basically satisfies the engineering application of aflatoxin detection. On account of it, spectral and image data should be collected at the same time, and in the process of application, it is necessary to fully consider the collection efficiency of hyperspectral images, especially those used in online sorting.

Now, we can answer the questions we proposed in the end of Section 4.1: (1) spectral information is more important than spatial information and using a few specific key bands selected by feature extraction can be good enough for aflatoxin-detected problems. (2) DL method can work well, and it is better than traditional recognition models.

REFERENCES

Ataş M, Yardimci Y, Temizel A. A new approach to aflatoxin detection in chili pepper by machine vision. *Computers and Electronics in Agriculture*, 2012, 87: 129–141.

Bao L, Zhang P, Lei Z, et al. Comparison between the post-column derivatization with bromine by HPLC and the fluorometric analysis for determination of aflatoxins in peanut. *Science of Inspection and Quarantine*, 2001, 11(5): 18–20.

Chang C C, Lin C J. LIBSVM: a library for support vector machines, 2011. http://www.csie. ntu.edu.tw/~cjlin/libsvm.

Chen Y, Lin Z, Zhao X, et al. Deep learning-based classification of hyperspectral data. *IEEE Journal of Selected Topics in Applied Earth Observations and Remote Sensing*, 2014, 7(6): 2094–2107.

Chu X, Wang W, Zhang L D, et al. Hyperspectral optimum wavelengths and fisher discrimination analysis to distinguish different concentrations of aflatoxin on corn kernel surface. *Spectroscopy and Spectral Analysis*, 2014, 34(7): 1811–1815.

Du Y, Chang C, Ren H, et al. New hyperspectral discrimination measure for spectral characterization. *Optical Engineering*, 2004, 43(8): 1777–1786.

Hinton G E, Osindero S, Teh Y W. A fast learning algorithm for deep belief nets. *Neural Computation*, 2006, 18(7): 1527–1554.

Hruska Z, Yao H, Kincaid R, et al. Fluorescence Imaging Spectroscopy (FIS) for comparing spectra from corn ears naturally and artificially infected with aflatoxin producing fungus. *Journal of Food Science*, 2013, 78(8): 1312–1320.

LeCun Y, Bottou L, Bengio Y, et al. Gradient-based learning applied to document recognition. *Proceedings of the IEEE*, 1998, 86(11): 2278–2324.

Malina W. Some multiclass Fisher feature selection algorithms and their comparison with Karhunen-Loeve algorithms. *Pattern Recognition Letters*, 1987, 6(5): 279–285.

Ministry of Health, PRC. *GB 2761-2011 The National Food Safety Standards of mycotoxins in food limited* (In Chinese). Beijing: China Standards Press, 2011.

Qiao X, Jiang J, Qi X, et al. Utilization of spectral-spatial characteristics in shortwave infrared hyperspectral images to classify and identify fungi-contaminated peanuts. *Food Chemistry*, 2017, 220: 393–399.

Samiappan S, Bruce L M, Yao H, et al. Support vector machines classification of fluorescence hyperspectral image for detection of aflatoxin in corn kernels. *Meeting Proceedings*, 2013: 1–4.

Silver D, Huang A, Maddison C J, et al. Mastering the game of Go with deep neural networks and tree search. *Nature*, 2016, 529(7587): 484–489.

Teena M, Manickavasagan A, Mothershaw A, et al. Potential of machine vision techniques for detecting fecal and microbial contamination of food products: A review. *Food and Bioprocess Technology*, 2013, 6(7): 1621–1634.

USDA. 2002. *Aflatoxin Handbook*. Washington, DC: USDA Grain Inspection, Packers, and Stockyards Administration. www.gipsa.usda.gov/publications/fgis/handbooks/afl_ins-phb.html.

Wiens T S, Dale B C, Boyce M S, et al. Three way k-fold cross-validation of resource selection functions. *Ecological Modelling*, 2008, 212(3): 244–255.

Yao H, Hruska Z, Kincaid R, et al. Development of narrow-band fluorescence index for the detection of aflatoxin contaminated corn. *Proceedings of the SPIE-The International Society for Optical Engineering*, 2011, 8027: 80270D.

Yao H, Hruska Z, Kincaid R, et al. Hyperspectral image classification and development of fluorescence index for single corn kernels infected with Aspergillus flavus. *Transactions of the ASABE*, 2013, 56(5): 1977–1988.

Yao H, Hruska Z, Kincaid R D, et al. Method and detection system for detection of aflatoxin in corn with fluorescence spectra. U.S. Patent Application 12/807,673, 2010.

Zhang N, Liu W, Wang W, et al. Image processing method of corn kernels mildew infection and aflatoxin levels. *Journal of the Chinese Cereals and Oils Association*, 2014, 29(2): 82–88.

5 Pixel-Level Aflatoxin Detection Based on Deep Learning and Hyperspectral Imaging

Aflatoxin is a kind of virulent and strong carcinogenic substance, and it is found widely in peanuts, maize, and other agricultural products. In order to detect aflatoxin in peanuts, we first built a hyperspectral imagery system using grating module, an SCOMS camera, and an electric displacement platform, and acquired 146 hyperspectral image cubes of 73 peanut samples before and after contaminated with aflatoxin. Then, we proposed a reshaped image method of pixel spectral for the convolutional neural network (CNN) method. By studying random selection data sets and comparing them with different identification models, we found that (1) reshape image established by the pixel-level spectral is good enough for aflatoxin-detected problems, and the overall recognition rate reached above 95% on pixel level. (2) Deep learning (DL) method is worked well, and it is better than traditional identification models, not only on the pixel level but also on the kernel identification. The recognition rate of above 90% on the kernel level can quickly be used in sorting machine's design.

5.1 INTRODUCTION

Aflatoxin – a toxic natural substance produced by fungi and molds found in certain food – is an ongoing threat to the global population. The toxin can be found in a number of food items but is most commonly found in nuts – particularly peanuts. Its level in food items has been limited strictly in the National Standards of P.R. China (Ministry of Health, PRC, 2011) and the United States (USDA, 2002), and the standard limits are 20 and 100 ppb (μg kg^{-1}) in the food grade and the feed grade, respectively. Research shows that the chronic intake of food infected with aflatoxin can increase the risk of dying from liver cancer by up to 66%. It is classified as a group 1 carcinogenic agent and is estimated to be 68 times deadlier than arsenic. Aflatoxin generally grows in damp environments such as storehouses that are not kept below a certain humidity level and can quickly spread once it develops, infecting other food and products. As it is colorless and tasteless, it can be extremely difficult to identify, while the substance can also withstand temperature up to 280°C, meaning it cannot be destroyed or removed by cooking or boiling. As a result, many traditional methods such as high-performance liquid chromatography (HPLC) have proven ineffective at detecting or removing aflatoxin (Bao et al., 2001).

On account of aflatoxin has the characteristic of surface or superficial distribution, in recent years, hyperspectral data have become a new method for aflatoxin detection, which has received widespread attention (Teena et al., 2013; Yao et al., 2013; Hruska et al., 2013; Zhang et al., 2014; Chu et al., 2014; Yao et al., 2010; Samiappan et al., 2013). Ataş et al. (2012) studied the hyperspectral imaging analysis technique of aflatoxin contamination in the cayenne pepper under two light sources (halogen lamp and ultraviolet lamp). They found that feature wavelength plays a key role in detection and the recognition rate reaches 85% under the 420 nm using leave-one-out and principal component analysis–support vector machine (PCA–SVM) method. Qiao et al. (2017) utilized the short-wave infrared hyperspectral images to classify and identify aflatoxin-contaminated peanuts. Using nonparametric weighted feature extraction (NWFE) and SVM as the classifier, the result shows that the classification accuracy of peanut pixels is above 90% in validation images. Hruska et al. (2013) studied the fluorescence imaging spectroscopy for comparing spectra from corn ears naturally and artificially infected with aflatoxin-producing fungus. Two image bands (437 and 537 nm) were used to calculate a ratio of the fluorescence, and a threshold of 0.5 was used to segment the contaminated ("hot" pixel) and good pixels. They found that the spectral analysis based on contaminated "hot" pixel classification showed a distinct spectral shift separation between contaminated and clean ears with fluorescence peaks at 501 and 478 nm, respectively. All inoculated and naturally infected control ears had fluorescence peaks at 501 nm that differed from uninfected corn ears. The above-mentioned researches were based on key wavelengths or key characters, and then they used a model like SVM (Chang and Lin, 2011; Zhongzhi and Limiao, 2018) to differentiate contaminated kernels from uncontaminated kernels.

In recent years, DL has become a superexcellent classifier in machine learning field, which is widely used in image classification which includes CNN, deep belief net (DBN), stacked autoencoder (SAE), and many models. In 1998, Lecun et al. proposed CNN as a classical algorithm, but the results of using this algorithm is not satisfactory. Hinton et al. (2006) proposed a DBN, which is a DL (including CNN) that made breakthrough progress. In 2016, Go program AlphaGo of DeepMind company beat the South Korea world Go champion Li Shishi. Hence, DL has attracted worldwide attention. In their article, Silver et al. (2016) use value networks and policy networks based on CNN and strategy of Monte Carlo tree search (MCTS). The aggregate score is 4:1. Recently, DL method was successfully used in hyperspectral data. Chen et al. (2014) used SAE and logistic regression for hyperspectral data and found it is better than linear SVM, radial basis function–support vector machine (RBF—SVM). On three datasets, the correct recognition rate is about 94%.

In this chapter, we propose a reshape image of pixel spectral as the input of the DL and use CNN as the DL model for aflatoxin detection based on hyperspectral images. The objectives of this study were to find (1) whether the pixel-level detection of aflatoxin contamination using reshape image can work well for DL method. In other words, whether the reshape method is satisfactory. (2) How DL method works when used in aflatoxin-detected problems, and whether it is good enough or better than traditional recognition models.

5.2 MATERIALS AND METHODS

5.2.1 Peanut Sample Preparation

Identification of peanuts containing aflatoxin in nature is difficult in a nondestructive way and requires professional knowledge, because it takes long time for *Aspergillus flavus* to grow in nature. Hence, most of the studies use aflatoxin solution to contaminate the healthy seeds. In the further experiment, contaminated peanuts in field could be used.

One breed of healthy peanut (Luhua 11#) was bought from a local market, which was purchased in 2016, provided by Shandong, China. Out of them, 73 peanut kernels had been selected as samples having good appearance and uniform size. The ratio of acetonitrile (from market) to aflatoxin B1 (from China national strain center) is 1:50. And then, we dropped the aflatoxin solution onto the peanut surface using a pipette (one drop of solution is about 1 μL, and it contains 0.02 μg aflatoxin). We divided these samples into three groups randomly, and then inoculated one, three, or five drops, respectively. One peanut seed weighs about 1 g (0.001 kg), which means that aflatoxin content in one peanut is about 20, 60, or 100 ppb (μg kg^{-1}). Although we divide the samples into three groups according to the degree of contamination, we do not distinguish them in the following test, because these three groups are all contamination kernels, which are used to investigate the generalization ability of our models.

5.2.2 Hyperspectral Imaging System and Image Acquisition

The system contains a grating spectrometer module (GSM, V8E, Dualix Spectral Imaging, CHN, 292–865 nm about 1 nm spectral bandwidth), an SCMOS CCD (ORCA-Flash2.8 C11440-10C, Hamamatsu, Japan), an electric displacement platform (HG-32-TA, Beijing UH Guano Dakota Co., Ltd., China), and a UV 365 nm LED illumination source (HRC-UV-4A, Shenzhen Weihai Lixin Technology Development Co., Ltd., China). Using these instruments, we establish a push-broom hyperspectral imaging system. Figure 5.1a and b is the real photo of experimental device and its structural representation.

The electric displacement platform was used for moving the sample step by step (0.002 cm), in order to take photos one by one. In each move, it outputs a 15 V electrical level which drives the CCD to take pictures. The exposure time of CCD was set to 0.1 s, and the image resolution was set to 1,200*1,600 pixels (about 2 MB). This picture is only one line image of the sample with a wavelength range of 292–850 nm. One peanut kernel is about 1.5 cm, and the electric displacement platform had to move 750 times. So, for one sample, we can acquire 750 images.

For each image (1,200*1,600 pixels), the 1,200 lines (total is) correspond to the wavelengths (292–865 nm) and the 1,600 columns refer to the pixels of one line about 0.002 cm on peanut surface. Using the 750 images of one peanut, we synthesized a hyperspectral image cube whose size is 1,200 pixels*1,600 pixels*750 pages. We rotated the cube to 1,600*750*1,200, and now the 1,200 pages (each image is 1,600*750) represented wavelength band image.

In order to compress the size of the cube, we only select peanut area about 500*375 pixels and compress the wave band to quarter (1/4). So, for each peanut, the cube was

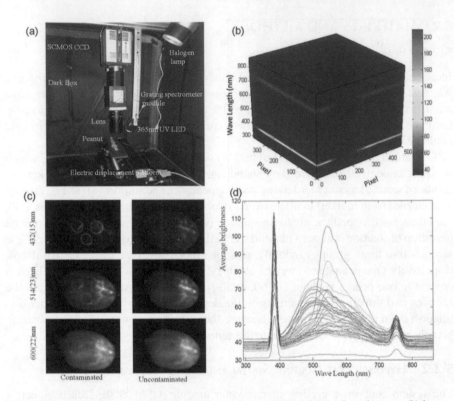

FIGURE 5.1 System integration and selected data: (a) experimental equipment, (b) hyperspectral image cube, (c) three band of a peanut before and after contamination, and (d) mean spectral of all peanuts.

turned to 500*375*300. Figure 5.1b illustrates the cube matrix of one of the samples, and all the hyperspectral cubes are the raw datasets that will be used in this study. Figure 5.1c shows images of three key bands. Because the reflectance is calculated with the brightness of the reflection, in Figure 5.1d, the y-axis represents the mean pixel brightness (according to the gray image, we set the pure white brightness value to 255 and the pure black to 0, and then add up all the pixel values in one picture to get the average brightness), and every single line represents one peanut.

In the experiment, we put the uncontaminated peanuts in line on the Teflon plate, and then took their images one by one. Next, we dropped aflatoxin solution onto the peanut surface using pipettes and captured their images once again. That is to say, we collected the hyperspectral images twice, before and after contamination. There were 73 samples (numbered 1–73), so we take photos 146 times totally. The size of raw data (146 hyperspectral images cube) is about 220 GB.

5.2.3 Hyperspectral Imaging Preprocessing

Comparing the two hyperspectral images taken before and after contamination of the peanut kernel, we can calculate the area of aflatoxin contamination. First, we obtained

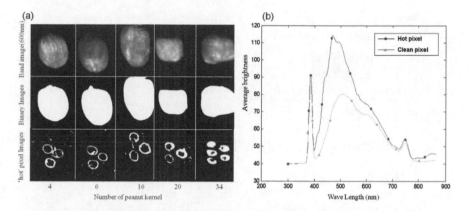

FIGURE 5.2 Image preprocessing: (a) "hot pixel" finding process and (b) mean spectral of "hot" and "clean" pixel.

the region of kernel through image binarization based on the visible band image (such as 600 nm). Then, for each pixel on the peanut area, we calculated the difference between the images before and after contamination, as a matter of experience, the images band is set to 430 nm. If the gray value ratio was greater than 1.2, we labeled these pixels as "hot" pixel (or contaminated pixels) on binary image; otherwise, they were labeled as "clean" (or uncontaminated) pixels. As illustrated in Figure 5.2a, there are the image preprocessing of some peanuts (numbered 4, 6, 16, 20, and 34). The first-line subimages are band images at 600 nm, the second-line subimages are the binary images of peanut's area, and the third are the aflatoxin-contaminated area. We summed up all of the "hot" and "clean" pixels' brightness. Figure 5.2b illustrates the mean waveform of these two kinds of pixels: the point line represents hot pixels and the triangle line represents clean pixels. We use average brightness value to represent the reflectance as shown in Figure 5.1d. It is obvious that there is a peak at 450 nm approximately, and the reflectance of contaminated pixels is higher than uncontaminated at 450–550 nm. For each hot and clean pixel, we got out each spectral as the sample for identification, and meanwhile, we recorded the position of each pixel.

5.2.4 CNN of Deep Learning Method

Here, CNN was used as the representative method of DL. The framework of CNN model is illustrated in Figure 5.3a. There are five hidden layers: the first layer is input layer, the second is convolution layer, the third is subsampling layer, the fourth is convolution layer, and the fifth is subsampling layer. The output layer is full connection layer. In our problem, the input is the spectral of each pixel of the peanut and the output is 0 or 1, which presents whether the pixel is been contaminated.

The flow chart of the method used for aflatoxin detection is illustrated in Figure 5.3b. In the first step, we collected the hyperspectral images by the instrument shown in Figure 5.1a, and then calibrated and extracted the spectral of "hot" or "clean" pixels in the region of interest (ROI) of the peanut. In the second step, we reshape each spectral to a gray image as the input for CNN. In the third step, we used

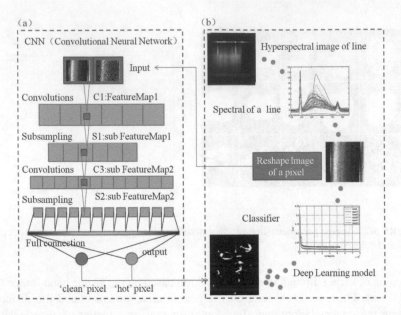

FIGURE 5.3 Deep learning method. (a) The structure of neural network and (b) the method flow chart of this paper.

different classifier to identify which pixel is really "hot" and labeled its position on the binary image. Here, these classifiers include traditional models (such as k-nearest neighbor (KNN), SVM, and back propagation artificial neural network (BP-ANN)) and modern models (such as CNN DL). Finally, we analyzed the performance of different datasets (for different kernels) and different models.

5.3 RESULTS AND DISCUSSION

5.3.1 Deep Learning for Training Kernels

Since the photo environment has not been changed before and after contamination by aflatoxin, the changes of pixels should only be caused by aflatoxin contamination. So, the changed pixels can be labeled "hot" pixels or contaminated pixels by aflatoxin, and the other pixels on the peanut skin can be labeled "clean" pixels. For each pixel, we extract the spectral wave from hyperspectral image cube and divide them into two sample classes according to the label of "hot" or "clean". Then for each pixel spectral (one-dimensional vector), we reshape it to a reshape image (two-dimensional matrix) as the input for CNN model. Here, the size of reshape image was set to 28*28 pixels.

The data of all hyperspectral images are huge (about 220 GB); in order to reduce the modeling time of DL and, at the same time, to establish the model using as less simple as possible (when the recognition rate is equal, the less simple the model use, the more practical the model is), here we selected six peanuts randomly as the dataset for training model which were numbered 9, 17, 30, 35, 41, and 42 and only contaminated kernels were used.

TABLE 5.1
Recognition Error by DL Model for Self-Sample

Sample Number	Pixel No.		Number of Epochs				
	Hot	Clean	1	2	5	50	100
9	7,473	74,833	0.0885	0.0885	0.0424	0.0315	0.0324
17	15,729	83,156	0.0711	0.0405	0.0259	0.0198	0.0220
30	12,561	36,611	0.2552	0.1968	0.0382	0.0081	0.0223
35	21,041	50,815	0.1220	0.0868	0.0330	0.0169	0.0189
41	18,446	44,251	0.1312	0.1222	0.0804	0.0601	0.0578
42	19,297	48,434	0.0152	0.0140	0.0114	0.0149	0.0110
Mean	–	–	0.1139	0.0915	0.0386	0.0252	0.0274

This CNN model is trained to discriminate whether a single pixel has been "polluted" or not, and the six peanuts we selected had 432,647 pixels in total. The recognition errors while using the CNN model of DL are listed in Table 5.1. As illustrated in the table, the total number of "hot" pixel sample is 94,547 and the "clean" number is 338,100. There are a large amount of data which are wide enough for classification task of DL. The sample size of general international handwritten dataset which was named MNIST (Liang et al., 2017) for DL is only 70,000. Each peanut sample basically satisfied the requirements of the amount of data size. Here, for each classification task, we only used the same kernel. For all pixels of the same peanut, we used 4/5 pixel spectral as training data and the rest 1/5 as testing data.

As illustrated in Table 5.1, with the increase of epochs, the error of DL declines, but the time consumed by the training model sharply increases, which is also illustrated in Figure 5.4a,b. As the number of epochs is changed from 1 to 100, the mean error declines from 11.39% to 2.74%, the time consumption increases from about 150′ to 15,000′. We should note that although training the model takes a long time, once the model is well trained, it takes short time to identify the pixels.

5.3.2 DEEP LEARNING FOR TESTING KERNELS

In Section 5.3.1, the pixels used in training set and testing set are the same for all kernel samples (from six peanuts). Although the recognition rate is high, we don't know its accuracy in identifying other peanuts. The performance of the model is determined by the model's ability to identify other kernels. Here, we selected five peanuts randomly as the testing kernels which were numbered 4, 6, 16, 20, and 34. First, we used the six kernel samples in Section 5.3.1 to establish the recognition model. In order to verify the scale of the model and reduce the time consumption of training models, 1/6 pixels of each kernel (whether it's "hot" or "clean") was selected randomly and combined as the input data of the training model. And then, the DL training program was run. Once the model is trained, this model can be used to identify the pixels on the testing kernels whether it has been contaminated or not. The result is illustrated in Table 5.2.

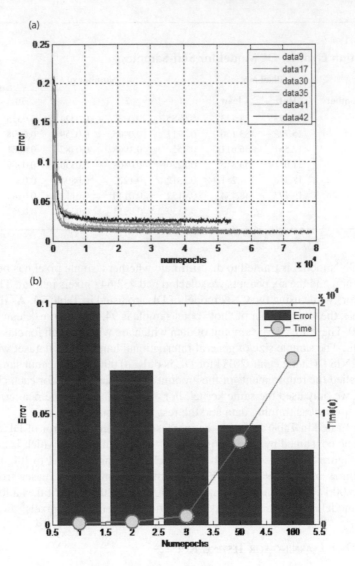

FIGURE 5.4 Error and time consumption with the epochs changed. (a) Neural network iteration process and (b) iteration error and elapsed time.

TABLE 5.2
Recognition Result by DL Model for Other Sample

	4 All	6 All	16 All	20 All	34 All	Mean
Pixel number	74,500	80,500	83,500	52,250	54,750	69,100
Elapsed time	99.5442	99.6846	106.7203	43.6803	48.5787	79.6416
Error	0.0132	0.0200	0.0687	0.0398	0.0412	0.03658

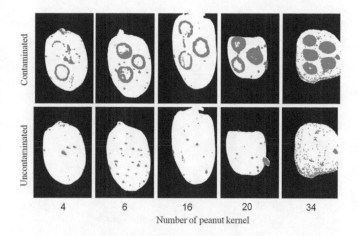

FIGURE 5.5 Pixel marker of aflatoxin for testing peanut samples.

From Table 5.2, it is easy to see that the average error of recognition rate is 3.66%, slightly higher than the recognition rate of the training set which is 2.74%. The mean elapsed time is 79.6416′ for 69,100 pixels, which is about 0.001′ for a pixel. This result is only about contaminated kernels.

Figure 5.5 illustrates DL-marked aflatoxin pixels (dark points), including before (bottom-line subimages) and after (top-line subimages) contamination of kernels. Since the aflatoxin solution was dropped by a pipette, and due to the effect of surface tension on liquid in natural air drying process, contaminated kernels presented a circular distribution, which conformed to people's experience. Uncontaminated kernels have little "hot" pixel, and it is randomly distributed. It may be caused by model recognition errors, or the kernel itself contains aflatoxin, because when selecting the kernel samples, they were not virus-free.

Figure 5.6 illustrates the percentage of area of predicted "hot" pixel and area of whole kernel, for all the training and testing kernels (totally 22 kernels). In this, the left bar shows the result of before contamination and the right bar shows the result of after contamination by aflatoxin. If we set the threshold to 0.05, it is easy to classify these kernels into two class (except No. 4 kernel). The total recognition rate is 91% (10/11). Since the model has been trained by combination pixels from six peanuts (each selected 1/6 pixels), if the size of sample for training model increased further, the accuracy of identifying if the kernel is contaminated will be further enhanced too.

5.3.3 MODELS COMPARED FOR ALL KERNELS

In this section, we use the kernel level to identify whether the peanut is polluted or not, all the 73 kernels before and after contamination (totally 146 hyperspectral image cubes), and we test the performance of DL. In order to verify the superiority of the algorithm, we also compared other models like KNN, Random Forest, RBF–SVM, and BP-ANN. When we identified it, we used the fivefold cross-validation method. That is, we randomly select 4/5 pixels as the training set and the remaining

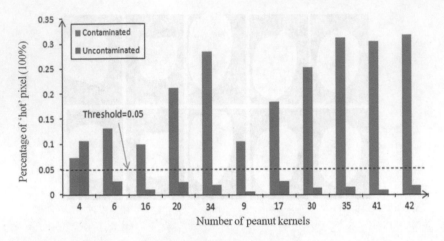

FIGURE 5.6 Identification result for all selected kernels.

1/5 as a test set, and then loop five times. The output of these models is two classes, that is, uncontaminated or contaminated (0 or 1). Then, we calculated the percentage of "hot" pixels and set 0.05 as the threshold to distinguish which kernel has been contaminated. Here, if identification was based on pixels, it was called pixel-level recognition, and if identification was based on kernels, it was called kernel-level recognition. The purpose of setting kernel-level recognition is to study the ability of identifying whether the kernel had been contaminated or not using pixels. The correct recognition rate results are illustrated in Table 5.3.

Overall, among the first four traditional models, KNN, RF, RBF–SVM, and BP-ANN, the recognition rate of BP-ANN is the best, and KNN is the worst. Furthermore, the DL is apparently higher than others. It is obvious that the method of DL can be used to improve the detection rate of aflatoxin kernels. This conclusion indicates that the DL is a new method for detection of agricultural products, and it is more effective than traditional methods.

When compared with kernel level, pixel level is better. Meanwhile, the samples for self are better than the samples for others. The reason may be that when taking multiple kernels, the environment has changed slightly in the long run (we used 3 weeks for sample testing). Therein, for long-time measurements, the changes of spectral variability caused by environmental factors should be taken into consideration.

TABLE 5.3
Recognition Results from Different Methods for All Kernels

	Sample	KNN	Random Forest	RBF–SVM	BP-ANN	DL
Pixel level	For self	92.27	94.14	94.25	95.51	97.26
	For others	89.30	92.79	89.88	94.25	96.34
Kernel level	For self	85.4	88.4	88.9	90.2	93.5
	For others	83.2	85.5	86.1	87.5	90.9

5.4 DISCUSSION

In traditional model of pattern recognition, the first step is feature extraction [using Fisher, successive projections algorithm (SPA), PCA, etc.], and the second step is using classifier (KNN, BP-ANN, SVM, etc.) for recognition. At present, as literature shows, most of the studies (Chang and Lin, 2011; Ataş et al., 2012; Hruska et al., 2013; Qiao et al., 2017) have adopted this traditional model. First, we select the key wavelength through feature extraction and then use classifier to identify whether the kernel is contaminated or not. However, in DL, there is no feature selection step, and the raw spectral data can be directly used as input data to the model (LeCun et al., 2015). Here, in this chapter, in order to have a fair comparison, all models have no feature extraction steps, which means that the original spectra were used as the input for all models. We believe that using full spectrum will provide *more* information than using only key wavelengths. Therefore, key wavelengths are not discussed in this chapter.

Recognition time is very important in the design of a sorting machine, because it directly affects the efficiency of the sorter. Besides this, the shorter the time is, the smaller the moving illegibility caused by the kernel movement. As illustrated in Section 5.3.2, it needs about 0.001' for a pixel. Here, we suppose that the size of one peanut kernel is about 1 cm*1 cm, it has 100*100 pixels, and using linear push scavenging camera, the time used to identify a line of pixels is 0.1'. It needs to be explained that this time is the calculation time obtained in the case of CPU, and if using GPU accelerates, it can at least shorten the time to the original's 1/100 (Campos et al., 2017). It is about 0.001'. According to the formula of gravity acceleration: $h = 1/2(g*t^2)$, the moving distance of kernel was 0.005 cm. At this time, the moving illegibility is only half a pixel and it can be ignored completely.

DL requires a large number of labeled data as the training dataset, but in practice, natural aflatoxin contamination kernels labeled at pixel level are difficult to be obtained. So in this chapter, we constructed labeled datasets using the changes before and after artificial contamination. Once a good model has been established, it can be used for identifying if the kernels are contaminated artificially or naturally. However, it should be pointed out that, limited by computing power of personal computer, the dataset constructed in this chapter is not big enough, and it does not take full advantage of DL, and further, we will expand the dataset and use accelerated computer by GPU, which will be the next step of our efforts.

The purpose of this chapter is to distinguish whether the kernel is contaminated with aflatoxin or not in the pixel level, so we don't discuss the degree of aflatoxin content. Further research may involve the statistics of contaminated pixels, the aflatoxin abundance in subpixel (Chen et al., 2017), etc. It is worth noting that environmental dust, especially organic matter like dandruff, may affect the robustness of test results, because these substances also have fluorescence; therefore, the detection environment must be clean. Soil dust has no fluorescence, so soil dust has little effect on the test results. Besides this, camera sensitivity and light intensity are two important factors for achieving the best result. We hope that the higher the UV sensitivity of the camera, the better and stronger the UV light source. For example, we can choose EMCCD camera (Kwon et al., 2017) or SiPM detectors (Bonanno et al., 2014) that are sensitive to fluorescence and use more stable and stronger UV laser as the light source. Finally, environmental stability is a problem that cannot be ignored, especially in practical application. It is necessary to keep the environment stable, and testing in a black box is a good choice.

5.5 CONCLUSIONS

Aflatoxin is a highly toxic, carcinogenic substance, which widely exists in peanuts and its products. In this chapter, first we established a hyperspectral imagery system using grating module, SCOMS camera, and an electric displacement platform, and then we acquired 146 hyperspectral image cubes of 73 peanut samples before and after being contaminated by aflatoxin.

In order to test the aflatoxin contamination, we proposed a reshape image method of pixel spectral for the CNN method. By comparing different identification models with our model, we found that (1) reshape image established by pixel-level spectral is good enough for aflatoxin-detected problems, and the overall recognition rate reached above 95%. (2) At pixel level or kernel level, DL method worked with high accuracy and better than traditional recognition models, and the recognition rate is higher than 96% at pixel level and higher than 90% at kernel level.

The conclusions of this chapter have positive significance for next sorter machine designing. According to the design idea of sorting machine, this model can be integrated into the upper computer system, which can quickly complete the design of sorting machine for aflatoxin testing.

REFERENCES

Ataş M, Yardimci Y, Temizel A. A new approach to aflatoxin detection in chili pepper by machine vision. *Computers and Electronics in Agriculture*, 2012, 87: 129–141.

Bao L, Zhang P, Lei Z, et al. Comparison between the post-column derivatization with bromine by HPLC and the fluorometric analysis for determination of aflatoxins in peanut. *Science of Inspection and Quarantine*, 2001, 11(5): 18–20.

Bonanno G, Marano D, Belluso M, et al. Characterization measurements methodology and instrumental set-up optimization for new SiPM detectors—Part II: Optical tests. *IEEE Sensors Journal*, 2014, 14(10): 3567–3578.

Campos V, Sastre F, Yagües M, et al. Distributed training strategies for a computer vision deep learning algorithm on a distributed GPU cluster. *Procedia Computer Science*, 2017, 108: 315–324.

Chang C C, Lin C J. LIBSVM: a library for support vector machines, 2011. www.csie.ntu.edu.tw/~cjlin/libsvm.

Chen L, Chen S, Guo X. Multilayer NMF for blind unmixing of hyperspectral imagery with additional constraints. *Photogrammetric Engineering & Remote Sensing*, 2017, 83(4): 307–316.

Chen Y, Lin Z, Zhao X, et al. Deep learning-based classification of hyperspectral data. *IEEE Journal of Selected Topics in Applied Earth Observations and Remote Sensing*, 2014, 7(6): 2094–2107.

Chu X, Wang W, Zhang L D, et al. Hyperspectral optimum wavelengths and fisher discrimination analysis to distinguish different concentrations of aflatoxin on corn kernel surface. *Spectroscopy and Spectral Analysis*, 2014, 34(7): 1811–1815.

Du Y, Chang C, Ren H, et al. New hyperspectral discrimination measure for spectral characterization. *Optical Engineering*, 2004, 43(8): 1777–1786.

Ministry of Health, PRC. *GB 2761-2011 The National Food Safety Standards of mycotoxins in food limited* (In Chinese). Beijing: China Standards Press, 2011.

Hinton G E, Osindero S, Teh Y W. A fast learning algorithm for deep belief nets. *Neural Computation*, 2006, 18(7): 1527–1554.

Hruska Z, Yao H, Kincaid R, et al. Fluorescence Imaging Spectroscopy (FIS) for comparing spectra from corn ears naturally and artificially infected with aflatoxin producing fungus. *Journal of Food Science*, 2013, 78(8): 1312–1320.

Kwon M, Ebert M, Young C, et al. Nondestructive fluorescence detection of hyperfine states of Rb using an EMCCD camera. *Bulletin of the American Physical Society*, 2017, 62: 130–110.

LeCun Y, Bengio Y, Hinton G. Deep learning. *Nature*, 2015, 521(7553): 436–444.

LeCun Y, Bottou L, Bengio Y, et al. Gradient-based learning applied to document recognition. *Proceedings of the IEEE*, 1998, 86(11): 2278–2324.

Liang T, Xu X, Xiao P. A new image classification method based on modified condensed nearest neighbor and convolutional neural networks. *Pattern Recognition Letters*, 2017, 94(15): 105–111.

Malina W. Some multiclass Fisher feature selection algorithms and their comparison with Karhunen-Loeve algorithms. *Pattern Recognition Letters*, 1987, 6(5): 279–285.

Qiao X, Jiang J, Qi X, et al. Utilization of spectral-spatial characteristics in shortwave infrared hyperspectral images to classify and identify fungi-contaminated peanuts. *Food Chemistry*, 2017, 220: 393–399.

Samiappan S, Bruce L M, Yao H, et al. Support vector machines classification of fluorescence hyperspectral image for detection of aflatoxin in corn kernels. *Meeting Proceedings*, 2013: 1–4.

Silver D, Huang A, Maddison C J, et al. Mastering the game of Go with deep neural networks and tree search. *Nature*, 2016, 529(7587): 484–489.

Teena M, Manickavasagan A, Mothershaw A, et al. Potential of machine vision techniques for detecting fecal and microbial contamination of food products: A review. *Food and Bioprocess Technology*, 2013, 6(7): 1621–1634.

USDA. 2002. *Aflatoxin Handbook*. Washington, DC: USDA Grain Inspection, Packers, and Stockyards Administration. www.gipsa.usda.gov/publications/fgis/handbooks/afl_insphb.html.

Wiens T S, Dale B C, Boyce M S, et al. Three way k-fold cross-validation of resource selection functions. *Ecological Modelling*, 2008, 212(3): 244–255.

Yao H, Hruska Z, Kincaid R D, et al. Method and detection system for detection of aflatoxin in corn with fluorescence spectra. U.S. Patent Application 12/807,673, 2010.

Yao H, Hruska Z, Kincaid R, et al. Development of narrow-band fluorescence index for the detection of aflatoxin contaminated corn. *Proceedings of the SPIE—The International Society for Optical Engineering*, 2011, 8027: 80270D.

Yao H, Hruska Z, Kincaid R, et al. Hyperspectral image classification and development of fluorescence index for single corn kernels infected with Aspergillus flavus. *Transactions of the ASABE*, 2013, 56(5): 1977–1988.

Zhang N, Liu W, Wang W, et al. Image processing method of corn kernels mildew infection and aflatoxin levels. *Journal of the Chinese Cereals and Oils Association*, 2014, 29(2), 82–88.

Zhongzhi H, Limiao D. Application driven key wavelengths mining method for aflatoxin detection using hyperspectral data. *Computers and Electronics in Agriculture*, 2018, 153: 248–255.

6 A Method of Detecting Peanut Cultivars and Quality Based on the Appearance Characteristic Recognition

The detection of peanut cultivars and quality is an important composition for peanut breeding and quality testing. In order to evaluate the feasibility of mass peanut seed detection via appearance characteristics, first we take pictures of 48 varieties and 6 different qualities of each variety with digital camera, and then we use the method of principle component analysis and artificial neural network to establish a seed recognition model which is made up of 49 distinct appearance characteristics with regard to their shape, texture, and color and optimize the model. The testing result indicates: after the model optimization, variety recognition rate and quality recognition rate reach 91.2% and 93.0%, respectively; the color characteristic plays an impactful role in the variety and quality detection; and the appearance characteristic is more helpful in detecting the quality than in differentiating between varieties. The detection method based on the machine vision possesses the cost and speed advantages, and it can be used in identification of peanut cultivars and evaluation of their quality.

6.1 INTRODUCTION

China is the biggest peanut producing and exporting country in the world. Seed facticity, purity, and quality are significant in testing, producing, marketing, importing, and exporting a new peanut variety. The detection of facticity and purity in the seeds of peanut mainly concerns the sort of the peanut variety, and the detection of peanut quality mainly scales trade quality especially the organoleptic quality. At present, the detection and research of peanut seed is carried out mainly through the method of manual work and bio-chemical tests. The workload is very heavy in identifying peanut seed by manual work. The manual method also calls for experience-rich people, and the method is fatigable. The biochemical identification calls for expensive equipment for detection, precise and complex technique for experimentation, and high cost. In order to cope with the demand in peanut breeding, further processing, and foreign trade, there is a compelling need to find a speedy and accurate method that identifies the peanut variety and quality.

The method of machine vision based on the seed image process is a new speedy, low-cost, non-fatigue method with no damnification, strong identification ability, high repetition, and capacity of detecting mass amounts. Using this method, people gained good results in maize (Hao et al., 2008; Yang et al., 2008; Zhu et al., 2007), rice (Sakai et al., 1996; Zhang and Zhang, 2007), and wheat (Dubey et al., 2006). The author has applied this method in detecting the variety and quality of only a small number of peanuts. Because it is seldom reported on peanut detection, and the varieties are limited in the test too, it is necessary to do some tests aiming at a large number of peanuts, so that we can evaluate the application value of machine vision method.

On the basis of our previous work (Han et al., 2007), here, we are going to collect a large number of pictures of peanut and design an experiment for recognition of peanut variety and quality. Then, we discuss the feasibility of recognition to peanut appearance characteristic. At last, we will analyze the role of appearance characteristic in machine recognition and the factors that affect the result of the detection.

6.2 MATERIALS AND METHOD

6.2.1 Materials for Test

We collect 50 peanut seeds with normal appearance characteristic from each of the 48 varieties cultivated in North China, and 50 peanut seeds of one variety with six different qualities (normal, with rubbing injury and damaged skin, cut-damaged, mildewed, germinated, and with impurities like stones) for the purpose of our research. The varieties selected basically cover the main production areas such as Shandong, Henan, Hebei, and Anhui.

6.2.2 Image Acquisition and Pretreatment

The established image acquisition system is shown in Figure 6.1. In order to avoid fluctuations in the daylight, we make a simple lamp box out of tin which is painted black to reduce the light reflection. In the next step, we take the profile pictures of the 50 peanut seeds of every variety or each quality of a selected variety, where the seeds are laid on the object stage of the lamp box, using a digital camera (Canon A610

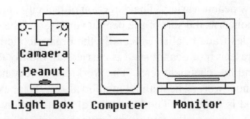

FIGURE 6.1 Image acquisition system.

1,728*2,304 pixels) set up at the "micro" function. Then, we transfer the pictures into a computer to prepare for the next step. The performance index of the computer we adopt is Lenovo Dual-Core Intel(R) CPU T1400@1.73 GHz, 1 GB Memory, Windows XP OS. The partial pictures we get from the image acquisition system are presented in Figures 6.2 and 6.3.

The image pretreatment stage includes image enhancement, de-noising, median filtering, edge detection, and morphology operation and color space conversion, and so on. All the image-processing procedures are based on self-programmed MATLAB® program.

FIGURE 6.2 Partial pictures of peanut in different varieties.

FIGURE 6.3 Partial pictures of peanut in different qualities.

6.2.3 The Appearance Characteristic Index of the Seed

The selected appearance characteristics (Wan, 2005) are divided into three classes:

1. Color: Twenty-four characteristic indexes: mean value, square deviation, skewness, kurtosis of RGB and HSV color spaces (R, G, B, H, S and V).
2. Morphological features: Fifteen characteristic indexes: seven characters describe the size, that is, length, width, major axis length, minor axis length, diameter and perimeter of the circumferential circle, and area of profile projection; eight characters describe the shape, that is, rectangular degrees, ovality, ratio between concave and convex, degree of circularity, ratio between minor and major axis lengths, compactness ratio, abscissa, and ordinate of the geometry centroid.
3. Textural features: Ten characteristic indexes: there are seven statistics moment invariants that describe the distribution characteristic of the grayscale value of the seed picture.

6.2.4 The Establishment of the Recognition Model

We establish a library of four pattern recognition features (as shown in Table 6.1): three of them are shape, texture, and color, and the other feature is the combination of these three. Obviously, there are too many feature variables in the feature library; so, there is redundancy and associativity among the variables. It will cost too much to use all these variables in recognition.

So, we should standardize and de-correlate the original data. Then according to the method of principle component analysis (Han et al., 2005), we call function "princomp" in the MATLAB and calculate the contribution rate of the variables. We choose the components that take up greater than 90% of the contribution rate for recognition as principle components. So, we can simplify and optimize the original feature to some extent.

At last, we call function "newff" in the MATLAB to construct forward neural network and set the number of the input, output, and hidden layer neurons, and then we call function "train" to do ANN training and the trained network to do the procedure of recognition.

The effect of recognition is evaluated by correct detection rate. The correct detection rate = the number of the correct recognized seeds/the total number

TABLE 6.1
Summary of Four Pattern Recognition Features

Model	Shape	Texture	Color	Combination
Number of characteristics	15	10	24	46
Principle component number	8	7	8	10
Accumulative contribution (%)	98.6	98.1	96.5	92.3

TABLE 6.2

Summary of the Recognition Results of Peanut Varieties

Model	Shape	Texture	Color	Combination
Number of cultivars correctly recognized	14	18	40	44
Correct recognition rate (%)	29.2	37.5	83.3	91.2

TABLE 6.3

Summary of the Recognition Results of Peanut Qualities

Model	Shape	Texture	Color	Combination
Normal	52	48	93	95
Knead damaged	71	84	91	93
Knives damaged	83	72	89	90
Moldy	59	76	94	96
Germinated	89	57	70	91
Impurities	64	73	92	93
Average	69.7	68.3	88.3	93.0

of the detecting seeds * 100%. We adopt the method called "cross validation" (Lachenbruch and Mickey, 1968) to test the recognition pattern. First, we consider the $n-1$ of the total n samples as a training set and use the set to train the neural network. Then, we consider the last 1 as testing set and test the network with it. In a similar way, alternate the training set and testing set, and train and test for n times. We have designed the testing method to analyze different varieties of peanut and different qualities of one peanut variety. The testing results are shown in Tables 6.2 and 6.3.

6.3 RESULTS AND ANALYSIS

6.3.1 The Result of Recognition on Peanut Varieties

In order to recognize the peanut varieties, first we should do some pretreatment for the 48 varieties which includes 50 seeds and detect the 46 appearance characteristic parameters for each seed, and then find the optimized principal components with the method of principal component analysis. Then, we distribute these principal components into distinguished characteristic classes as the input of the neural network, train the network, and detect the peanut parameters. The gathered summary results are shown in Table 6.2. For the sake of chapter length, we just list the summary results instead of the detection rate of each variety.

6.3.2 THE RESULT OF RECOGNITION ON PEANUT QUALITIES

In order to recognize the peanut qualities, first we should do some pretreatment for the 50 seeds of each variety with different qualities. The qualities include normal, with rubbing injury and damage in the red skin, cut-damaged, mildewed, germinated, and with impurities like stones. Also, we detect the 46 appearance characteristic parameters for each seed and find the optimized principal components with the method of principal component analysis. Then, we distribute these principal components into distinguished characteristic classes as the input of the neural network, train the network, and detect the peanut parameters. The gathered summary results are shown in Table 6.3.

6.4 ANALYSIS OF THE RESULTS OF RECOGNITION AND DETECTION

We simplify the peanut characteristic parameters with the method of principal component analysis and reduce the original 15 shape characters, 10 textural characters, 24 color characters, and 46 combined characters to 8, 7, 8, and 10, respectively. There are much redundancy and associativity in the different characters of peanut. By abstracting the chief components without affecting the recognition effects, the method of principal component analysis largely reduces the characteristic data and increases the detecting speed.

The recognition rates of the 50 seeds of the 48 peanut varieties based on the four features, that is, shape, texture, color, and combined features, are 29.2%, 37.5%, 83.3%, and 91.2%, respectively. The results indicate that the shape and texture of different varieties of the testing peanuts are very close, and these characteristics of different varieties overlap each other; the recognition rate of color character is higher indicating that it is adequate to distinguish the varieties by color to some extent, and there is less overlap between different varieties; and the combined parameters can raise the recognition rate further while enlarging the cost on data processing.

We can draw a conclusion from the quality detection in peanuts that the color and combined characters are superior to the shape and textural characters. It is similar to the conclusion of variety recognition. Additionally, we discover that the detection results are different between the seeds having different qualities with identical feature library and the seeds having identical qualities with identical feature library. And the roles different seed characters play in detection are different. For example, the shape characters are more sensitive in germinated peanut, while it is less sensitive in distinguishing normal and mildew ones; the textural characters are more sensitive in peanuts having red skin breakage and rubbing injury, while the color characters are less sensitive in germinated ones. It is consistent with the judgment based on our experience.

With the four-feature library, the effect of quality detection is superior to the effect of variety detection. We can conclude that the discrimination validity of different qualities based on peanuts' appearance is higher than the discrimination validity of different varieties. Especially based on the shape and textural characters, the quality detection is more obvious than variety detection. But the detection rate is generally low with shape and textural characters, overall below 60%. Although the shape and texture of seeds are steady peanut production factors, they are heavily influenced by the ecological environment, cultivation conditions, and soil quality (Lan et al., 1968), which influences also the correctness of detection results.

6.5 DISCUSSIONS

From the detection results, we can draw some conclusions: (1) each of the seed appearance characteristics plays its unique role in detection, and these characteristics can be kept in the order of shape < texture < color character based on detection results; (2) the color character of the seeds is very obvious in detection, and it improves the detection result by adopting other character parameters; and (3) the discrimination validity of appearance characteristics in variety detection is higher than that in quality detection. The method of image processing is more adequate in peanut quality detection.

It should be pointed out that if there are more varieties and more overlaps between characteristics, discrimination validity becomes worse, which influences the recognition results (Sapirstein and Kohler, 1999). The number of seeds reflexes the representativeness to the variety. In order to eliminate interference of the random difference between the seeds and improve the detection results, we reduce the sample size to as low as 400.

The seed's inner qualities such as taste, nutrient component, and protein content are not discussed in this chapter. These characteristics could hardly be detected by visible light images, but it could also be detected by imaging techniques in other waveband such as near-infrared spectroscopy analysis.

We are just concerned about the appearance characteristics of the peanut seed, except for the appearance characteristics of the peanut legume. Besides the size and color characters, the appearance characteristics of peanut legume include kernel number, observability for the waist, thickness of the reticulation, the projection of the peanut mouth, and legume type. The recognition rate can be increased further if the above-mentioned factors are also considered.

REFERENCES

Dubey B P, Bhagwat S G, Shouche S P, Sainis J K. Potential of artificial neural networks in varietal identification using morphometry of wheat grains. *Biosystems Engineering*, 2006, 95: 61–67.

Han Z-Z, Huang H-M, Ye H-T, Preprocessing of speech recognition in intensive environment noises based on ICA/BSS. *Journal of Yunnan University*, 2005, 44: 255–259 (in Chinese with English abstract).

Han Z-Z, Kuang G-J, Liu Y-Y, Yan M, Quality identification method of peanut based on morphology and color characteristics. *Journal of Peanut Science*, 2007, 36: 18–21 (in Chinese with English abstract).

Hao J-P, Yang J-X, Du T-Q, Cui F-Z, Sang S-P. A study on basic morphologic information and classification of maize cultivars based on seed image process. *Scientia Agricultura Sinica*, 2008, 41: 994–1002 (in Chinese with English abstract).

Lachenbruch P A, Mickey M A. Estimation of error rates in discriminant analysis. *Technometrics*, 1968: 1–10.

Lan J-H, Li X-H, Gao S-R, Zhang B-S, Zhang S-H. QTL analysis of yield components in maize under different environments. *Acta Agronomica Sinica*, 2005, 31: 1253–1259 (in Chinese with English abstract).

Lan S J, Henderson L M. Uptake of nicotinic acid and nicotinamide by rat erythrocytes. *Journal of Biological Chemistry*, 1968, 243(12): 3388–3394.

Sakai N, Yonekawa S, Matsuzaki A. Two-dimensional image analysis of the shape of rice and its application to separating varieties. *Journal of Food Engineering*, 1996, 27: 397–407.

Sapirstein H D, Kohler J M. Effects of sampling and wheat grade on precision and accuracy of kernel features determined by digital image analysis. *Cereal Chemistry*, 1999, 76: 110–115.

Wan S-B. Quality study of peanut. *Agriculture Press of China*, 2005 (in Chinese with English abstract).

Yang J-Z, Hao J-P, Du T-Q, Cui F-Z, Sang S-P. Discrimination of numerous maize cultivars based on seed image process. *Acta Agronomica Sinica*, 2008, 34: 1069–1073 (in Chinese with English abstract).

Zhang C, Zhang H. Rice figure edge detection based on mathematical morphology. *Journal of the Chinese Cereals and Oils Association*, 2007, 22: 139–141 (in Chinese with English abstract).

Zhu S, He G, Li Z. Whole kernel detection: determination of amylose content of maize with near-infrared transmittance spectroscopy. *Journal of the Chinese Cereals and Oils Association*, 2007, 22: 44–49 (in Chinese with English abstract).

7 Quality Grade Testing of Peanut Based on Image Processing

The quality of peanut kernels is referred to every aspect of the profit of supply and marketing. A back propagation (BP) neural network model of quality grade testing and identification is built based on 52 appearance features such as the shape, texture, and color using the technology of computer image processing. A comprehensive test is carried out in 1,400 grains to find out unsound kernel, mildewing, impurity, hetero-variety, and other aspects with the aim of achieving results with the accuracy rate of 95.6%. According to the national standards, a method is designed for testing peanut kernels' grade based on their specification and quality, in which 100 peanut grains are tested with results of the comprehensive test reaching the accuracy rate of 92%. The methods that are discussed this chapter to test the quality and distinguish the grade of peanuts based on their appearance and specifications can produce results with high accuracy rate, which can have a positive impact on the peanut industry's productivity and development.

7.1 INTRODUCTION

Peanut is one of the prime crops for the peasants in China that helps increase their income from agriculture. But the country's international trade price of peanut is only 80% of the average market price. The level of automation in quality testing of peanut kernels is low, and most of the work is done manually. Hence, the workload is heavy which may increase workers' burden and need them to have ample testing experience. In addition, it makes the testing more costly and time-consuming. Because of the upward trend in the trade through import and export of peanuts, finding solutions to this contradiction becomes more and more crucial.

The testing based on image processing and machine vision is a new method which is speedy and repeatable with high distinguishing rate, less depreciation, low cost and manual work, and it can also be used in batch test. This method is used in appraisal of peanut kernels based on their quality. Many scholars made a series of meaningful explorations in its application on maize (Zhao et al., 2009), rice paddy (Sakai et al., 1996), wheat (Dubey et al., 2006), and other general crops, and obtained good results. However, achievements in research are very limited in testing the quality of peanut kernels. In the few available works (Xiong et al., 2007a,b; Chen et al., 2007; Han and Zhao, 2009), sample set and feature set are commonly small and there is lack of universality; the testing items and standards of quality are given by people without any quantizing criterion.

This research is intended to build a set of integrated quantizing system for grading peanut kernels based on their appearance quality using image processing and machine vision. The quantizing model of peanut kernels' appearance quality is built based on the international data, trade standards, and expert knowledge. It has three main objectives. The first one is to build larger sample and feature database breaking through the limitations of former testing which is based on small sample set and feature set. The second one is to build pertinently testing-item system on account of the national and other trade standards. The third one is to build an integrated image testing program and a method for assessing peanut kernels' quality based on some common features and grade specifications in their appearance.

7.2 MATERIALS AND METHOD

7.2.1 MATERIALS FOR TESTING

The sample is chosen from primitive peanut kernels harvested by one farmer of Junan, Shangdong, which has not been chosen by hand, and the main variety of which is the big grain of hua 37. Grains of different qualities are subjected to investigation.

Images of chosen peanut kernels were collected; the image-collection system is shown in Figure 7.1. In order to take full images of peanut kernels, two-eye vision system is built in which a camera is set on top and a scanner is set at the bottom to view both top and bottom parts of peanut kernels. In order to avoid the instability of natural light, the image collection is processed in a hermetic lamp house in which the internal walls are covered with black light absorption cloth to reduce reflection. Also a layer of transparent diffusion reflection rubber sheet is added to the scanner's surface in order to lessen the glisten phenomenon, which occurs in the process of taking a photo. Peanut kernels of different qualities are laid evenly on the scanner in order to get the top and bottom images, which will be stored in a computer through a transmission line for later process.

The equipment used in the process of testing are as follows: Olympus E300 (1,728*2,304 pixels) camera of microspur pattern, CanoScan 8800F flat plate CCD scanner of 600 dpi resolution, and the ThinkPad SL300 computer with an image processing program based on the MATLAB® R2008a software installed.

Parts of the images collected in the process are shown in Figure 7.2. The preprocessing of images includes reinforcement, de-noising, median filtering, edge

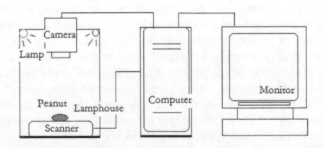

FIGURE 7.1 Graphical representation of the system of image collection.

FIGURE 7.2 Images of peanuts of different qualities.

detection, morphological operation, transition of the color space of image, and other routine pretreatment methods (Han and Zhao, 2009). Confined to the space, no more methods will be pointed out.

7.2.2 MAINTAINING THE INTEGRITY OF THE SPECIFICATIONS

The peanut kernels exported from China needs to be long-oval, light-red coated without variegation, of no crack, of uniform color, of tidiness, and handsomeness. And the size, purity, and other specifications are all regulated clearly. The quality quantizing figures are shown in Table 7.1, which are made according to the national, traditional, and international standards.

Integrating the national and traditional data, aiming at different purposes, and allowing for expert knowledge, the measurement of peanut's overall quality can be divided into three steps: the first one is to identify the five types of defect items that are having unsound kernels, mildewed, impure, hetero-variety, and whole semi-grain; the second one is to identify normal peanut kernels and divide them into five categories based on their size; and the third one is to test for their each specification and divide them into five grades based on quality.

TABLE 7.1

Quantizing Requirement of Peanut Kernels' Appearance Quality

Standard of Size			Standard of Grade		Standard for Other Restrictions	
Size Specification	(Grains/ Ounce)	Grade		Purity Rate (%)	Standard	Specification
24/28	23.51~27.5	1		96.0	Hetero-variety	≤5%
28/32	27.51~31.5	2		94.0	Whole semi-grain	≤10%
34/38	33.51~37.5	3		92.0	Impurity	≤1%
38/42	37.51~41.5	4		90.0	Unsound Kernel	≤4%
No-grading	XX~401.0	5		88.0	Mildewing	≤0.03%

Note: Referring to the national standard of peanut kernels of the People's Republic of China (GB1533-1986), the national standard of peanut kernels of the People's Republic of China (GB/T1532-2008), national agricultural traditional standard of green food of peanut (NY/T420-2000), and the national standard of non-polluted peanut (DB34/T 252.4-2003).

7.2.3 Model of Quality Recognition

It is obvious that the problem arises in the first step of recognizing accurately the normal kernels, unsound ones (such as worm-eaten one, one with disease speck, gemmiparous one, crushed one, immature one, whole semi-grain one, or hulled little one), mildewy ones, impurities (clod, stone, tile, glass, hull), and hetero-variety ones.

The peanut kernel samples of 14 specifications should be divided into two groups: one group is used to train the recognizing model, and the other group is to be tested. Each group contains 100 grains counting up to 2,800 grains. The 54 features selected from the images of both sides of peanuts are used as inputs of the image recognition, the output of which is four-bit coded of thermometer method. A three-layer BP neural network needs to be constructed, in which the numbers of input and output nodes are 104 and 4, respectively, and the hidden layer is identified as 11 neurons through many experiments. First, the network model needs to be trained by the feature set of the training group, the result of which is to be tested by the testing group.

The 54 features can be divided as follows: 24 features reflect the color such as mean, variance, skewness, and kurtosis of the components of the color spaces are R, G, B, H, S and V; 7 features reflect the size such as length, width, length of the major and minor axes, circumference, diameter of the equilateral circle, and side projected area; 8 features reflect the shape such as rectangle degree, ovality, concave–convex proportion, circularity, ratio between the lengths of the major and minor axes, compact ratio, and ordinate and abscissa of the opposite centroid; 15 features reflect the texture – of which, 7 are statistical invariant moments that reflect the frequency of the distribution feature of the gray value of the grain image, and 8 reflect features such as mean of gray-level image, variance, smoothness, third-order moments, consistency, and entropy. The definition of these features and the neural network model are referred in the corresponding literatures (Zhao et al., 2009; Sakai et al., 1996).

7.2.4 Method of Grading

To recognize the types of the grains is the premise of the quality grading. The next step is to distinguish the specification and grade of grains that are tested as normal. The feature of the area of the image is used mostly in distinguishing the specification. A regression equation can be used, where Y is the area and X is the weight of a grain's image; hence, the regression curve is: $Y = 0.978X - 0.0248$, the correlation coefficient (R^2) is 0.9841, and there is a positive correlation between the specification and weight of the grain as shown in Table 7.2.

By measuring the area of the pixel, we can get the specification of the grain. In the national standard, there is no relative regulation that is made about the percentage of grains having corresponding specification in the total number. Allowing for quantizing request and expertise knowledge, we make a provision that if the percentage of grains having corresponding specification amounts to above 95% of the total

TABLE 7.2

Relationship between the Specification of Peanut Kernel and the Area of Pixel

Specification of Size	(Grains/Ounce)	Gram/Grain	Area of the Pixel
24/28	23.51~27.5	0.829–0.970	2,704–3,154
28/32	27.51~31.5	0.970–1.111	3,155–3,605
34/38	33.51~37.5	1.182–1.323	3,831–4,281
38/42	37.51~41.5	1.323–1.464	4,282–4,732
No grading	XX~401.0	Others	Others

number, then whole lot can be assumed to conform to quality standards; otherwise, a prompt should be given that these grains should be filtered mechanically.

After the specification of the grains is given, the purity level needs to be calculated in order to categorize them into different grades based on their corresponding specification. The purity level given in the national standard means the percent of weight that the pure peanut kernels account for the total peanut kernels. Seeing that all the grains filtered mechanically are of the same size mostly, the mean weight of each seed can be substituted by the average weight of all seeds, namely, the weight index can be substituted by the number of seeds having corresponding specification. So, the approximate calculation formula can be as follows:

$$P_{\text{rate}} = \left(1 - \frac{\dfrac{1}{2}N_1 + \dfrac{1}{4}N_2 + N_3 + N_4 + N_5}{\displaystyle\sum_{i=1}^{5} N_i}\right) \times 100\% \qquad (7.1)$$

In the above formula, N_1 represents the number of unsound kernels, N_2 represents the number of whole semi-grain ones, N_3 represents the number of hetero-variety ones, N_4 represents the number of mildewy ones, and N_5 represents the number of impure ones. The grade of the grains can be distinguished according the national standard (Table 7.1) after the purity level is calculated. Details: two whole semi-grain grains are calculated as one whole grain, allowing for the whole semi-grain kernels' edible value, it's reasonable for it to be calculated as unsound kernels.

7.3 RESULTS AND ANALYSIS

7.3.1 ANALYSIS OF RECOGNITION RESULTS OF GRAINS' QUALITY

The results of the judging method of neural network on grains of different specifications are shown in Table 7.3. In the table, 100 (97+2A+1P) represents that there are 100 grains in this grade, of which 97 are the same as the number gets by hand; +2A represents 2 grains are judged as worm-eaten ones labeled as D; and +1H represents 1 grain is judged as normal seed labeled as P.

TABLE 7.3

Ten Parcels of Peanut Kernels' Detailed Testing Conclusion

Number	Test Conclusion by Hand	Test Conclusion by Machine	Test Conclusion of Specification	Test Conclusion of Grade
1	24/28, grade 1	24/28, grade 1	T	T
2	24/28, grade 2	24/28, grade 3	T	F
3	34/38, grade 1	34/38, grade 1	T	T
4	24/28, grade 3	28/32, grade 3	F	T
5	38/42, grade 2	38/42, grade 2	T	T
6	24/28, grade 4	24/28, grade 4	T	T
7	Irregulars	Irregulars	T	T
8	24/28 grade 1	24/28, grade 1	T	T
9	28/32 grade 5	28/32, grade 5	T	T
10	38/42 grade 1	No grading	F	F

The conclusion of using Neural network are as follows:

- The accuracy rate of the integrated judging result on grains of 1,400 specifications averages out at 95.6%, and the accuracy rates of the judging result on unsound and impure grains average out at 95.3% and 96%, respectively.
- Taking edible value in to consideration, unsound kernels are still fit for consumption, the impurities have no value, the mildewy kernels can only be used as organic fertilizer; the quality of peanut kernels can be classified as unsound ones, impure ones, mildewy ones, hetero-variety ones, and normal ones; the whole semi-grain kernels are classified as unsound ones; the accuracy rates of recognition of unsound and impure kernels are, respectively, 97.1% and 96.8%.
- Although the accuracy rates of recognition of hetero-variety and normal kernels are, respectively, 96% and 97%, the integrated accuracy rate of the two classes is 100%. If only at the range of the national standard, it will not have effect on the trade of peanut kernels.
- The rate of recognition of the kernels with disease speck and mildewy ones is higher. There are three mildewy kernels recognized as ones with disease speck and six as clod, three kernels with disease speck as clod and seven as mildewy ones. The prime reason for the above phenomenon is that grains with these specifications are similar on the face. Fortunately, clod and mildewy kernels do not have any value, and the ones with disease speck have little value. So, this low accuracy rate will have little effect on the test of grade.

7.3.2 Analysis of the Result of Specification and Grading

Totally, 100 parcels of peanuts of different specifications are sampled to be tested, and due to lack of space, only 10 parcels of peanut kernels' test results are sampled (Table 7.3); the total accuracy rate is shown in Table 7.4.

TABLE 7.4

The Test Accuracy Rate of 100 Parcels of Peanut Kernels

Item	Wrong Only in Testing Specification	Wrong Only in Testing Grade	Wrong in Both Specification and Grade	Right in Both Specification and Grade
Ratio (%)	1	5	1	93

It can be seen from the above test results that the total accuracy rates of testing specification and grade are higher that it reaches up to 93% meeting the qualification of use in the main. The inaccuracy rate of testing grade is four times more than that of testing specification, the reason of which is that it is more complex, there are more items to be tested, and it is more complex to quantize in testing grade, and that it is so easy to test specification that the good quantizing result can be obtained from only one feature (the area of image). The inaccuracy rates of testing grade and specification are all lower that it reaches up to 1%. The system has good performance of generalization.

7.4 CONCLUSIONS

The method of image processing used in testing peanut kernels' quality is researched in the chapter; a series of quantizing designs are made referring to the data of national standard. The testing results are as follows: the accuracy rate of the integrated test-ing result of grains of 1,400 specifications averages out at 95.6%, and the accuracy rate of testing specification and grade of the 100 grain samples reaches up to 93%. It can be seen that the method of image processing used in testing peanut kernels' quality has higher performance, and it can be used in the productive practice after been improved.

There is one point worthy to be described in this chapter that peanut kernels' qual-ity is researched based only on the specification and quantizing quality of big grains of peanut kernels, without referring to small grains. The big and small grains are only different in size, so the method researched in this chapter can also be referred to in the process of the latter. Moreover, the quality of peanut kernels includes not only external qualities but also the internal ones such as texture, nutrient component, protein con-tent, moisture of the grain, and production area. The internal quality is not researched in this chapter because of the limitation of machine vision. It is difficult to test the features by visual light image; instead, near-infrared analysis and other band image technology can be used. Moreover, the method used in the current work is designed for testing the grade of peanut kernels, which can be used in categorizing peanuts based on their grades and other production processes after being little improved.

REFERENCES

Chen H, Xiong L, Hu X, et al. Identification method for moldy peanut kernels based on neural network and image processing. *Transactions of the Chinese Society of Agricultural Engineering*, 2007, 23: 158–161 (in Chinese with English abstract).

Dubey B P, Bhagwat S G, Shouche S P, et al. Potential of artificial neural networks in varietal identification using morphometry of wheat grains. *Biosystems Engineering*, 2006, 95: 61–67.

Han Z, Zhao Y. A cultivar identification and quality detection method of peanut based on appearance characteristics. *Journal of the Chinese Cereals and Oils Association*, 2009, 24: 123–126 (in Chinese with English abstract).

Sakai N, Yonekawa S, Matsuzaki A. Two-dimensional image analysis of the shape of rice and its application to separating varieties. *Journal of Food Engineering*, 1996, 27: 397–407.

Xiong L, Chen H, Zhang J. The test of peanut integrity based on machine vision. *Cereals and Oils Processing*, 2007a, 3: 74–72 (in Chinese with English abstract).

Xiong L, Ren Y, Xiao R. The test of peanut size based on machine vision. *Hubei Agricultural Sciences*, 2007b(3), 46: 464–465 (in Chinese with English abstract).

Zhao C, Han Z, Yang J, et al. Study on application of image processing ear traits for DUS testing in maize. *Acta Agronomica Sinica*, 2009, 42: 4100–4105 (in Chinese with English abstract).

8 Study on Origin Traceability of Peanut Pods Based on Image Recognition

In order to investigate variety-specific characteristics reflected by peanuts from different origins, we use a scanner to capture the images of the same species (Huayu 22) from three different regions. Each variety includes one front and two side images of 100 peanuts. For each image, we have acquired 50 characteristics including shape, color, and texture. We build an artificial neural network (ANN) model for identification based on these characteristics and those optimized by principal component analysis (PCA). Result shows that for species of different origins, the maximum rate of detectability reaches 100%. The methods used in this chapter have positive significance to the distinctness, uniformity and stability (DUS) testing of peanut.

8.1 INTRODUCTION

Seed is an important organ of crops, and the same seed varieties have some species-specific properties that are different from other varieties in nature such as genetic specificity, consistency, and stability, which are the main contents of seed DUS. As influenced by the external environment, a variety of crops of the same seeds from different areas reflect different characteristics. These features can only be identified based on rough judgments from previous experience. These methods are costly and time-consuming, and require considerable experience. They cannot judge the origin based on quality quickly, which restricts the efficiency of seed test.

In recent years, machine vision inspection based on computer-based digital image processing is a new fast, effective method which is able to identify highly repetitive, high-volume samples. Zhao et al. (2009) studied the application of image processing technology in ear traits for DUS testing in *maize*, which confirmed the bright prospects of the technology in quality inspection of corn seeds. Our team (Han and Zhao, 2009, Han and Yang, 2010) and Chen et al. (2007) studied the role of peanut seed kernel in the identification of species, which provides a guideline to further explore the key features carried by the organism itself in species recognition; image processing method is also used in other grains such as wheat (Sakai et al., 1996) and rice (Dubey et al., 2006).

Research of image processing in the origin recognition is less. This method is used less for testing peanut seeds. Peanut is also an important organ: will the peanuts from different origins have specific appearance? This is investigated in this study.

8.2 MATERIALS AND METHOD

8.2.1 Test Materials

The peanut variety chosen for the test is Huayu 22 (Table 8.1); it is collected from three different areas: the seed stations, farmers' field, and breeder. These sample varieties are produced in eastern Shandong Yantai, Qingdao, and Laiyang. For each species, we select 100 normal peanuts without any damage.

In the process of obtaining pictures, we fully consider the shooting specification (Yu, 2008) and requirement characteristics of new peanut varieties' DUS test photographs. We capture images with a scanner, and the seeds were placed in a fixed order and direction on the scanner. In order to scan the background, the scanner's lid is kept fully open, and the 100 peanuts are uniformly placed on the scanner. In order to obtain more comprehensive pictures, we take images of the front and two sides of the peanuts. There's no back image because it is difficult to obtain. One of the varieties of seed is shown in Figures 8.1 and 8.2.

The scanner used in experiment is Cano Scan 8800F, Flatbed CCD scanner, with an optical resolution of 4,800 dpi*9,600 dpi, a maximum resolution of 19,200 dpi, and a scan range of 216*297 mm; the computer used in the experiment is idea Centre Kx 8160; CPU is Intel Core Duo 2 Quad Q8300 2.50 Hz, memory DDRIII40; Flash 10, HDD 5000; Windows XP operating system.

TABLE 8.1
Materials Source

No.	Variety	Source	Year of Gathering
001	Huayu 22	Farmers of Laiyang	2011
010	Huayu 22	Seed station of Yantai	2011
011	Huayu 22	Breeder of Laiyang	2010

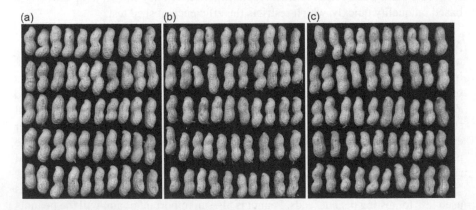

FIGURE 8.1   Scanned images of three sources.

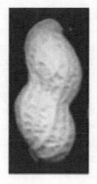

FIGURE 8.2 Scanned images of one variety.

TABLE 8.2
Statistical Characteristics

Category	Amount	Characteristics
Dimension	7	Length, width, long- and short-axes length, perimeter, equal-area diameter, side projected area
Shape	8	Squareness, ovality, bump ratio, circularity, short-/long-axis ratio, compactness, the relative mass of the vertical and horizontal axes, etc.
Color	24	Mean, variance, skewness, and kurtosis of R, G, B, H, S, and V components
Texture	15	Seven statistical invariant moments, mean, variance, smoothness, third-order moments, consistency, entropy, etc. of gray image

8.2.2 Methods

Picture preprocessing includes image enhancement, de-noising, median filtering, edge detection, morphological operations, image color space conversion, and other conventional pretreatment methods.

The selected appearance features are divided into four broad categories of 54 characteristics (Table 8.2), in which the color, shape, and texture features, respectively, are captured from the color image, binary image, and texture image. The definition of these characteristics refers to the relevant documents (Hao et al., 2008; Yang et al., 2008). Image preprocessing and feature extraction process are based on MATLAB® R2008a software.

8.2.3 The Method Used to Optimize

Statistical method used is the PCA, and the recognition model is back propagation (BP) neural network algorithm.

PCA is an effective way to analyze statistical data, and its purpose is to find a space in the dataset of vectors as possible data can be projected from the original

R-dimensional space down to the M-dimensional space projection ($R > M$), and then save the main information after dimensionality reduction to make the data more manageable. PCA is a minimum mean square error of optimal dimensionality reduction method. PCA is used to find some mutually orthogonal axes along the largest variance direction of the dataset.

BP neural network is trained by error BP algorithm of multilayer feed forward networks, which is one of the most widely used neural network models. BP network can learn and store a lot of input–output model mapping, without having to reveal the mathematical description of the mapping equation in advance. Its learning rule is the steepest descent method, which adjusts the weights and threshold by BP network to continuously make the smallest error sum of squares of network. BP neural network model topology includes input layer, hidden layer, and output layer.

8.3 RESULTS AND ANALYSIS

Image recognition is based on the characteristics and statistical methods, using different statistical indicators to get a series of statistics which we call the statistical features that often include shape, color, and texture characteristics (Table 8.2).

During species identification, each statistical characteristics has a certain ability to judge. In the use of pattern classification system for diagnosis, the more features used, in theory, the higher recognition accuracy achieved. But if the more characteristics are detected, the cost is greater. Therefore, an effective amount of a suitable feature will become very meaningful. However, as statistical indicators increase, the dimension of statistical characteristics also needs to make the necessary dimension reduction. PCA will be a good choice to make the statistical properties extracted. Not only for data dimensionality reduction, it can also find the optimized combination of characteristics suitable for all classes.

Figure 8.3 shows the recognition rate curve using the statistical characteristics of the original features and characteristics to identify in the circumstances of

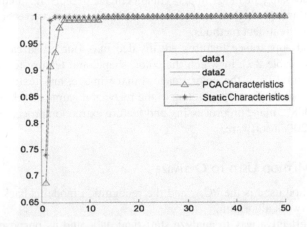

FIGURE 8.3 Recognition rate graph under accumulative characteristics.

TABLE 8.3

Recognition Rate of Part Characteristics

	Number of Features	Single Characteristics (%)	Cumulative Recognition Rate
Statistical characteristics	1	71.2	70.1
	2	68.3	91.0
	3	68.6	95.4
	4	84.1	95.7
PCA	1	74.0	69.7
	2	70.0	89.7
	3	75.7	95.3
	4	81.7	97.9

different numbers. It can indicate that (1) the recognition effect of the PCA-optimized feature is better than the raw statistic features in the case of having less feature in the characteristics of No. 1, 2, 3, and 4; the recognition rate increased by 22.5%, 15.5%, and 22.1%; and as the number of characteristic dimension increases, this advantage becomes less obvious. (2) In the feature recognition of original features and statistical characteristics in the first three or four features, the recognition rate has reached 100%. So, four features have been fully capable of effective identification of peanuts from three different sources.

Table 8.3 details the statistics of the single-feature recognition rate and the cumulative average recognition rate, which are ten times the first four statistical and PCA features. It shows the different recognition effects of different statistical features such as in the first four features, the fourth has the highest recognition rate which reaches 84%, and the cumulative feature of the first four characteristics reaches 95.7%.

8.4 DISCUSSIONS

There is a high positive correlation between heritability and recognition rate (Yang et al., 2011). However, due to different growing conditions, it showed a different appearance. Plant organs depend on cell division and growth; the shape depends on cell differentiation; color r is closely related to the metabolism of substance; and texture is the ultimate form of expression of cell division, growth, differentiation, and metabolic interactions.

This chapter does not consider the recognition effect of a single feature when recognizing a feature of original statistics (described in Chapter 9), but, overall, because PCA is the optimal combination of the second-order statistical sense of the above 54 characteristics, the features optimized by the PCA do not consider the difference.

It should be noted that this chapter only is only about the recognition of origin of peanut pods from a variety of different sources, and the reflections of different species of origin are not the same [Department of Agriculture testing new varieties of plants (Guangzhou) sub-centers]. Identification of origin is an important research content of traceable origin technique. We need to build a broader sample of the source for wider origin recognition.

The images collected in this chapter are taken by manually placing the peanuts on the scanner, but the shapes of the peanut pods are not regular, so they cannot be fully structured when placed, which leads to some error in the recognition results. In addition, as the initial weights of neural network are given randomly by the system, it makes the results of the training vary to some extent. In the experiment, the recognition results are the average of results obtained by ten repeated experiments. In addition to the number of hidden-layer neurons, there is no theoretical guidance, and after many experiments, we selected the N-29-8-3 neural network structure.

8.5 CONCLUSION

Traceability of seed origin is an important part of seed quality testing. In the process of producing automatic traceability, correctly identifying the origin of the seed is the basic premise in order to validate the feasibility of automatic traceability of the origin of varieties of peanut pod on image recognition technology. We bring up a peanut pod origin identification method based on physical characteristics of peanut and BP-ANN classification algorithm.

We have chosen a peanut variety (Huayu 22), taken 50 images of each origin by cameras, and measured 54 appearance features of 4 categories to study. We study combinatorial optimization methods of the PCA features. And further study is made about the optimization problem of feature dimensions in the process of neural network pattern classification, which automatically sets up a peanut origin traceability model. The results show that correct identification rate of the model for three different areas of the peanut pod reaches 100%, the recognition rate of the first four statistical features and PCA characteristics reached 95.7% and 97.9%, respectively. PCA feature recognition rate performance increased by about 2% overall than the original features. The other two features have some differences in the recognition, but little effect overall. Building a neural network model of 4-29-8-3 made the overall recognition rate reach 95%, which is far higher than man-made judgments and can greatly improve the recognition efficiency. Methods and conclusions of this chapter have a positive meaning to the study of automatic traceability of origin of peanuts.

REFERENCES

Chen H, Xiong L, Hu X, et al. Identification method for moldy peanut kernels based on neural network and image processing. *Transactions of the CSAE*, 2007, 23(4): 158–161.

Dubey B P, Bhagwat S G, Shouche S P, et al. Potential of artificial neural networks in varietal identification using morphometry of wheat grains. *Biosystems Engineering*, 2006, 95(1): 61–67.

Han Z, Yang J. Detection of embryo using independent components for kernel RGB images in maize. *Transactions of the CSAE*, 2010, 26(8): 222–226

Han Z, Zhao Y. Image analysis and system simulation on quality and variety of peanut. *Journal of the Chinese Cereals and Oils Association*, 2010a, 25(11): 114–118.

Han Z-Z, Zhao Y-G. A cultivar identification and quality detection method of peanut based on appearance characteristics. *Journal of the Chinese Cereals and Oils Association*, 2009, 24(5): 123–126.

Han Z-Z, Zhao Y-G. Quality grade detection in peanut using computer vision. *Scientia Agricultura Sinica*, 2010b, 43(18): 3882–3891.

Hao J-P; Yang J-Z; Du T-Q. A study on basic morphologic information and classification of maize cultivars based on seed image process. *Acta Agronomica Sinica*, 2008, 41(4): 994–1002.

Sakai N, Yonekawa S, Matsuzaki A. Two-dimensional image analysis of the shape of rice and its application to separating varieties. *Journal of Food Engineering*, 1996, 27: 397–407.

Yang J, Hao J, Du T, et al. Discrimination of numerous maize cultivars based on seed image process. *Acta Agronomica Sinica*, 2008, 34(6): 1069–1073.

Yang J, Zhang H, Hao J, et al. Identifying maize cultivars by single characteristics of ears using image analysis. *Transactions of the CSAE*, 2011, 27(1): 196–200.

Yu S. Chinese peanut varieties and their genealogy. *Shanghai Science and Technology Press*, Shanghai, 2008, 12.

Zhao C, Han Z, Yang J, et al. Study on application of image process in ear traits for DUS testing in maize. *Acta Agronomica Sinica*, 2009, 42(11): 4100–4105.

9 Study on the Pedigree Clustering of Peanut Pod's Variety Based on Image Processing

Using image recognition, images of 20 peanut varieties are collected through scanner. For each variety, the images of 100 peanut seeds' positive and two sides are collected; from each image, 50 characteristic features related to shape, color, and texture are obtained. A cluster analysis model based on these features and another cluster analysis model with principal component analysis (PCA) data optimizing are built. In the next step, we have got a pedigree clustering tree for 20 peanut varieties. The first 17 principal components of PCA features have been able to fully simulate the 50 statistical features. There is almost no difference between the classification categories of peanut when the cumulative contribution rate is 85% or more.

9.1 INTRODUCTION

China is the largest country for peanut industry, and the peanut is one of the world's most economically important oil crops. New varieties of peanut are coming soon. Pod is an important organ of peanut, and the distinctness, uniformity, and stability (DUS) testing according to the observation and analysis of its pod shape and size is a very important aspect for breeding a new peanut variety. In normal circumstances, plant breeders measure a large number of plant resources and identify the shape and size of appropriate variety, so as to give identification indicators for new variety, which includes traits, size, color, texture, and so on of the pods. The China's peanut DUS Test Guide (the guide for the briefness, Sub-center of New Plant Variety Test of Department of Agriculture, 2010) specifies the morphological characteristics of pod as the main character, and it can be an effective variety testing and identification based on these indicators. But currently these data collection mainly relies on visual grading and manual measurement. There will be a lot of problems such as slow rate, low accuracy, and poor objectivity. Moreover, as the number of registered varieties increases, there is a need to add new characters to distinguish between similar varieties. In addition, rise in the number of indicators increases the difficulty and complexity of statistics and analysis of the problem. There is a contradiction that as we keep add observed indicators to complete the research process, the presence of too many indicators creates confusion among people. The indicators often have some relevance. In practice, we want to use smaller number of indicators to reflect more original information, and want to find an more effective methods for data reduction.

Image processing is carried out by a new plant DUS testing technology – a highly automatic, effective collection technique, including biochemical techniques and molecular techniques – approved by International Union for the Protection of New Varieties of Plants (UPOV). Image processing technology is used to collect data of seeds' appearance features and then identify the different varieties based on these features. Zhao et al. (2009) have picked up dozens of features for corn seed based on its appearance image and have confirmed that the image processing technology is efficient in corn seed testing. Han and Yang (2010) have picked up the key feature of corn embryo using independent component analysis method. It provides an idea for further exploring the role which the key features of the organism itself played in variety identification. The image process methods are also used in other crops such as wheat (Dubey et al., 2006) and rice (Sakai et al., 1996). However, it is less used in peanut seed testing. The author has explored in depth that the appearance of peanut seed plays some role in identification of its variety (Han and Zhao, 2009, 2010a,b), and obtained a better conclusion. But fewer reports are available about using image processing to study similarities between variety and lineage relationships. Appearance of peanut pod, as a reproductive organ, plays an important role in plant taxonomy and identification of crops. It is an important aspect in DUS testing of a new variety. This chapter discusses the study of peanut pod lineage relationships and variety clustering method, and the influence of PCA characteristic dimension reduction for variety clustering.

9.2   MATERIALS AND METHOD

9.2.1   EXPERIMENTAL MATERIALS

There are 20 peanut breeds chosen for this experiment, all from the farmers' retained seeds, and the samples are the main peanut breeds from major peanut-producing areas of Hebei, Shandong, Rizhao, Weifang, Qingdao, and Laiyang (see Table 9.1). We choose 100 peanut pods from each breed without damage.

We take images of all the 100 pods on a scanner placing them in a fixed order and direction. In order to get a black background, we open the scanner lid fully. In order to obtain complete images, we take two side images and one positive image of the pod as shown in Figure 9.1. There is no back image because it is difficult to scan its back.

We use Canon scanner Canon Scan 8800F, flatbed CCD scanner, with an optical resolution of 4,800 dpi*9,600 dpi, a maximum resolution of 19,200 dpi, and a scan range of 216*297 mm; and computer lenovel idea Centre Kx 8160, Intel Core Duo 2 Quad Q8300 2.5 GHz, memory DDRIII4G; Flash 1G, HDD 500GB; Windows XP operating system. We use MATLAB® R2008a to program for image preprocessing and extracting features from captured images.

TABLE 9.1

Experimental Materials

No.	Variety	Producing Area	No.	Variety	Producing Area
1	Jihua 2	Hebei	11	Ai 2	Laiyang
2	Jihua 4	Hebei	12	Lainong 13	Laiyang
3	Jihua 5	Hebei	13	Luhua 11	Qingdao
4	Zhongnong 108	Hebei	14	16-2	Qingdao
5	Tianfu 3	Hebei	15	Xiaobaisha	Rizhao
6	Weihua 8	Hebei	16	Luhua 9	Rizhao
7	Huayu 22	Laiyang	17	p12	Weifang
8	Huayu 25	Laiyang	18	Unknown 1	Weifang
9	Qinghua 6	Laiyang	19	Unknown 2	Weifang
10	Peanut 101	Laiyang	20	Unknown 4	Weifang

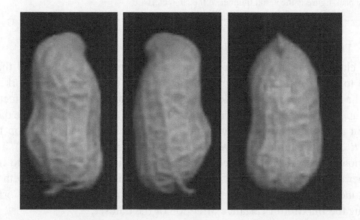

FIGURE 9.1 Scanned image of one variety (Jihua 2).

9.2.2 METHODS

Image preprocessing and feature extraction: Image preprocessing includes conventional pretreatment methods of image enhancement, noise removal, median filtering, edge detection, morphological operations, and image color space conversion. Image recognition is based on the statistical characteristics. We use different statistical indicators to obtain a series of statistical results, and we call them statistical characteristics that include shape, color, and texture. The appearance features we extracted are 50 totally and are divided into three categories (see Table 9.2). The color, shape, and texture features were obtained from the color image, the binary image, and the texture image. The definition of these features are available in the literature (Yang et al., 2008; Zhao et al., 2009).

TABLE 9.2

Statistical Characteristics and Numbers

Class	Characteristic (Feature)
Shape	There are eight characteristics of size, they are area length of long axis, length of short axis, length, width, circumference, diameter of equilateral circle, and convex area. There are five characteristics of shape: they are ellipticity, rectangle ratio, circularity, compact ratio, and concave convex ratio.
Color	The characteristics of the three components of RGB color space are mean, variance, and kurtosis. Their serial number are, respectively, 14–16, 17–19, 20–22, and 23–25. The characteristics of the three components of HSV color space are mean, variance, and kurtosis. Their serial numbers are 26–28, 29–31, 32–34, and 35–37.
Texture	The characteristics of gray-level image include mean 38, variance 39, smoothness 40, third-order moments 41, consistency 42, and entropy 43. There are seven statistically invariant squares that reflect the number distributing characteristic of seeds image's gray value. Their serial numbers are 44–50.

Note: The back figure is the serial number of the characteristic.

9.2.2.1 PCA Algorithm

Directly clustering a sample of more indicators is more complex, and the results are hard to analyze; however, dimension reduction can reduce the workload. So, we use the PCA approach to reduce the dimensionality. PCA is an effective data analysis approach in statistics. Its purpose is to find a set of vectors in the data space to explain the variance of the data, and the data dimensional projection from the original R-dimensional space down to the M-dimensional space ($R>M$). The main information from the data is saved after the dimensionality reduction to make the data easier to handle. The PCA is an optimal dimension compression method under minimal mean square error. The PCA method is to find some mutually orthogonal axes along the maximum variance direction of the dataset.

9.2.2.2 Clustering Algorithm

Clustering method is a multivariate statistical method that classifies the samples or variables according to the similarity degree of nature. It can classify the multi-index data. In this chapter, we use the shortest distance method. The idea is that we see every sample as one class and then calculate the distance between the classes and merge the closest pair to a new class. Then, we extract the corresponding property values of the new class, recalculate the distance between classes, and repeat merging until the stop conditions are met.

9.3 CONCLUSION AND ANALYSIS

9.3.1 Statistical Characteristics Clustering

The samples we collected are great, and directly using these samples to cluster is not appropriate because of the large amount of data, and the clustering results are

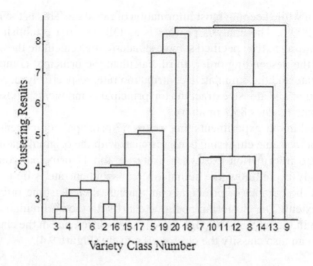

FIGURE 9.2   Clustering pedigree chart of statistical characteristics.

more difficult to analyze. So in our experiment, we only use the mean value of the corresponding characteristics of each type to clustering, and the pedigree shown in Figure 9.2 can be broadly classified into three categories. Obviously, the No. 9 Qinhua 6 peanut belongs to only one category, because this variety is crossbred from Baisha 1016 and 99D1, and it belongs to pearl beans type of little grain peanut varieties. In addition, the No. 7, 10, 11, 12, 8, 14, and 13 are clustered into the second category, and the other is divided into the third category. The peanuts in the second category are mainly from the production areas of Laiyang and Qingdao. Maybe the external environment leads to the common characteristics of appearance. The No. 18 and 20 in the third category are more different from other varieties. After, we investigated the germplasm resources and found that the No. 18 is a typical variety of single-fruit peanut, and the No. 19 is a typical multi-fruit variety of peanut. Such peanuts are mainly produced in Hebei. It can be seen that the type of fruit and their production area have great influence on the result of clustering process.

Moreover, the clustering pedigree can reflect the genetic relationship between varieties to some degree (Zhao et al., 2009). For example, there are close kinships between No. 3 and No. 4, No. 2 and No. 16, and No. 11 and No. 12. The new peanut variety, Jihua 4, is a hybrid developed by pedigree method with 88-8 (Jinhua 5) as female parent and 8609 (7924 × Lainong 4-4) as male parent by Cereal and Oil Crops Institute, Hebei Academy of Agriculture and Forestry Sciences. Lainong 13 is developed with Jinhua 28 as female parent and 534-211 as male parent. There are close kinship among these varieties.

9.3.2   PCA CLUSTERING

In many practical problems, there is a certain relationship between the multiple variables. On the basis of the correlation analysis, multiple variables interrelated originally can make up some less independent comprehensive indicators by linear transformation while keeping most information of raw data. This is the PCA method (Yang et al., 2008). The analysis process is as follows: first establish sample data matrix R according to the specified sample standard; then calculate the characteristic roots of R in the descending order; third, calculate the principal component of the contribution rate and the cumulative contribution rate, respectively, according to the existing formula; and at last, extract the top principal components whose cumulative contribution rate reaches 85% or above.

It was found in our experiments that we use 17 principal components for clustering and obtain the same clustering pedigree chart with the original features. It shows that there is no information loss when using the top 17 principal components for clustering analysis, and now, the cumulative contribution rate is up to 100%. With a decrease in the number of principal components, there is some information loss to a certain extent. The clustering pedigree chart with the accumulation rate above 85% is shown in Figure 9.3. Although there is some difference in the classification of subclasses, it can also classify the classes effectively (Figure 9.4).

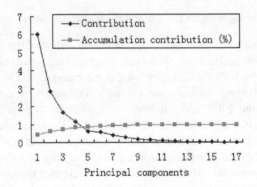

FIGURE 9.3   The contribution and accumulation rate of the top 17 principal components.

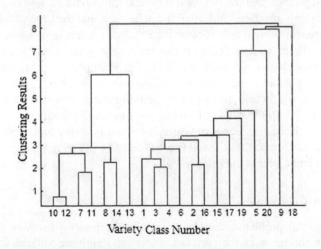

FIGURE 9.4   Clustering effect using five principal components.

9.4 DISCUSSIONS

The principal components after PCA are independent of each other. The effect of analysis depends on the correlation between the original indicators. If the correlation is very strong, then the PCA results will be good, and the vice versa is not good. The indicators selected in this chapter generally have some sort of common or related characteristics. Therefore, the actual results obtained cannot really eliminate the impact of the multiple correlations of the original variables. Artificially adopting useless variables and unilaterally reinforcing the importance of some variables can lead to severe information overlap, while the contribution rate of the principal components will change greatly, which seriously affects the objectivity of the analysis results. Therefore, before the PCA, variables must be carefully selected in order to obtain more objective results of the analysis (Yang et al., 2011).

Directly clustering the samples with more indicators is complex, and the results are hard to analyze; however, dimension reduction can reduce the workload. So, we first use the PCA to reduce the dimensionality, and then take eigenvalues whose cumulative contribution rates are greater than 85% and their corresponding eigenvectors to calculate the principal component. Instead of the original variables, we use the selected principal components for clustering analysis. At last, the classes for clustering are determined, and the result of clustering is appraised.

The fruit type of normal peanut pods is different because of different varieties and producing areas, and this is the interaction of dietary and environmental factors in genetics which are expressed in appearance by different characteristics such as shape, color, and texture. Clustering according to the characteristics can bring great convenience to the follow-up statistical analysis and research. The objective of clustering is to find the model that can describe and distinguish data class and concepts and can also be used to predict the class of objects whose varieties are unknown (Han et al., 2009). System clustering, as an effective method to solute this kind of problems, is a multivariate statistical method, which is to classify samples or variables according to the qualitative intimacy similarity (Han and Zhao, 2010a).

The size of plant organs depends on the cell division and growth, the shape depends on cell differentiation, the color is closely related to the metabolism of chromogenic material, and the texture is the final reflection of interaction between the cell's cleavage, growth, differentiation, and metabolism. Therefore, we can generally agree that the traits of these four kinds of properties are independent of each other. In addition, there is a high degree of positive correlation between heritability and recognition rate (Hao et al., 2008). The study of peanut pod's image characteristics in this chapter is based on the above conclusions.

Using image processing methods to measure plant morphology not only replaces part of the manual measurement and obtains more precise result, but can also find more pedigrees of peanut variety through clustering analysis and distinguish unknown varieties.

PCA method reduces parameters' dimension of variety discrimination and pedigree clustering to some degree, and accelerates the speed of system clustering and enhances the system's maneuverability and practicability.

It should be said that although the result of this study is suitable to genealogy clustering, it is only the initial result in this realm. The gathered peanut seeds are all from the peasants' that are reserved for planting, so the seeds' purity is distrustful and the collected seeds are picked randomly, and the optimization of the chosen seeds' intrinsic characteristics is not performed. Because of the above-mentioned reasons, the pedigree relationship of peanut's new variety cannot be fully sketched. So before using it formally in pedigree clustering and variety identification, it is necessary to increase the number of varieties for test and further survey the diversities of varieties, consistence of the variety, and yearly stability of producing area.

9.5 CONCLUSIONS

We have taken one positive and two side images of peanut pods for 20 varieties using a scanner, and 100 peanut pods have been chosen from each variety. We have got 50 features for three categories of shape, color, and texture of each image, using image recognition method. We have set up a cluster analysis model based on these features and another cluster analysis model with PCA data optimizing. In the next step, we have got a clustered tree for 20 peanut varieties. The first 17 principal components of PCA features have been able to fully simulate the 50 statistical features. There is almost no difference in the classification categories of peanut when the cumulative contribution rate reaches 85% or more. In this chapter, the indicators of characteristics are measured which can represent the peanut variety. Their cluster analysis and principal components are performed by a program based on MATLAB software. This can provide reference not only for further classification of peanut shape and statistical analysis, but also for the research of peanut pedigree. This also provides reliable basis for the test of peanut's distinctness, uniformity, and stability (DUS test).

REFERENCES

Dubey B P, Bhagwat S G, Shouche S P, et al. Potential of artificial neural networks invari-
    etal identification using morphometry of wheat grains. *Biosystems Engineering*, 2006,
    95(1): 61–67.
Han Z, Yang J. Detection of embryo using independent components for kernel RGB images
    in maize. *Transactions of the CSAE*, 2010, 26(8): 222–226.
Han Z, Zhao Y. A cultivar identification and quality detection method of peanut based on
    appearance characteristics. *Journal of the Chinese Cereals and Oils Association*, 2009,
    245: 123–126.
Han Z, Zhao Y. Image analysis and system simulation on quality and variety of peanut.
    Journal of the Chinese Cereals and Oils Association, 2010a, 25(11): 114–118.
Han Z, Zhao Y. Quality grade detection in peanut using computer vision. *Scientia Agricultura
    Sinica*, 2010b, 43(18): 3882–3891.
Hao J-P; Yang J-Z; Du T-Q. A study on basic morphologic information and classification of
    maize cultivars based on seed image process. *Acta Agronomica Sinica*, 2008, 41(4):
    994–1002.
Sakai N, Yonekawa S, Matsuzaki A. Two-dimensional image analysis of the shape of rice
    and its application to separating varieties. *Journal of Food Engineering*, 1996, 273:
    97–407.

Sub-center of New Plant Variety Test of Department of Agriculture (Guangzhou). The shooting specifications of photos of characteristics in DUS testing of new varieties of peanuts. Agriculture Press of China, Beijing, 2010, 6.

Yang J, Hao J, Du T, et al. Discrimination of numerous maize cultivars based on seed image process. *Acta Agronomica Sinica*, 2008, 34(6): 1069–1073.

Yang J, Zhang H, Hao J, et al. Identifying maize cultivars by single characteristics of ears using image analysis. *Transactions of the CSAE*, 2011, 27(1): 196200.

Zhao C, Han Z, Yang J, et al. Study on application of image process in ear traits for DUS testing in maize. *Acta Agronomica Sinica*, 2009, 42(11): 4100–4105.

10 Image Features and DUS Testing Traits for Identification and Pedigree Analysis of Peanut Pod Varieties

DUS (distinctness, uniformity, and stability) testing of new peanut varieties is an important method for peanut germplasm evaluation and identification of varieties. In order to verify the feasibility of DUS testing for identification of peanut varieties based on image processing, 2,000 pod images of 20 peanut varieties were obtained by a scanner. First, six DUS testing traits were quantified successfully using a mathematical method based on image processing technology, and then, size, shape, color, and texture features (totally 31) of peanut pods were also extracted. On the basis of these 37 features, the Fisher algorithm was used as feature selection method to select "good" features to expand DUS testing traits set. Then, support vector machine (SVM) and K-means algorithm were used as model of varieties recognition and clustering analysis method, respectively, to study the problems of varieties identification and pedigree clustering comprehensively. It is found that, by the Fisher feature selection method, a number of significant candidate features for DUS testing have been selected, which can be used in the DUS testing further; using the top half of these features (about 16) after ordered by the Fisher discrimination, the recognition rate of SVM model is more than 90% in identification of varieties, which is better than unordered features. Besides this, a pedigree clustering tree of 20 peanut varieties has been built based on K-means clustering method, which can be used in in-depth study of the genetic relationship between different varieties. In summary, the results of this chapter may provide a novel reference method for future DUS testing, peanut varieties identification, and study of peanut pedigree.

10.1 INTRODUCTION

Peanut is one of the most economically important oil crops. China is the first peanut industry superpower in the world. Correct identification of peanut varieties is the most important aspect for genetic breeding and DUS testing.

Traditionally, manual measurement was used as a major method in practice for DUS testing in germplasm resources research, such as measuring the average weight, thickness, length and width of kernel, bulk density, volume weight, and density of the quantitative traits (features). These features were always evaluated by hand in field tests generally. This manual approach brings some problems such as low efficiency and high cost, which limits the efficiency of seed testing. In recent years, digital image processing has arisen as a new detection method with high speed, strong identification ability, and repeatability, and the burdensome traditional testing methods have been paid very less attention. As a highly automated intelligence technology, similar to biological, chemical, and molecular technology, image processing has become a new technology for plant DUS testing and has been approved by The International Union for the Protection of New Varieties (UPOV, 2002). Image processing method can not only measure crop morphology automatically, instead of partial manual labor, but it also has some new DUS candidate properties with higher precision.

Using image processing technology, Yang et al. (2008) and Hao et al. (2008) extracted dozens of features from corn kernel and confirmed the importance of this technology in corn seed inspection. Furthermore, Zhao et al. (2009) studied ear traits of maize using DUS testing based on image processing, quantified seven DUS testing traits successfully, and found that image processing is a useful tool for gathering and quantifying maize ear DUS and several other traits. Han and Zhao (2010, 2012) used the qualities of peanut kernel in recognition of the seed varieties and obtained good results. Using pod beak and constriction, Deng et al. (2015) analyzed DUS traits quantitatively based on Freeman chain code, and extracted and quantified three DUS testing traits successfully using image processing techniques. Besides this, image processing methods were also applied for wheat, rice (Sakai et al., 1996; Dubey et al., 2006), and other crops (Lootens et al., 2007; Keefe, 1999). However, in terms of inspection of peanut pods, there is not much relevant research.

DUS testing traits of peanut include pod, kernel, and plant (root, stem, and leaf) (GB/T 1532-2008; DB34 T 252.4-2003). As an important peanut organ, peanut pods play an important role in plant taxonomy and crop breeding. It is an important aspect of DUS testing for new varieties. Some genetic and environmental factors are found to be reflected in the appearance of peanut pods; thus, the appearance of peanut pods can be used to identify different varieties. In this chapter, we first propose quantitative methods of six DUS testing traits stated by Agriculture Department of China (2010) and evaluate four features of peanut pods (size, shape, color, and texture) using statistical calculation. Then, we use the Fisher method (Chu et al., 2014) as feature selection method to find potential DUS testing features and use SVM (Samiappan et al., 2013) as classifier to identify these peanut varieties. At last, we investigate the cluster analysis and pedigree relations for different peanuts using K-means (Aristidis et al., 2003) method.

10.2   MATERIALS AND METHOD

10.1.1   Materials

10.1.1.1   Peanut Samples

Twenty varieties of peanuts (100 pods of each variety), with typical and consistent appearance, were selected manually. These samples were collected from Hebei province and Shandong province (Qingdao, Weifang, and Laiyang), and were harvested in 2011. They were mainly large peanut varieties in the north of China (see Table 10.1).

TABLE 10.1
Variety Identification Experimental Materials

No.	Varieties	Area	No.	Varieties	Area
1	Yihua 2#	Hebei	11	Ai 2	Laiyang
2	Yihua 4#	Hebei	12	Lainong 13#	Laiyang
3	Yihua 5#	Hebei	13	Luhua 11	Qingdao
4	Zhongnong 108	Hebei	14	16-2	Qingdao
5	Tianfu 3#	Hebei	15	Xiaobaisha	Qingdao
6	Weihua 8#	Hebei	16	Luhua 9#	Qingdao
7	Huayu 22	Laiyang	17	p12	Weifang
8	Huayu 25	Laiyang	18	Unknown name 1	Weifang
9	Qinghua 6 #	Laiyang	19	Unknown name 2	Weifang
10	Huasheng	Laiyang	20	Unknown name 3	Weifang

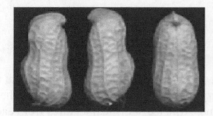 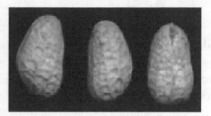

FIGURE 10.1   Scanned images of two varieties (Luhua 11# and Qinghua 6#).

10.1.1.2   Image Acquisition

We used a flatbed scanner (CanoScan 8800 F) for acquiring peanut pod images, which is a plate-type CCD scanner with an optical resolution of 4,800*9,600 dpi, a maximum resolution of 19,200 dpi, and a scanning range of 216*297 mm. For scanning, 100 pods were placed side by side on the glass panel of the scanner. In order to make the background black, the cover was opened fully. In order to acquire more complete images for each pod, the images of front and back sides were scanned. Figure 10.1 illustrates the images of two varieties (Luhua 11# and Qinghua 6#). Besides this, the computer used in the experiment was Lenovo idea Centre Kx 8160 with Intel Core CPU@ 2.5 GHz, 4 GB RAM, 1 GB Flash memory, and Windows 7 operating system. Image preprocessing and feature extraction were based on MATLAB® 2008a integration environment.

10.1.2   METHODS

10.1.2.1   Feature Extraction

Image preprocessing included traditional methods such as image enhancement, de-noising, median filtering, edge detection, morphological operation, and image color

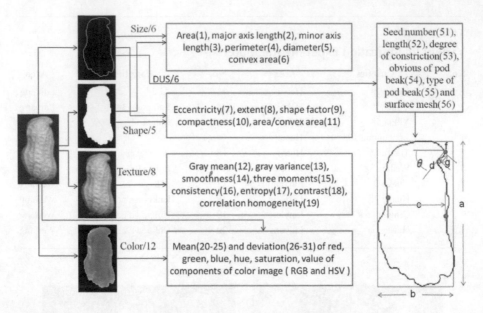

FIGURE 10.2 Quantification of DUS testing traits and extraction of statistic features.

space transformation. After preprocessing, we obtained four kinds of images, that is, color image (RGB/HSV), gray image, binary image, and edge image. Based on these images, we quantified DUS testing traits and extracted statistical features from each image as illustrated in Figure 10.2. Besides surface texture, the mean value of two side images was also used.

1. Extraction of statistical features of images

   The statistical features can be divided into four categories (as shown in Figure 10.2): size, shape, color, and texture. Therein, size and shape features were extracted from edge and binary images, texture features were extracted from gray image, and color features were extracted from RGB and HSV images. The numbers of size, shape, texture, and color features are, respectively, 6, 5, 8, and 12. All these statistical features were illustrated in the following, and they were defined in our previous related researches (Han et al., 2010; Han and Zhao, 2012).

   Size/6: Area (1), major axis length (2), minor axis length (3), perimeter (4), diameter (5), and convex area (6)

   Shape/5: Eccentricity (7), extent (8), shape factor (9), compactness (10), and area/convex area (11)

   Texture/8: Gray mean (12), gray variance (13), smoothness (14), three moments (15), consistency (16), entropy (17), contrast (18), and correlation homogeneity (19)

   Color/12: Mean (20–25) and deviation (26–31) of red, green, blue, hue, saturation, and value of components of color image (RGB and HSV).

2. Quantification of DUS testing traits

The quantification includes six DUS testing traits that were stated in peanut varieties DUS testing standard of China (Agriculture Department of China, 2010). They are grain number, length, degree of constriction, obvious of pod beak, type of pod beak, and surface mesh. Among these, three traits (length, degree of constriction, and pod beak) have already been quantified, particularly in our previous study (Deng et al., 2015). The quantified process is illustrated in the subimage at the right bottom corner of Figure 10.2.

Grain number (32): Defined as the number of the waist minus one, here, the number of the waist was already calculated in degree of constriction below

Length (33): Demonstrated as a, it is the length of minimum bounding rectangle

Degree of constriction (34): Defined as $1 - c/b$, where b is the width of minimum bounding rectangle, c is the width of pod's waist (the horizontal ordinate difference between left and right waists), here b and c can be calculated by boundary Freeman chain code (Deng et al., 2015);

Obvious of pod beak (35): Defined as $g*\cos(\theta)$, here, g is the distance between d and f, θ is the angle between the horizontal line and the line connecting points d and f, in which, d and f are, respectively, the protruding point ($p'(z) = 0$ and $p''(z) > 0$) and the sunken point ($p'(z) = 0$ and $p''(z < 0)$) of the pod beak

Type of pod beak (36): The absolute value of curvature at the position of the pod beak protruding point (denoted as f)

Surface mesh (37): The number of mesh holes after binarization of pod image and extracting skeleton of pod area.

10.1.2.2 Analysis and Identification Model

Fisher: Effective feature selection aims to seek for a subset (M) to represent the full features from the original features set (R). The subsets of features ($M < R$) include as much information as possible of the original features and are equal or more efficient because dimensionality reduction of original data makes the identification less time-consuming. The Fisher algorithm (Samiappan et al., 2013) is used as a method for feature selection to identify the most significant features in this chapter, which can also be used in the development of the image identification system. The features that have stronger ability for discrimination will be good features or traits for DUS testing. With the reduced features, it will be possible to construct a simple machine vision system for peanut identification and detection. In this chapter, the Fisher discrimination ability will measure the importance of each features.

SVM is a method based on Vapnik-Chervonenkis (VC) dimensional theory in statistical learning and the principle of minimizing the structure risk, and was first proposed by Cortes C and Vapnik V (Cortes and Vapnik, 1995). It has been widely used in many fields and can solve both linear and nonlinear multivariate calibration problems. Rather than a quadratic programming (QP) problem, a set of linear equations were used to get the support vectors (SVs). Here, we utilize SVM as the classifier. The radial basis function (RBF) is used as the kernels function in consideration of its excellent performance. In this study, the optimal parameter values of kernels function are calculated by grid search, and they are calculated by free IBSVM toolbox (v2.91) (Chang and Jin, 2011) in MATLAB 2008a.

K-means is one of the most commonly used unsupervised clustering methods (Aristidis et al., 2003). First, it chooses K objects discretionarily as the initial clustering center from N objects, and the rest of the objects will be divided to K classes according to the similarity (or distance) to clustering center. Then, calculate the clustering center (mean of all objects clustering) of the new class after each clustering step. Repeat this process until the standard measurement function begins to convergence; here, the mean square error is used as a standard measure function typically. After clustering is completed, the distance within one class is as small as possible, and the distance between classes is as large as possible. In this chapter, it is to measure the genetic relationship between different peanuts by calculating the distance between classes.

10.3   RESULTS AND ANALYSIS

10.3.1   Feature Selection by Fisher

The Fisher discrimination of each feature will be used to evaluate the importance of peanut identification and DUS testing. Here, we calculated the Fisher discrimination of above 37 features including size, shape, texture, color features, and quantified DUS testing traits. The result is illustrated in Figure 10.3, in which the discrimination ability of each single feature has been drawn as histogram in the bottom, the average discrimination ability of each kind is shown above (shadow height), and the numbers in the brackets is the average value.

It is found that different features have different discrimination abilities. Among the five kinds of features, DUS testing traits have the best discrimination ability with mean 7.68 and the color features have the worst discrimination ability. The reason may be that the DUS testing traits are designed by agronomy experts based on years

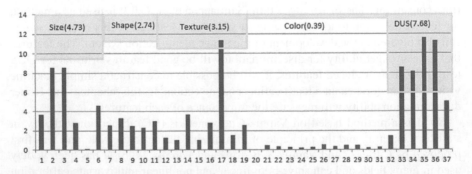

FIGURE 10.3   The Fisher discrimination ability of different features.

of experience and they have been tested for a long time. Moreover, the color of peanut pods is similar to that of dry soil as per DUS testing, but the main color of peanut pods of the samples we selected is principally yellow, so the color difference is not obvious. The size, shape, and texture features have medium discrimination ability; however, the ability of size and texture features is better than that of shape features. It may be the fact that for these three types of features, the genetic factor plays a greater role than the environmental factor. Besides this, the shape features are different from the other two kinds of features, and they are more likely to be affected by the hardness of the earth. Even so, according to the Fisher discrimination, all statistical features can be sorted in descending order. Also, we can select some good features for seed identification and DUS testing from these statistical features. Features such as major axis length (2), minor axis length (3), convex area (6), and entropy (17) are better than grain number (32), which can be seen as candidate features for DUS testing. These "good" features can be regarded as being extended to future DUS testing traits. Grain number is a "bad" feature, for which the reason may be that most of the peanuts we chose are double kernel peanuts, and the difference is not significant in grain number. Besides this, for area (1), extent (8), smoothness (14), consistency (16), area/convex area (11), perimeter (4), eccentricity (7), and shape factor (9), their discrimination ability is high, and they can also be used as a potential trait for DUS testing in the future. Here, we have chosen these 12 features for DUS testing as candidate features.

10.3.2 Variety Identification by SVM

Accurate peanut recognition is the basis for identification of new varieties, identification of fake seeds, seed testing, and DUS testing. Theoretically, with an increase in the number of features, the recognition rate increases accordingly. However in the actual pattern classification, a large number of features will increase the burden of computer identification and consume longer time. Among these features, there often exists redundancy information, which may lead to decrease of recognition efficiency.

Here, SVM algorithm and fivefold cross-validation method (Wiens et al., 2008) were employed for peanut recognition. For each variety, 80 of the 100 samples were used as training set, and the rest were used for testing. There are 20 varieties of peanuts; each variety has 100 pods; and for each pod, 37 features have been extracted. So, the size of feature matrix is 20*100*37.

Table 10.2 illustrates the identification results using different feature sets. The left column is the identification rate of size, shape, texture, color features, and DUS traits. The right column is the identification rate of top six, 7–14, 15–21, 22–28, and last nine features after the features ordered according to the weight of the Fisher discrimination. It is found that different classes of features have different identification abilities. In terms of recognition rate, the order will be DUS traits > size features > texture features > shape features > color features. This conclusion is consistent with the Fisher discrimination. After ordered based on discrimination ability, the strongest discriminative features have been sorted to the first and the top seven features have the best accuracy (above 70%), which is higher than that of DUS traits (60%–70%). This also reflects that only using six traits stipulated by the

TABLE 10.2
Contribution of Different Classes of Features by SVM

Feature Set	Training Set	Testing Set	Arrange Set	Training Set	Testing Set
DUS	**68.4**	**60.2**	**Top 6**	**76.6**	**71.2**
Size	58.3	55.2	7–14	48.0	41.8
Shape	40.7	38.0	15–21	40.6	34.7
Texture	50.1	48.3	22–28	33.5	28.6
Color	28.3	24.5	Last 9	21.3	20.6

Note: Bold values represent the highest recognition rate.

national DUS test guidelines is not effective in identifying new varieties, and it needs to properly expand the features set of the guideline.

In accordance with the weight of the Fisher discrimination, putting the features that have better discrimination ability on top can effectively improve the efficiency of identification of varieties, and using fewer features, higher recognition rate can be achieved. As illustrated in Figure 10.4, the diamond line shows the increase of recognition rate with the increase of features arranged by their serial numbers in Section 10.2.2.1 and the square line shows the increase of recognition rate with the increase of features according to the discrimination ability by Fisher. When only half (18) of the features are used (including 6 DUS testing traits and 12 candidate features), the recognition rate has reached more than 90%.

The recognition results of each variety obtained using the top half (18) features are illustrated in Table 10.3, in which the figure in the header line of horizontal and vertical coordinates is the peanut variety number, and the value in the table is the identification number of horizontal variety which is recognized as vertical variety. It is found that the overall recognition rate has reached 92.5%. There is a big difference between the recognition rates of different categories such as the 02, 03, 04, and 16

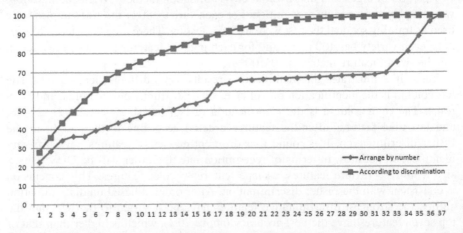

FIGURE 10.4   The effect on SVM model when the quantity of features increased.

TABLE 10.3
Identification Results of Different Varieties Based on SVM Model

No.	01	02	03	04	05	06	07	08	09	10	11	12	13	14	15	16	17	18	19	20
01	97	0	2	1	0	0	0	0	0	0	0	0	0	0	0	0	0	0	0	0
02	0	77	0	0	0	3	0	0	0	0	0	0	0	0	0	20	0	0	0	0
03	2	0	82	13	0	3	0	0	0	0	0	0	0	0	0	0	0	0	0	0
04	0	0	15	85	0	0	0	0	0	0	0	0	0	0	0	0	0	0	0	0
05	0	0	1	0	97	0	0	0	0	0	0	0	0	0	1	0	1	0	0	0
06	1	2	0	1	0	94	0	0	0	0	0	0	0	0	0	2	0	0	0	0
07	0	0	0	0	0	0	96	0	0	2	1	1	0	0	0	0	0	0	0	0
08	0	0	0	0	0	0	0	95	0	1	1	0	0	3	0	0	0	0	0	0
09	0	0	0	0	0	0	0	0	100	0	0	0	0	0	0	0	0	0	0	0
10	0	0	0	0	0	0	2	1	0	97	0	0	0	0	0	0	0	0	0	0
11	0	0	0	0	0	0	1	0	0	0	88	11	0	0	0	0	0	0	0	0
12	0	0	0	0	0	0	1	0	0	0	10	87	1	1	0	0	0	0	0	0
13	0	0	0	0	0	0	0	0	0	0	0	0	99	1	0	0	0	0	0	0
14	0	0	0	0	0	0	0	4	0	0	0	1	0	95	0	0	0	0	0	0
15	0	0	0	0	1	0	0	0	0	0	0	0	0	0	94	0	3	0	1	1
16	0	21	0	0	0	0	0	0	0	0	0	0	0	0	0	78	1	0	0	0
17	0	0	0	0	2	0	0	0	0	0	0	0	0	0	5	0	93	0	0	0
18	0	0	0	0	0	0	0	0	0	0	0	0	0	0	0	0	0	99	1	0
19	0	0	0	0	0	0	0	0	0	0	0	0	0	0	0	0	2	0	98	0
20	0	0	0	0	0	0	0	0	0	0	0	0	0	0	0	0	0	1	0	99

peanut varieties. This may be because these species are similar in appearance and closely related to each other. In the next section, we will further study the genetic relationship between peanut varieties based on pedigree clustering method.

10.3.3 Paternity Analysis by K-Means

Due to the large number of samples (about 2,000 samples) directly used for data clustering, the clustering result will be very difficult to analyze. So in this experiment, only the mean of features of each variety was used for clustering. K-means clustering method was used in this experiment, and the dendritic diagram is shown in Figure 10.5.

Figure 10.5 illustrates that peanuts are roughly classified into three categories. It is obvious that Qinghua 6# (numbered 9) is divided into a separate class. By investigating its germplasm resources, we found that this variety is the hybrid of Baisha 1016 and 99D1, and it belongs to Spanish-type small peanut variety which is different from other big peanut varieties. Besides this, the varieties numbered 7, 10, 11, 12, 8, 14, and 13 were gathered to the second major category, and the rest were divided into the third category. The varieties of the second category were mainly harvested from Qingdao, China, and it may be due to external environmental factors that make this category exhibit the same characteristics in appearance. The varieties numbered

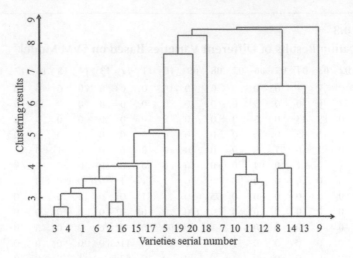

FIGURE 10.5   Varieties clustering tree.

18 and 20 of the third category have big difference from other varieties. By inves-tigating its germplasm resources, we find that the variety numbered 18 is a typical single kernel peanut, and the variety numbered 19 is multi-kernel peanut. This kind of peanut is mainly distributed in Hubei province, China. It is obvious that different pod types and different origination influenced the clustering results. The varieties numbered 3 and 4, the varieties numbered 2 and 16, and the varieties numbered 11 and 12 may have a close genetic relationship. In the cluster tree, they gather together first, which further illustrates that they are easily confused (Table 10.3).

10.4   DISCUSSIONS

10.4.1   Biological Basis for Seed Testing with Appearance

Using ten seeds of inbred maize, Hall et al. (2016) compared the identification abilities based on the dataset of morphology, metabolism, ribonucleic acid (RNA), and single-nucleotide polymorphisms (SNPs), and the results indicated that the morphological features ranked only second to SNP. By calculating the morpho-logical similarity matrix of inbred maize and the coefficient of genetic relationship, Babić et al. (2012) compared the similarities between morphology and gene, and found that description of morphology was very helpful for estimation of genetic similarity and the morphological information had great significance for maize breeding, especially in the treatment of a large number of germplasm resources or in the case of unknown germplasm. In a work published in *Science*, Cortijo et al. (2014) found that plants showed significant morphological differences, and these differences could be passed to the next generation. Based on genetic data spanning on seven generations, some epigenetic traits of plants can be stably inherited for at least 20 generations. Nagy et al. (2003) compared random amplified polymorphic

DNA (RAPD), simple sequence repeat (SSR), and morphological markers, and the results showed that these first two markers could only partially reflect the genetic relationship between maize varieties, but when combined with morphological data, it could provide close association of cluster analysis and formation of pedigree data. At the same time, using morphological data for classification is economic, intuitive, simple, and rapid, which is still a basic method for studying germplasm resources, and morphological data is one of the important basic gist which are indispensable to classification of different varieties.

The morphological appearance of crop traits is the result of interaction between genetic and environmental factors in genetics. In appearance features, the size of plant organs depends on cell division and growth, the shape depends mainly on cell metabolism, the color is closely related to plant pigmentation, and the texture is the reflection of cell division, growth, differentiation, and metabolism of interaction. So, it can be assumed that these four characteristic features are mutually independent (Yang et al., 2010). In this study, identification and DUS testing of peanut pods by image technology are carried out based on the above ideas.

10.4.2 Finding Candidate Features for DUS Testing

Image processing method can not only measure plant morphology with higher precision, instead of partial manual measurement, but can also provide new DUS candidate properties. This study found that peanut pods' image can provide more valuable candidate properties and can identify peanut pods effectively. The Fisher discrimination ordering can put "good" features on top and makes better performance recognition than previous sorting methods.

In DUS testing, whether a feature is suitable or not, besides labor force, mainly depends on its ability to discriminate between varieties. To a certain extent, the level of discrimination indicated the distinction between different varieties (Yang et al., 2011). If the recognition rate of a feature is high, its ability to discriminate between varieties is high naturally, and if the recognition rate is minimum 0, it indicates that there is no difference between the varieties and the feature does not have any discrimination; the maximum 1 indicates that the character has the biggest distinction ability. The same is true for a set of features.

Among these five kinds of features, DUS testing features have the highest discrimination. The reason may be that these features combine years of experience of breeders and are excellent in practice. Color features have the worst discrimination ability, which may be because the germplasm resources we selected are almost the same. The germplasm resources are relatively narrow, and the theme color is earthy yellow. In addition, based on the clustering results, the morphological characteristics are not effective in cultivar identification, probably due to the different habitats, soil environment, and other factors affecting the development of peanut pod. Since the collected peanut varieties were reserved seeds from local farmers, after planting many years, these seeds are inevitable mixture, which leads to impure varieties and cause the deviation of recognition result. More testing using breeder's seeds will make the results of the research work more convincing.

The pod type of normal peanuts showed difference because of differences of varieties, habitats, and external environment. Genetically, it is the result of interaction between genetic and environmental factors; in the case of appearance, it is different in size, shape, color, and texture (Agriculture Department of China, 2010). According to the clustering of characteristics, the study of genetic relationship and pedigree distribution can provide great convenience for subsequent statistical analysis and research; at least, it can provide an objective method for selection of breeding varieties.

It should be explained that the results of this study can be applicable to DUS testing. However, this is only a preliminary result in this field; so, before using this method for DUS testing, it is necessary to further expand the number of varieties to further investigate the candidate traits among the varieties of diversity, variety, origin, and stability in future.

10.5 CONCLUSIONS

The DUS testing of varieties is an important method for breeding new peanut variety and trading its seeds. Accurate identification of seed varieties is the premise. In order to verify the feasibility of image processing method for variety identification and DUS testing of peanut pods, 5 categories of 37 features of peanut pods have been quantified and extracted using the Fisher feature selection, SVM varieties identification, and K-means cluster analysis method. A series of problems in the selection of candidate features for variety DUS testing, establishment of varieties identification model, and clustering of varieties pedigree have also been studied.

This study selected a number of meaningful alternative features for DUS testing, and it is found that, with these optimal features, the recognition rate of SVM model can achieve high recognition performance with accuracy above 90%. The results from further investigation of 20 peanut cultivars pedigree clustering tree are consistent with the results of SVM model. Finally, the meaning and limitation of the genetic basis of DUS testing using image processing have also been discussed. This study provides a new idea for peanut DUS testing in the future and has certain reference value for peanut germplasm identification.

REFERENCES

Agriculture Department of China. New varieties of plants test sub-centre (Guangzhou), photos standards of new peanut varieties DUS test. *China's Agriculture Press*, Beijing, 2010, 6 (in Chinese).

Aristidis L, Nikos V, Verbeek J. The global k-means clustering algorithm. *Pattern Recognition*, 2003, 36(2): 451–461.

Babić M, Babić V, Prodanović S, et al. Comparison of morphological and molecular genetic distances of maize inbreds. *Genetika*, 2012, 44(1): 119–128.

Chang C C, Lin C J. LIBSVM: A library for support vector machines. *ACM Transactions on Intelligent Systems and Technology (TIST)*, 2011, 2(3): 27.

Chu X, Wang W, Zhang L-D, et al. Hyperspectral optimum wavelengths and Fisher discrimination analysis to distinguish different concentrations of aflatoxin on corn kernel surface. *Spectroscopy and Spectral Analysis*, 2014, 34(7): 1811–1815.

Cortes C, Vapnik V. Support-vector networks. *Machine Learning*, 1995, 20: 273–297.

Cortijo S, Wardenaar R, Colomé-Tatché M, et al. Mapping the epigenetic basis of complex traits. *Science*, 2014, 343(6175): 1145–1148.

DB34 T 252.4-2003. National standards of the People's Republic of China: Pollution-free peanut IV: Peanut (kernel). *Standards Press of China*, Beijing, 2003 (in Chinese).

Deng L, Du H, Han Z. Identification of pod beak and constriction and quantitative analysis of DUS traits based on Freeman Chain Code. *Transactions of the Chinese Society of Agricultural Engineering (Transactions of the CSAE)*, 2015, 31(13): 186–192 (in Chinese with English abstract).

Dubey B P, Bhagwat S G, Shouche S P, et al. Potential of artificial neural networks in varietal identification using morphometry of wheat grains. *Biosystems Engineering*, 2006, 95(1): 61–67.

GB/T 1532-2008. National standards of the People's Republic of China: Peanut. *Standards Press of China*, Beijing, 2008 (in Chinese).

Hall B D, Fox R, Zhang Q, et al. Comparison of genotypic and expression data to determine distinctness among inbred lines of maize for granting of plant variety protection. *Crop Science*, 2016, 56(4): 1443–1459.

Han Z, Zhao Y. Quality grade detection in peanut using computer vision. *Scientia Agricultura Sinica*, 2010, 43(18): 3882–3891 (in Chinese with English abstract).

Han Z, Zhao Y. Variety identification and seed test by peanut pod image characteristics. *Acta Gronomica Sinica*, 2012, 38(3): 535–540 (in Chinese with English abstract).

Han Z, Zhao Y, Yang J. Detection of embryo using independent components for kernel RGB images in maize. *Transactions of the CSAE*, 2010, 26(3): 222–226 (in Chinese with English abstract).

Hao J, Yang J, Du T, et al. A study on basic morphologic information and classification of maize cultivars based on seed image process. *Acta Agronomica Sinica*, 2008, 41(4): 994–1002 (in Chinese with English abstract).

Hsu C-W, Chang C-C, Lin C-J. A practical guide to support vector Classi_cation [EB/OL], April 15, 2010. www.csie.ntu.edu.tw/~cjlin

Keefe P D. Measurement of linseed seed characters for distinctness,uniformity and stability testing using image analysis. *Plant Varieties & Seeds*, 1999, 12(2): 79–90.

Lootens P, van Waes J, Carlier L. Evaluation of the tepal colour of Begonia×tuberhybrida Voss for DUS testing using image analysis. *Euphytica*, 2007, 155(1–2): 135–142.

Nagy E, Gyulai G, Szabó Z, et al. Application of morphological descriptions and genetic markers to analyse polymorphism and genetic relationships in maize (Zea mays L.). *Acta Agronomica Hungarica*, 2003, 51(3): 257–265.

Sakai N, Yonekawa S, Matsuzaki A. Two-dimensional image analysis of the shape of rice and its application to separating varieties. *Journal of Food Engineering*, 1996, 27: 397–407.

Samiappan S, Bruce L M, Yao H, et al. Support vector machines classification of fluorescence hyperspectral image for detection of aflatoxin in corn kernels. *Meeting Proceedings*, 2013: 1–4.

UPOV. General introduction to the examination of distinctness, uniformity and stability and the development of harmonized descriptions of new varieties of plants (TG/1/3). The International Union for the Protection of New Varieties of Plants, Geneva, 2002: 11.

Wiens T S, Dale B C, Boyce M S, et al. Three way k-fold cross-validation of resource selection functions. *Ecological Modelling*, 2008, 212(3): 244–255.

Yang J, Hao J, Du T, et al. Discrimination of numerous maize cultivars based on seed image process. *Acta Agronomica Sinica*, 2008, 34(6): 1069–1073 (in Chinese with English abstract).

Yang J, Zhang H, Hao J, et al. Identifying maize cultivars by single characteristics of ears using image analysis. *Transactions of the CSAE*, 2011, 27(1): 196–200 (in Chinese with English abstract).

Yang J, Zhang H, Zhao Y, et al. Quantitative study on the relationships between grain yield and ear 3-D geometry in maize. *Scientia Agricultura Sinica*, 2010, 43(21): 4367–4374 (in Chinese with English abstract).

Zhao C, Han Z, Yang J. et al. Study on application of image process in ear traits for DUS testing in maize. *Acta Agronomica Sinica*, 2009, 42(11): 4100–4105 (in Chinese with English abstract).

11 Counting Ear Rows in Maize Using Image Processing Method

Ear rows is an important agricultural feature in *maize* (*Zea mays* L.). In order to examine the feasibility of ear rows counting method by machine vision, a new method was created. This method is based on edge marker and discrete curvature. Seventy-eight digital images of four maize cultivars were scanned from two sides of maize fragment face. Based on these images, we find that the number is mostly 12–18, the absolute error of detection is 0.103, and the relative error rate is 0.66%. The accuracy rate is above 90%. Machine vision has the advantages of low cost and high speed when compared to manual or biochemical detecting methods, and it is feasible to be applied to identification of numerous maize cultivars.

11.1 INTRODUCTION

Ear row is one of the important features in maize. There are some differences in the number of ear rows between different varieties, and even the growth conditions can have impacts on the number of ear rows in maize (Guo et al., 2004). So, it is very important to count the ear rows exactly in the scientific research and practices such as breeding, cultivation, and new varieties of distinctness, uniformity, and stability (DUS) testing in maize. The method to count the ear rows in maize traditionally depends on manual work, but there are inherent defects in the manual counting method. The machine vision counting is a fatigue-free new method based on the computer digital image processing with high testing speed, high ability to identify, high repeatability, and detection of large quantity. People used this testing method for recognition of crop seeds gained good results in corn (Hao et al., 2008; Yang et al., 2008), peanuts (Han et al., 2007), rice (Sakai et al., 1996), wheat (Dubey et al., 2006), and lentils (Venora et al., 2007). People also gained some preliminary results in testing corn seedlings (Zhang and Yang, 2005; Liang and Yang, 2006) and growth in the period of adult phase. However, there is no report on the recognition of corn ear traits using image processing method.

This research intends to make an intelligent recognition and analysis of the corn cross sectional morphology using the corn cross-sectional image obtained by a scanner, depends on a new counting method combining edge marking and discrete curvature, gains a distribution character, and makes an analysis of the impact of factors during the counting.

TABLE 11.1
Fringy Linage Characteristics of Eight Maize Breeds

Maize Cultivars	Breeding Academe	Row Numbers
Zha 918	Nei-meng	10–12
Jing 24	Bei-jing	12
H21	Lai-nong	12–14
P128	Zhong-nong	14
Luyuan 92	Shan-dong	14–16
478	Lai-zhou	16–18
Dan 340	Dan-dong	18–20
Danhuang 02	Dan-dong	20–22

11.2 MATERIALS

We take 78 cross-sectional images of four varieties of corn cultivated in 2007. Here are the four varieties: Nongda 108, Zhengdan 958, Jinhai No. 5, and Tianta No. 5.

11.3 PROCEDURE

11.3.1 IMAGE OBTAINING

The scanned cross-sectional images of corn are taken using a flat-panel scanner with a resolution of 600 dpi.

11.3.2 CHARACTERISTIC INDICATORS OF CORN VARIETIES

The traditional corn variety recognition method based on image processing is largely based on the characteristics of shape, color, and texture of the corn seed. But the related characteristic of corn fringy linage is ignored, although the fringy linage is an important characteristic of a variety. Table 11.1 lists parameters and corn fringy linage characteristic of eight representative varieties. From this table, we can see clearly that the fringy linage is an important character of a variety.

11.3.3 PRETREATMENT

First of all, in order to count the ear rows in maize, we need to do some pretreatment.

We transfer the image data into a computer, convert them to grayscale image (Figure 11.1a), enhance image quality (Figure 11.1b), treat binary image with median filter (Figure 11.1c), fill inner holes (Figure 11.1d), remove the internal interference with a dark circle (Figure 11.1e), and get the profile of the image by edge detection method (Figure 11.1f).

11.3.4 CONSTRUCTION OF THE COUNTING MODEL

As an example image region R (Figure 11.2a), first, we should calculate the acreage as A and figure out the center of the image as $(\overline{x}, \overline{y})$ with Formula (11.1).

$$A = \sum_{(x,y)\in R} 1, \bar{x} = \frac{1}{A} \sum_{(x,y)\in R} 1, \bar{y} = \frac{1}{A} \sum_{(x,y)\in R} 1 \qquad (11.1)$$

Then, we start to find the edge and figure out the distribution of the function $f(\cdot)$ of the edge radius r with the method of edge marking. First, scan the coordinates x, y in the pixel by row and by column. If we find coordinates $(x1, y1)$ whose value equals 1, then restrict three searching directions (right, downright, and down), as shown in Figure 11.2b, and figure out the radius $r1$; then, go on and find the second coordinates $(x2, y2)$ whose value equals 1 and figure out the radius $r2$; then, search the connected eight directions as Figure 11.2c, to find two points $(x1, y1)$ and $(x2, y2)$ whose value equals 1, and then abnegate the point $(x1, y1)$ and figure out the radius $r3$ of the coordinates $(x3, y3)$; and at last, evaluate $x1 = x2$, $y1 = y2$, $x3 = x2$, $y3 = y2$, and do the calculation until the searched point is the first pixel.

The scratched shape of the corn is shown in Figure 11.3, and the point marked by "*" stands for the center of the area $(\bar{x}, \bar{y})$; following the direction of searching, we get the graph of function $f(\cdot)$ which is the function of the edge point radius r. If we use three-order polynomial to approximate the outline, we can conclude the function of discrete curve $k(t)$ as Formula (11.2), in which $b1$, $b2$, $c1$, $c2$ are taken from Zhang and Yang (2005). We get the protuberance of the outline with discrete curve distribution and get the number of the curve go through 0, so we can easily get the row number of the corn cross section. The number in this example is 14, and it is counted by self-programmed MATLAB® software automatically.

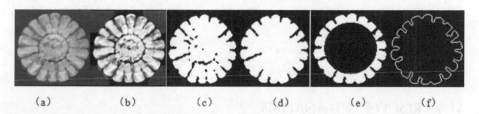

FIGURE 11.1 Picture pretreatment: (a) grayscale image, (b) enhancement image, (c) binary image, (d) hole filling image, (e) remove center image, and (f) edge image.

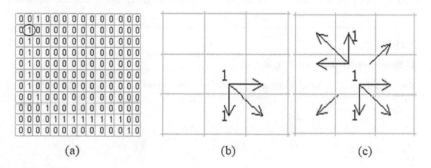

FIGURE 11.2 Borderline search sketch map: (a) image edge sketch, (b) starting pixel, and (c) intermediate pixel (eight connection directions).

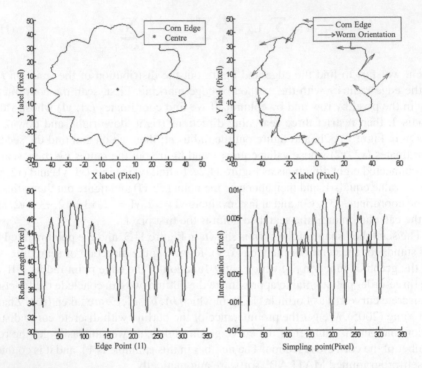

FIGURE 11.3 Transverse section linage discrimination: (upper left) center and edge, (upper right) forward direction, (lower left) distance of edge pixel to center, and (lower right) number of zero crossing.

$$k(t) = \frac{\left(r'f'' - r''f'\right)}{\left(r'^2 - f'^2\right)^{3/2}} = 2\frac{c_1 b_2 - c_2 b_1}{\left(c_1^2 + c_2^2\right)^{3/2}} \tag{11.2}$$

11.4   RESULTS AND ANALYSIS

We adopt comparison method to test the counting model, which numbers the 78 corn pictures from 1 to 78 and compares the row number counted by computer with that counted by people; the results are shown in Table 11.2. If we consider the number

TABLE 11.2
Detection Results of Ear Rows

Row Numbers	Total Numbers	Miscarriage Rows	Miscarriage Rate (%)
12	2	0	0
14	29	3	10.345
15	1	0	0
16	33	2	6.061
17	1	0	0
18	12	2	16.667
Average	15.46	0.103	0.663

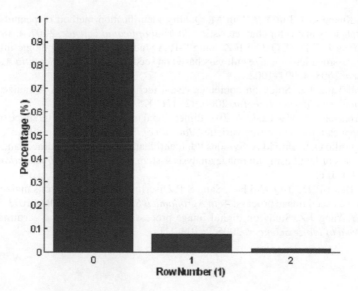

FIGURE 11.4   Miscarriage of justice probability.

counted by people as frame of reference, then the correct rate of the number counted by computer reaches as high as 91.03%. The correct rate is figured as the following formula: Correct rate (%) = The correct number counted by computer/the total number which the corn are tested × 100% (Figure 11.4).

We can draw some conclusions from the above results: (1) the number of corn ear row often comes to even and takes the percentage of 97.44; (2) the number of corn ear row often comes to 14, 16, and 18 rows and takes the percentage of 94.87; (3) the correct rate of the number counted by computer reaches as high as 91.03% if we consider the number counted by people as criterion; and (4) the major reason of error counting is drawn by mildew seed and severe seed deformity (Figure 11.4).

11.5   CONCLUSIONS

There are different ear row numbers in different varieties of corn, and with this characteristic, we can distinguish different varieties to some extent. If we consider the number counted by people as a criterion, the deviation in the data offered by the computer is 0.103 rows, the relative error rate is 0.66%, and the correct rate reaches as high as 91.03%, which is comparatively fine. So, we can conclude that it is feasible to achieve ear rows in maize counted automatically by the method of machine vision.

REFERENCES

Dubey B P, Bhagwat S G, Shouche S P, Sainis J K. Potential of artificial neural networks in varietal identification using morphometry of wheat grains. *Biosystems Engineering*, 2006, 1: 61–67.
Guo Q-F, Wang Q-C, Wang L-M. Cultivation science of maize. *Shanghai Science and Technology Publishing House*, Shanghai, 2004, 12: 786–787.

Han Z-Z, Kuang G-J, Liu Y-Y, Yan M. Quality identification method of peanut based on morphology and color characteristics. *Journal of Peanut Science*, 2007, 4: 18–21.

Hao J-P, Yang J-Z, Du T-Q, Cui F-Z, Sang S-P. A study on basic morphologic information and classification of maize cultivars based on seed image process. *Scientia Agricultura Sinica*, 2008, 4: 994–1002.

Liang S-M, Yang J-Z. Study on computer visual technique application in maize. *Chinese Agricultural Science Bulletin*, 2006, 12: 471–475.

Sakai N, Yonekawa S, Matsuzaki A. Two-dimensional image analysis of the shape of rice and its application to separating varieties. *Journal of Food Engineering*, 1996, 27: 397–407.

Venora G, Grillo O, Shahin M A, Symons S J. Identification of sicilian landraces and canadian cultivars of lentil using an image analysis system. *Food Research International*, 2007, 40: 161–166.

Yang J-Z, Hao J-P, Du T-Q, Cui F-Z, Sang S-P. Discrimination of numerous maize cultivars based on seed image process. *Acta Agronomica Sinica*, 2008, 6: 1069–1073.

Zhang X-Y, Yang J-Z. Study on digital image process application in the germination test. *Journal of Maize Sciences*, 2005, 2: 109–111.

12 Single-Seed Precise Sowing of Maize Using Computer Simulation

In order to test the feasibility of computer simulation in field maize planting, the selection of the maize single-seed precise sowing method is studied based on the quadratic function model $Y = A \times (D - Dm)^2 + Ym$, which depicts the relationship between maize's yield and planting density. The advantages and disadvantages of two planting methods under the condition of single-seed sowing are also compared: method 1 is optimum density planting, while method 2 is the ideal seedling emergence number planting. It is found that the yield reduction rate and yield fluctuation of method 2 are all lower than those of method 1. The yield of method 2 increased by at least 0.043 ton hm^{-2} and showed more advantages over method 1 with higher yield level. Further study made on the influence of seedling emergence rate on the yield of maize finds out that the yield of the two methods are all highly positively correlated with the seedling emergence rate and the standard deviation of their yield are all highly negatively correlated with the seedling emergence rate. For the study of the breakup problem of sparse caused by the single-seed precise sowing method, the definition of seedling missing spots is put forward. The study found that the relationship between the number of hundred-dot spots and field seedling emergence rate is the parabola function $y = -189.32x^2 + 309.55x - 118.95$ and the relationship between the number of spot missing seedling and field seedling emergence rate is the negative exponent function $y = 395.69e^{-6.144x}$. The results may help in guiding the maize seeds' production and single-seed precise sowing to some extent.

12.1 INTRODUCTION

The method of single-seed precise sowing is frequently adopted in the process of simplified cultivation of maize, which can greatly reduce the quantity of seeds used and save the trouble of thinning and supplementary planting; thus, less labor force is consumed than that of traditional cultivation management (GB 4404.1-2008). The crop area of maize in China is more than 500 million mu, and the percentage ratio of single seed sowing is less than 40% according to incomplete statistics (Li et al., 2007; Yang et al., 2015). As large quantities of migrant workers move to the cities, study on the simplified cultivation of single-seed precise sowing has great realistic significance (Dong and Liu, 2011; Wei et al., 1985; Yang et al., 2013). To find the optimum seed density for maximum yield or maximum profit, breeders carried out a series of studies (Steve and Meredith, 2014).

Traditional seeding mode is carried out by placing several seeds in one hole to ensure full stand. For example, if the missing seedling rate is less than 1%, then the number of seeds in one hole is $\geq -2/\log(1 - $ field seedling emergence rate). According to the national norm of seeds of 2008 (GB 4404.1-2008) that regular seeds' germination rate is 85% (assume that the field seedling emergence is equal to the germination rate), the number of seeds n is calculated as $n \geq -2/\log(1 - 0.85) = 2.4$ and $n \approx 3$. Namely, only by placing three seeds in one hole, the planting requirement that field missing seedling rate is less than 1% be met. The missing seedling rate of 1% means that the field seedling emergence rate is 99%. In the case of single-seed precise sowing, the national norm seems too low. Single-seed sowing realizing full stand seems too far to be true. Today, more and more farmers are adopting the method of single-seed sowing, which will soon take the place of traditional seeding mode. Yet, there is not enough theoretical guidance with formidable persuasive power in single-seed sowing mode.

Full, even, and strong seedling is the foundation of maize prolificacy (Figure 12.1). The single-seed precise sowing and incomplete seedling emergence can bring about

FIGURE 12.1 Full and even seedling and sparse breakups: (a) corn seedlings grow well and vigorously and (b) shortage of seedlings and ridge breaking.

the problem of sparse breakups, leading to reduction of yield to some degree. Thus, in order to get the greatest complex economic benefit, certain field seedling emergence rate must be guaranteed. Then, by adopting the single-seed precise sowing, what will the influence of field seedling emergence rate on the yield be like? Different planting methods will influence the yield, then in the case of single-seed precise sowing, which planting method is the best? Planting density will influence the yield, then how much will the influence be? Different planting densities can lead to the yield compensation because of sparse breakups, then how much will the compensation be? Traditionally, study and data collection of the above-mentioned problems are all from field work, which is time consuming and costly. One data collection requires the actual measurement of an entire maize growth period (always 1/2 to 1 year), which is too slow and inefficient. As the field experiment is always influenced by the factors such as soil, fertilizer, moisture, sunshine, and weather, the experiment results lack universality. Sometimes each experiment has different results, and results of some experiments are even opposite; thus, it is difficult to get objective laws. Study results lack universality because of specific field and climate. Moreover, in maize hybridization breeding process, a specific breed is always planted in different densities empirically to look for its density-tolerance parameter. Thus, differences exist in the measurement results of the same field in different years.

Theoretically, whether seeds will germinate is random, so it can be calculated from probability as the field seedling emergence rate. The rows and ridges where maize are planted can be simplified and represented as one grid matrix in computer, and the areas of missing seedling are represented as one series of connected regions in the computerized binary image. So, computer analog simulation can be used to look for a better planting and cultivation method under the simple model. Using computer, the planting method's field seedling emergence rate, planting density, and compensation range of seed yield can be simulated precisely and the results are precise and objective. Then, the results will be further determined by a field experiment and used to enhance the efficiency of maize cultivation study. Based on this thought, this chapter studies the advantages and disadvantages of the two methods (method 1: number of sowing seeds per mu = recommended planting density; method 2: number of sowing seeds per mu = recommended planting density/field seedling emergence rate) which are regularly used in the single-seed precise sowing production. Further study is made on the comparison of the number and stability of the two methods' yield, field seedling emergence rate's influence on the number and stability of yield, missing seedling's yield compensation mode, and the distribution rule of the seedling missing spots (number of hundred-dot spots and number of spot missing seedlings).

12.2 MATERIALS AND METHODS

12.2.1 MATHEMATICAL DEPICTION OF THE PROBLEM

Under the same exterior conditions (such as soil, fertilizer, moisture, sunshine, and weather), the planting density of maize may influence its yield. According to the field experiment results of the past years (GB 4404.1-2008; Li et al., 2007), the relationship between maize's yield and density can be depicted as a singlet curve, and curve fitting with quadratic function model can get perfect fitting results (Yang et al., 2015).

It can be assumed that maize's yield is the quadratic function of planting density:

$$Y = A \times D^2 + B \times D + C \qquad (12.1)$$

In this formula, Y represents the yield and D represents the planting density. In the parabola model, there are three parameters, namely, A, B, and C. The first two, A and B, respectively, the maximum yield and the optimum density, embody explicit biologic and physical meaning. The last one, C, is the quadratic term's coefficient, which has no explicit biologic and physical meaning but explicit geometric meaning. The vertex formula of quadratic function (1) is $Y = A \times (D - Dm)^2 + Ym$. The vertex $P(Dm, Ym)$ of the parabola represents the maximum yield that can be acquired at the optimum density with $Dm = -B/2 \times A$ and $Ym = (4 \times A \times C - B^2)/4 \times A$.

According to the vertex formula model ($Y = A \times (D - Dm)^2 + Ym$) of the parabola, yield is the derivative to density:

$$dY/dD = 2 \times A \times (D - D_m), A < 0 \qquad (12.2)$$

The derivative is the change rate of yield function, which can precisely express the variation of yield brought about by density. When density deviates left or right from the optimum density for one plant/m^2, there is $dY/dD = \pm 2 \times A$. They are the slope of the two tangents which are across the departure point P and Q on the yield curve. Density sensitivity can be defined as the deburrers from tangent P(with the slope > 0) to line Q(with the slope < 0) as shown in Figure 12.2.

Each variety has an optimum planting density M (individual plant/mu) under certain conditions, and the Ms for different varieties vary. Now M is known. Under ideal state, the number of seedling emergence D^* equals M (individual plant/mu). But the actual number of seedling emergence D^S is influenced by ρ, which makes the actual number of seedling emergence $D^S < D^*$. In the case of single-seed precise sowing, $D^S = D^* \rho$. Then,

Method 1: Planting according to the optimum density, the planting number is D^*, then the actual number of seedling emergence is $D^S = D^* \rho$, and the nutrition area of each individual plant is the reciprocal of actual planting number D^S, $1/D^S$.

Method 2: Planting according to the ideal number of seedling emergence, the planting number is D, then the actual number of seedling emergence is D^* and $D \times \rho = D^*$, and the nutrition area of each individual plant is the reciprocal of actual planting number D, $1/D$.

Problem: In the case of single-seed precise sowing, among the two methods, methods 1 and 2, which one is better? And how does the area of missing seedling distribute?

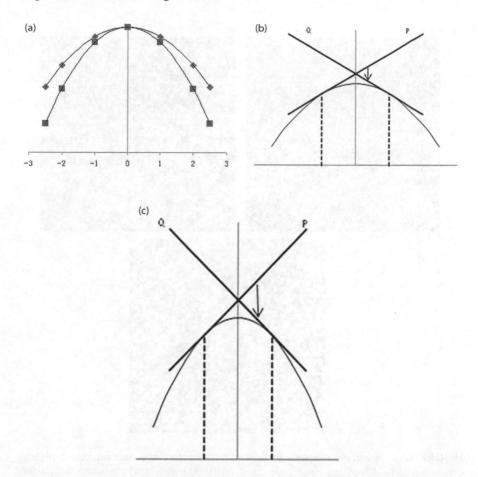

FIGURE 12.2 Relationships between yield and planting density: (a) relationship schematic of parabola, (b) schema of low sensitivity, and (c) schema of high sensitivity.

12.2.2 Method of Analog Simulation

The simulation platform of the experiment is IBM-SL3000 computer, XP operating system, and MATLAB® 2008a software.

12.2.2.1 Computer Simulation of Planting Method

Seedling emergence methods are automatically generated by computer, and then the random number of 0-1 distribution is generated. The number 0 represents no seedling emergence, and 1 represents seedling emergence. In the binary image, seedling emergence is expressed as black dot, while no seedling emergence is expressed as white dot. The generated 0-1 distribution matrix by controlling the emergence probability of 0 and 1 can be used to simulate the seedling emergence status, control the filed seedling emergence rate, and control the planting number of each mu, namely planting density by controlling the matrix's rows and columns. This simulation

FIGURE 12.3 Simulation images of field seedling emergence under different planting densities and field seedling emergence rates: (a) 10,000 individual plant/hm^2 with a field seedling emergence of 85%, (b) 3,600 individual plant/hm^2 with a field seedling emergence of 85%, and (c) 3,600 individual plant/hm^2 with a field seedling emergence of 50%.

process is realized with the binornd function of MATLAB, and data are calculated with Excel. The formula used is R = BINORND (N,P,MM,NN), in which $N = 1$ represents the random number of 0-1 distribution; $1 - p$ represents field seedling emergence rate; MM and NN represent the planting rows and columns of each mu, respectively. Thus, MM*NN represents planting strains per mu, i.e. planting density.

Figure 12.3 shows the simulation images of field seedling emergence in different conditions. These are the seedling emergence status under the conditions of 10,000 and 3,600 individual plants/hm^2 with a field emergence rate of 85%, and 3,600 individual plant/hm^2 with a field emergence of 85% and 50%.

12.2.2.2 Seedling Missing Spots and Missing Seedling Compensation

Seedling missing spots are defined as the successive missing seedling area in a successive area. As maize around the missing seedling area has more soil area, growing space, and the advantages of ventilation and shadow, there is certain increase in

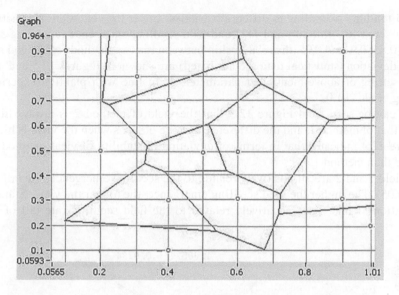

FIGURE 12.4 Voronoi diagram. (The dark circle indicates lack of seedlings.)

the yield of individual plant. This is the so-called yield compensation effect. The compensation effect is related to the floor space of individual plant. Then, how is the floor space determined? Seedling missing spot is shared by maize around it, and its area is determined by Voronoi chart (Daniel, 2011; Liu et al., 2014; Li et al., 2012; Bock et al., 2010).

Another problem is how to estimate individual plant's yield according to its floor space. Many trials have proved that the relationship between maize's yield and planting density is parabolic. The result of the parabola equation divided by planting density is the numeric equation of relationship between individual plant's yield and planting density. Planting density and individual plant's floor space are reciprocal to each other. Thus, this relationship equation can be used to calculate the compensation effect.

According to the parabola model, yield is the parabola function of density, that is, $Y = F(x)$, then individual plant's floor space is the reciprocal of density $1/x$, and the yield of individual plant is $Yd = F'(1/x)$. The compensation strategy of sprout deficiency: When there is no sprout deficiency, the yield of individual plant should be determined by the planting density, and when there is sprout deficiency, the yield of individual plant should be determined based on the per plant area by Voronoi diagram (Figure 12.4).

12.3 RESULTS AND ANALYSIS

12.3.1 COMPARISON OF THE TWO PLANTING METHODS' YIELD

In the simulation, the row spacing is set at 0.6 m; under the condition of optimum planting density being 4.5, 6, 7.5, 9, 10.5, and 12 individual plant m^{-2}, planting density is adjusted by changing the planting space, which can be calculated according to row space and density. The maximum yield is set at 5.25, 7.5, 9.75, 12, and

14.25 ton hm^{-2} according to different varieties. Under the condition of sensitivity being set at 16°, 48°, 80°, and 112° and field seedling emergence rate being set at 0.75, 0.85, 0.9, and 0.95, the relationship between the two methods' yield and standard deviation (statistical data are not listed) are shown in Figures 12.5 and 12.6. The setting of above-mentioned parameters is in line with planting experience (Yang et al., 2013).

It can be seen from Figure 12.3 that the yield of method 2 is always higher than that of method 1, and the difference is more obvious when the yield is higher. Figure 12.4 indicates that in terms of deviation of stability of yield, method 2 is lower than method 1.

Yield 1 and 2 are all positively related to the seedling emergence rate (all of P is <0.05). The higher the filed seedling emergence, the higher the yield. Standard deviations 1 and 2 are negatively related to the field seedling rate (all of P is

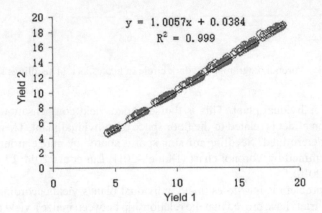

FIGURE 12.5   Comparison of two policies' yield.

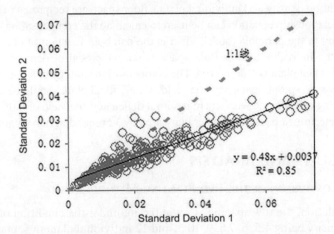

FIGURE 12.6   Comparison of two policies' standard deviation of output.

<0.05). The larger the standard deviation, the lower the stability and the larger the production hazard. The reason of fluctuation is the change in the number of individual plants and field distribution pattern of seedling emergence. The lower the field emergence rate and the larger the change in the number of individual plant and field distribution pattern, the bigger the change of yield, namely, the larger the standard deviation. Thus, in the production of single-seed precise sowing, more attention should be given to the quality of seed, planting bed, and seedling quality, and various measures should be adopted to enhance the field seedling emergence rate with a goal to attain high and stable yield (Figures 12.5 and 12.6).

12.3.2 INFLUENCE OF FIELD SEEDLING EMERGENCE RATE ON YIELD

Table 12.1 shows the linear regression coefficient of influence of different methods' yield reduction on the field seedling emergence rate.

The results indicate that the higher the field seedling emergence rate, the less the reduction of yield (the regression coefficient is negative); the speed of yield reduction in method 2 is always lower than that in method 1 (the former's absolute value of regression coefficient is lower than the latter's).

High field seedling emergence rate is one of the main reasons for the high yield of single-seed precise sowing. Under the condition of single-seed precise sowing, to ensure planned number of individual plant, it is recommended that it should be sowing based on the field seedling emergence rate is recommended. Method 2 can increase the yield for at least 0.043 ton hm^{-2}, and the amount of increased production ascends proportionally with the yield level. Based on the data mentioned above, the yield of method 2 is higher than that of method 1.

TABLE 12.1

Linear Regression Coefficient of Influence of Different Methods' Yield Reduction on the Field Seedling Emergence Rate

Optimum Density[a]	Method 1[b]	Method 2[b]
4.5	−5.483	−3.86
6	−10.142	−7.445
7.5	−14.236	−10.912
9	−18.952	−14.875
10.5	−26.118	−20.857
12	−33.604	−27.539

[a] Sensitivity =112.
[b] Results are all similar at random sensitivity.

TABLE 12.2

Multiple Regression Analysis of Yield Reduction to Sensitivity and Field Seedling Emergence Rate

Optimum Density	Method	Intercept	Sensitivity	Sensitivity*Rate	R-Square	Maximum Yield -Simulated Yield
4.5	1	−0.033	0.048	−0.049	0.959	0.31
4.5	2	−0.059	0.035	−0.034	0.97	0.238
6	1	−0.011	0.088	−0.091	0.938	0.518
6	2	−0.084	0.067	−0.066	0.97	0.401
7.5	1	0.044	0.123	−0.127	0.958	0.617
7.5	2	−0.047	0.097	−0.097	0.989	0.498
9	1	0.016	0.165	−0.169	0.979	0.584
9	2	−0.044	0.132	−0.133	0.994	0.486
10.5	1	0.122	0.225	−0.233	0.959	0.79
10.5	2	0.004	0.184	−0.186	0.995	0.657
12	1	0.157	0.289	−0.3	0.931	0.888
12	2	0.026	0.242	−0.246	0.991	0.749

12.3.3 INTERACTIONS BETWEEN SENSITIVITY AND FIELD SEEDLING EMERGENCE RATE

Density sensitivity is moderately and positively correlated to the maximum yield (Yang et al., 2015). Under different planting densities of 4.5, 6, 7.5, 9, 10.5, and 12, the reduction of yield line (maximum yield minus simulated yield), the multiple regression analysis of the two policies' sensitivity, and field seedling emergence rate are shown in Table 12.2.

When sensitivity is 112, the multivariate equation is simplified as a linear equation. When sensitivity is set at other values, things are similar. The common character is that the regression coefficients are all negative, which indicates that no matter what the values of sensitivity are, the higher the field seedling emergence rate, the less the yield reduction.

12.3.4 SEEDLING MISSING SPOTS AND ITS DISTRIBUTION RULE

Seedling missing spots are defined as the successive missing seedling area in a successive area which is represented as a region of successive 0 and four series connection region is selected. In the simulation of seedling missing spots under the case of single-seed precise sowing, a matrix model of 60 rows and 120 columns is adopted to simulate a piece of land, the simulated row spacing is two times of plant spacing, and the field emergence rate is set, respectively, at 0.95, 0.925, 0.9, 0.85, 0.8, 0.75, and 0.7. Considering the length of the chapter, only the means of 1,000 times simulation for each field seedling emergence rate are listed (Table 12.3).

TABLE 12.3

Number of Missing Seedlings and the Distribution Rule of Missing Seedlings Spots

Field Seedling Emergence Rate	Number of Seedling Missing Spots	Number of Spots	Number of Hundred-Dot Spots	Number of Spots Missing Seedling
0.95	360.8	293.3	4.1	1.2
0.925	541.5	393.4	5.5	1.4
0.9	720.7	465.6	6.5	1.5
0.85	1,080.5	538.8	7.5	2
0.8	1,438.7	533.4	7.4	2.7
0.75	1,799.1	468.7	6.5	3.8
0.7	2,159.4	368	5.1	5.9

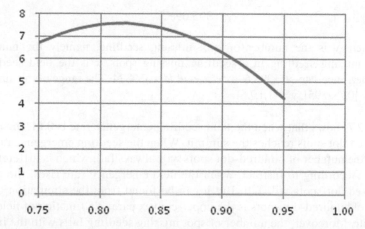

FIGURE 12.7 Relationship between the number of hundred-dot spots and field seedling emergence rate.

Rules acquired from the table are shown as follows:

1. The number of hundred-dot spots is the parabola function of field seedling emergence rate (Figure 12.7).

$$y = -189.32x^2 + 309.55x - 118.95$$

where y is the number of hundred-dot spots (i.e., number of seedling missing spots in each 100 planting points) and x is the field seedling emergence rate. The range of raw data is $x = [0.7, 0.95]$, $y = [3.5, 8.2]$; $R^2 = 0.9499$ and extreme point (0.8175, 7.5834).

2. The number of spot missing seedling is the negative exponent function of field seedling emergence rate (Figure 12.8).

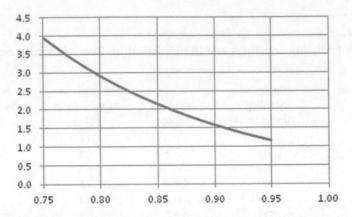

FIGURE 12.8 Relationship between the number of spot missing seedling and field seedling emergence rate.

$$y = 395.69e^{-6.144x}$$

where y is the number of spot missing seedling, namely the number of missing seedling in a seedling missing spots, x is the field seedling emergence rate, $y = 395.69e^{-6.144x}$, and $R^2 = 0.9824$. The range of raw data is $x = [0.75, 0.95]$, $y = [1.1, 7.6]$.

Figure 12.7 shows that when the field seedling emergence rate is 0.81, the number of hundred-dot spots reaches the summit. When the seedling emergence rate rises or falls, the number of hundred-dot spots will always fall, which is different from instinct. According to instinct, when the field emergence rate rises, the number of hundred-dot spots will fall. But it can be found from the simulation that the number of hundred-dot spots is the upper-convex parabola function of field emergence rate. Moreover, the number of spot missing seedling falls with the increase of field seedling emergence rate, but this trend is not linear and the number of spot missing seedling is the negative exponent function of field emergence rate (Figure 12.8).

The influence factors of maize yield are very complex. It involves variety (Peng et al., 2014), seed vigor (Li et al., 2013), seeding quality, and other aspects (Xiao et al., 2014; Drury, et al., 1999; Zhu et al., 2013), and planting density is only one of the key factors (Chu et al., 2014; Duncan, 1958). Our previous studies have shown that Monte Carlo (Han et al., 2015) is an effective simulation method in maize planting.

12.4 CONCLUSION

In this chapter, computer simulation technology is used to simulate field experience, realizing the concept of planting maize in computer. Preliminary study is made on the selection of maize planting density and planting mode. The influence of field seedling emergence rate on maize yield is inspected. Further discussion is made on the problem of sparse breakups because the field emergence rate cannot reach 100% under the case of maize's single-seed precise sowing. The relationship between maize yield model and dominant factor, and the definition of seedling missing spots are put forward. The study found that under the condition of single-seed precise sowing of maize, to ensure planned number of individual plant, the method recommended is that it should increasing the number of planting seed according to the field seedling emergence rate correspondingly. High field seedling rate is one of the main reasons for the high yield of single-seed precise sowing; the number of hundred-dot spots is the upper-convex parabola function of field emergence rate; and the number of spot missing seedling is the negative exponent function of field seedling emergence rate. The results have guiding significance to the maize seeds production and single-seed precise sowing to some degree.

REFERENCES

Bock M, Tyagi A K, Kreft J U, et al. Generalized Voronoi tessellation as a model of two-dimensional cell tissue dynamics. *Bulletin of Mathematical Biology*, 2010, 72(7): 1696–1731.

Chu J, Lu H, Xue J, et al. Field experiment and effect of precise mechanical sowing of maize based on wide-narrow row deep rotation and no-tillage technology. *Transactions of the CSAE*, 2014, 30(14): 34–41 (in Chinese with English abstract).

Daniel R. The geometric stability of Voronoi diagrams with respect to small changes of the sites. *Proceedings of the 27th Annual ACM Symposium on Computational Geometry (SoCG 2011)*, 2011, pp. 254–263.

Dong X, Liu R-T. Study on region of space partition algorithm based on Voronoi diagrams. *Journal of Harbin University of Commerce (Natural Sciences Edition)*, 2011, 27(6): 867–880.

Drury C F, Tan C S, Welacky T W, et al. Red clover and tillage influence on soil temperature, water content, and corn emergence. *Agronomy Journal*, 1999, 91(1): 101–108.

Duncan W G. The relation between corn population and yield. *Agronomy Journal*, 1958, 50(2): 82–85.

GB 4404.1-2008. State standard of the People's Republic of China: Quality standard of crop seed. *China Standards Press*, Beijing, 2008, p. 12.

Han Z, Cao H, Gao H, et al. Monte-Carlo simulation of yield effect under singular seeding strategy in maize based on Voronoi diagrams. *Transactions of the CSAE*, 2015, 31(13): 17–21 (in Chinese with English abstract).

Li D, Ban X, Zhou C, et al. Quality standards for maize seeds used in planting one seed per hill. *Seed World*, 2013, 12: 35–36 (in Chinese with English abstract).

Li F-H, Zhou F, Wang Z-B. Study on the optimal density in the different maize varieties. *Seed*. 2007, 26(2): 77–80.

Li H, Li K, Kim T, et al. Spatial modeling of bone microarchitecture. *IS&T/SPIE Electronic Imaging. International Society for Optics and Photonics*, 2012: 82900P-1–82900P-9.

Liu S, Wu S, Wang H, et al. The stand spatial model and pattern based on Voronoi diagram. *Acta Ecologica Sinica*, 2014, 34(6): 1436–1443 (in Chinese with English abstract).

Peng Y, Yang F, Zhao X, et al. Deep-sowing tolerant characteristics in maize inbred lines. *Agricultural Research in the Arid Areas*, 2014, 32(1): 25–33, 51 (in Chinese with English abstract).

Steve B, Meredith B. Crop insights: Corn seeding rate considerations for 2014[R/OL], 2014 [2014-4-7]. www.pioneer.com/home/site/us/agronomy/library/corn-seeding-rate-considerations/2015-5-23 verified.

Wei L, Dai J, Liu J. Studied on leaf layer structure and light distribution in high-yield corn population and their effects on yield. *Journal of Shenyang Agricultural College*, 1985, 16(2): 1–8.

Xiao J, Sun Z, Jiang C, et al. Effect of ridge film mulching technique and furrow seeding of spring corn on water use and yield in semi-arid region in Liaoxi area. *Scientia Agricultura Sinica*, 2014, 17(10): 1917–1928 (in Chinese with English abstract).

Yang J-Z, Zhang H-S, Du J-Z. Meta-analysis of evolution trend from 1950s to 2000s in the relationship between crop yield and plant density in maize. *Acta Agronomica Sinica*, 2013, 39(3): 515–519.

Yang J-Z, Zhao Y-M, Song X-Y. Sensitivity of crop yield to plant density with an example in maize. *Journal of Biomathematics*, 2015, 30(2): 243–252.

Zhu J, Tao H, Gao Y, et al. Effects of sowing irrigation on plant establishment, water consumption and yield of summer maize. *Journal of China Agricultural University*, 2013, 18(3): 34–38 (in Chinese with English abstract).

13 Identifying Maize Surface and Species by Transfer Learning

In this chapter, we study the feasibility of using transfer learning to identify the embryo surface and variety of maize seeds. First, the original image is segmented, and region of interest (ROI) is extracted; second, the extracted image is imported into a model that is pretrained by image-net data set; and finally, the recognition task can be completed efficiently through the local fine-tuning of the model. In order to comprehensively evaluate the productivity of a maize species, we test and identify its embryo surface, non-embryo surface, and longitudinal cut surface. The model we use here has also been applied to and achieved good results in recognition of peanut pod. Experimental results show that the transfer learning can be used for image recognition and has good prospects of application. More maize varieties need to be studied in future to expand the application of transfer learning to image recognition.

13.1 INTRODUCTION

Maize is a widely distributed grain crop in China; in 2015, China's area of maize cultivation was 38,119.31 hectares, and the total harvest reached 22,463.16 million tons. Seed is an irreplaceable essential material in agricultural production, which is an internal cause and a key factor for determining and achieving high yield and good quality of crops, and a core carrier of various technical measures. Therefore, it is necessary to propose an effective strategy for breeding process, especially to improve the speed and accuracy of identification and classification of target seeds of maize to the maximum extent.

The image processing technology has been widely used in the inspection and grading of agricultural products such as cereals, vegetables, and fruits based on their color, shape, defects, and many other external appearances (Liao et al., 1994; Yarnia et al., 2012). It is of great significance in sowing process of agriculture to judge the position and germination of maize embryo. At present, image processing methods are adopted to identify the position of maize germ surface (Wang et al., 2014).

Methods for seed identification include morphological methods (Sánchez et al., 1993) electrophoretic detection of proteins, DNA molecular markers (Ye et al., 2013), and genetic markers (Wang et al., 2011). However, the methods are expensive and time-consuming, and involves manual work (Mahesh et al., 2015), which may lead to loss in the accuracy of detection. In order to avoid this loss, some people use machine vision technology to achieve accurate detection (Boelt et al., 2011).

Because the shape of grains in all maize varieties is very similar, it is difficult to distinguish the maize species by image processing. Instead, some people use the color of grains to identify the maize species (Yan et al., 2010). We created a model combining computer vision and transfer learning in order to achieve better results than image processing method.

In this chapter, we first identify the seed surface, embryo surface, non-embryonal surface, and longitudinal section of different maize varieties, and compare the accuracy of identification in all cases. It is very important to select the best identification scheme for practical application. Few people identify the variety of maize through longitudinal cut surface. We make attempts and hope to make new discoveries.

To sum up, we use deep learning method to identify maize varieties. However, in many practical applications, reorganizing training data and rebuilding models are expensive or even impossible. It is great to reduce the need and workload of reorganizing training data. In this case, knowledge transfer or transfer learning is required between task domains. As big data repositories become more and more popular and use relevant but not identical existing data sets, interest in the target domain makes transfer learning of solutions an attractive approach. Transfer learning has been successfully applied to image classification (Duan et al., 2012; Kulis et al., 2011; Zhu et al., 2011), human disease detection (Tan et al., 2018), text emotion classification (Wang et al., 2011), software defect classification (Nam et al., 2017), multilingual text classification (Carneiro et al., 2015; Prettenhofer and Stein, 2010; Zhou et al., 2014), plant disease identification (Ramcharan et al., 2017), and human activity classification (Harel and Mannor, 2011). The transfer learning method greatly reduces the workload of image acquisition, enabling the small data sets to achieve higher recognition accuracy (Razavian et al., 2014; Donahue et al., 2014; Yosinski et al., 2014). The model is pretrained using image-net data set, and the data are imported into the model. The last layer of the model is replaced with the full connection layer for local fine-tuning to improve the recognition accuracy (Figure 13.1).

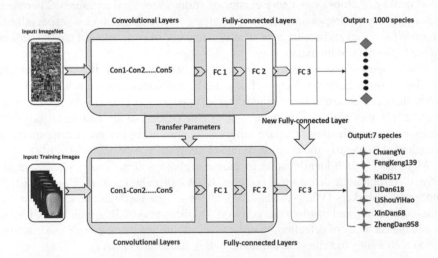

FIGURE 13.1 Network model structure diagram.

13.2 MATERIALS AND METHOD

13.2.1 IMAGE CHARACTERISTICS

The seeds used in the experiment were obtained from the experimental field of Qingdao Agricultural University, which guaranteed the purity of the breed without any hybridization. We chose 800 grains from each maize variety for identification of embryo and non-embryo surfaces. In this study, we attempted to identify seven maize varieties; hence, we collected 270 seeds from each variety, totally 1,890 seeds. The identification was based on the appearances of embryo surface, non-embryo surface, and longitudinal cut surface.

We used a scanner (Canon 8800F) to take the images of maize seeds. The seven maize varieties include ChuangYu, FengKen139, KaDi517, LiDan618, LiShouYiHao, XinDan68, and ZhengDan 958 (Figure 13.2).

13.2.2 WORKFLOW DIAGRAM

In order to expand the data set, 1,890 collected images were processed. The first step was to conduct horizontal and vertical processing for the image, and then each image was rotated once every ten degrees. Finally, we obtained 204,920 images. The number of training set images and the number of test set images were in the ratio of 7:3. The specific data set is divided as shown in Figure 13.3. The main work of the experiment is shown in Figure 13.4.

13.2.3 CONVOLUTIONAL NEURAL NETWORK

Deep learning constitutes a recent, modern technique with promising results and high potential for image processing and data analysis. As deep learning has been successfully applied in various domains, it has recently also entered the domain of

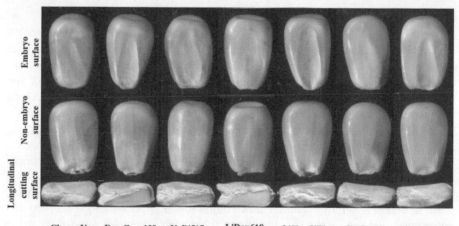

ChuangYu   FengGeng139   KaDi517   LiDan618   LiShouYiHao   XinDan68   ZhengDan958

FIGURE 13.2 Maize images.

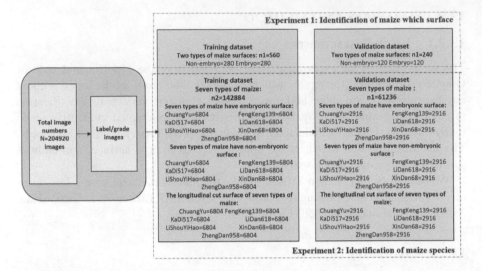

FIGURE 13.3 Experimental data collection diagram.

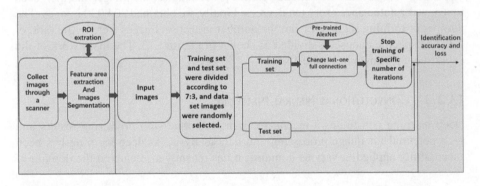

FIGURE 13.4 Experimental flowchart.

agriculture (Kamilaris and Prenafetaboldu, 2018). Meanwhile, convolutional neural network (CNN) has achieved good results in extraction of information about cultivated land (Lu et al., 2017), classification of plant species (Lee et al., 2015), and classification of plant diseases by leaf images (Sladojevic et al., 2016). Therefore, we carried out transfer learning on the basis of CNN and achieved excellent results.

In the experiment, a convolutional neural network, AlexNet (Krizhevsky et al., 2012), was used to extract the features of maize images and identify maize species. The network is comprised of eight layers of network, containing five convolutional layers and three full connection layers (Figure 13.5). Previously, the standard activation functions used tanh or sigmoid functions, but with gradient descent training, these saturated (saturating) nonlinear activation functions would cause gradient disappearance and reduce the training speed. This problem can be avoided by using unsaturated nonlinear activation function ReLu. Overlapping pooling improves accuracy and reduces the probability of overfitting. Dropout is adopted to prevent overfitting in full connection layer. The network adopts small-size convolutional layer with three to five layers, because the small-size convolutional layer is more nonlinear than a large-size convolutional layer, making the function more accurate and reducing the number of parameters (Liao et al., 2019). Because the depth of the network and the small size of the convolution kernel help to achieve implicit regularization, the network starts to converge after a few iterations. Therefore, we selected this network for accomplishing the identification task of maize seeds.

13.2.4 Transfer Learning

Transfer learning is not training a completely blank network; the model can recognize the distinguishing features of a specific category of images, much faster and with significantly fewer training examples and less computational power (Kermany et al., 2018).

The AlexNet trains the ImageNet data set sufficiently. Although the neural network cannot recognize the maize variety before training, it provides good initial values for the maize recognition network. Good starting values are critical to network training. We keep all layers before the last output layer and connect these layers to a new layer for the new classification problem. We transfer the last layer to the new classification task by replacing them with a layer of full connection, a softmax layer, and a classification output layer; specify options for the new layer of full connection based on the new data; and then, set the total layer of connection to the same size as the number of classes in the new data. In order to learn faster in the new layer than in the transfer layer, the weight learning rate factor and the bias learning rate factor values of the fully connected layer were increased.

13.3 PERFORMANCE OF THE MODEL

The curve of recognition accuracy and cross loss (Figure 13.5) is drawn to illustrate the performance of our model. The curve of the training process was processed by smoothing with a smoothing factor of 0.6, and the variation trend was easily observed. We also add a part of the image with high noise interference in the training set to avoid the occurrence of overfitting phenomenon, which is conducive to the generalization of the classification model. When drawing the precision curve, training data set is the solid line and verification data set is the dashed line. When the cross-entropy loss is plotted, the training data set is solid line and the verification data set is dashed line.

The model training process proceeded for 250 iterations on the problem of recognizing maize varieties based on their embryo and non-embryo surfaces, and the recognition accuracy was 99.58%.

For the recognition model, the training set and test set are divided as in Table 13.1.

In order to explore which angle is more effective to identify the maize species, experiments were conducted in both embryo and non-embryo surfaces. The numbers of training and test sets of each angle are the same, and the number of iterations of the training process is the same to guarantee the fairness of the experiment. After 2,500 iterations of the model, the recognition accuracy of two angles was obtained. The accuracy rate of embryo surface recognition is 91.61% and that of non-embryo surface recognition is 91.25% (Figure 13.5a and c).

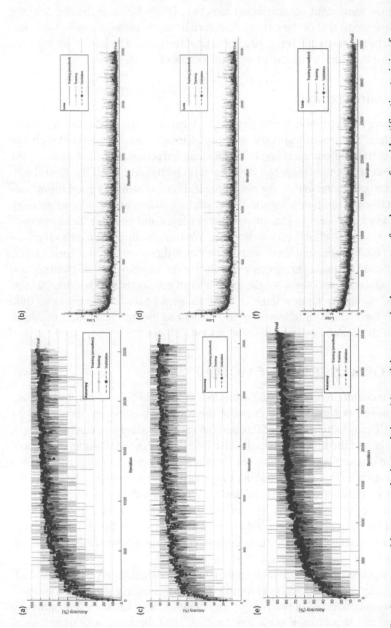

FIGURE 13.5 Accuracy and loss curve during training: (a) precision curve in the process of embryo surface identification training for seven kinds of maize, (b) loss curve in the process of embryo surface identification training for seven kinds of maize, (c) precision curve in the process of non-embryo surface identification training for seven kinds of maize, (d) loss curve in the process of non-embryo surface identification training for seven kinds of maize, (e) precision curve in the training process of longitudinal cut surface identification of seven kinds of maize, and (f) loss curve in the training process of longitudinal cut surface identification of seven kinds of maize.

TABLE 13.1

Statistics of Maize Embryo and Non-Embryo Surfaces

Items	Number of Total Images	Number of Trained Images	Number of Verified Images	Number of Model Identification	Accuracy Rate (%)
Non-embryo surface	400	280	120	119	99.16
Embryo surface	400	280	120	120	100
Total	800	560	240	239	99.58

We have further identified the species of maize by longitudinal cut surface. In the previous maize species identification experiments we know, no work has been done on identification of maize species by longitudinal cut surface using deep learning. Due to more interference factors in longitudinal cut surface, the number of iterations of model training increased to 3,500 times to obtain better experimental results. The model shows exciting results in the aspect of longitudinal cut surface recognition, with a recognition accuracy of 87.36% (Figure 13.5e). The experimental results provide feasible support for the identification of maize species using longitudinal cut surface.

The results of three recognition models are shown in Figure 13.6.

13.4   PERFORMANCE OF LIMITED DATA MODEL

To further evaluate the performance of our model, we reduced the number of images in the data set and verified the stability of the model with the reduced data set. We collected 270 images from three angles for each variety of maize, and the classification of specific data sets is shown in Table 13.2.

At the same time, we identified the three aspects mentioned above, and the training iterations of the model were 2,500 times.

The accuracy rates of three angle recognition of embryo surface, non-embryo surface, and longitudinal cut surface are 92.59%, 95.24%, and 81.05%, respectively.

13.5   COMPARISON WITH MANUAL METHOD

We use a separate seven kinds of maize seed images, a total of 210 images, to compare the identification abilities of our model and manual method. In order to standardize the process of artificial recognition by our model, the time of recognition is fixed. Within an interval of 5 min, 30 images must be recognized. Manual classification was done by comparing the appearance with sample images, and then, the number of matching images is counted to further calculate the accuracy. At the same time, the model also has recognized 30 images at the set interval, and the results are statistically analyzed.

Based on the observation in the experiment, we obtained surprising results: the performance of the transfer learning model is far better than that of the manual

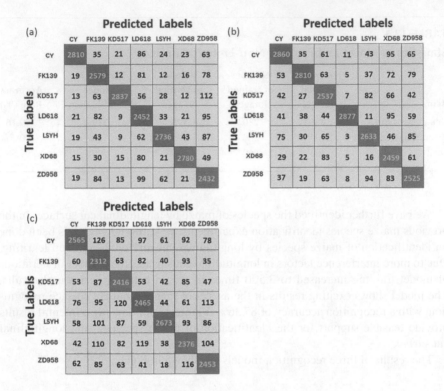

FIGURE 13.6 Confusion matrices for identification of (a) embryo surface, (b) non-embryo surface, and (c) longitudinal cut surface of seven kinds of maize.

method, the accuracy of the manual average recognition is generally low, and the highest accuracy of the artificial optimal average is 45.36%. The accuracy rate of the model in this experiment reached 92.38% (Figure 13.7). In the manual classification process, there was interference of many factors such as artificial subjectivity and fatigue degree, which had a great impact on the recognition results. In contrast, the recognition performance of the model is more stable.

13.6 EXPANDING APPLICATION

In order to further prove the universality of the model, we also applied it to the identification of peanut pod variety and achieved good results. We identified 12 varieties of peanut pods, for which we collected 300 images of each variety from two sides and the front (Figure 13.8). Thus, the number of images in each variety's training and test sets became 70 and 30, respectively.

After 2,500 iterations, the recognition accuracy of the method in identifying the peanut species based on the images from three angles of pod was calculated: the right side of the peanut pod was 89.99%, the left side of the peanut pod was 86.34%, and the front side of the peanut pod was 90.35% (Figure 13.9). Based on these results, our model is better in identifying peanut species using the front image of the pod. At the same time, this experiment also proves that our model has good generalization.

TABLE 13.2
The Data Set of Limited Data Model

	Embryo Surface			Non-Embryo Surface			Longitudinal Cut Surface	
Items	Number of Training Set	Number of Test Set	Items	Number of Training Set	Number of Test Set	Items	Number of Training Set	Number of Test Set
ChuangYu	63	27	ChuangYu	63	27	ChuangYu	63	27
FengGeng139	63	27	FengGeng139	63	27	FengGeng139	63	27
KaDi517	63	27	KaDi517	63	27	KaDi517	63	27
LiDan618	63	27	LiDan618	63	27	LiDan618	63	27
LiShouYiHao	63	27	LiShouYiHao	63	27	LiShouYiHao	63	27
XinDan68	63	27	XinDan68	63	27	XinDan68	63	27
ZhengDan958	63	27	ZhengDan958	63	27	ZhengDan958	63	27
Total	441	189	Total	441	189	Total	441	189

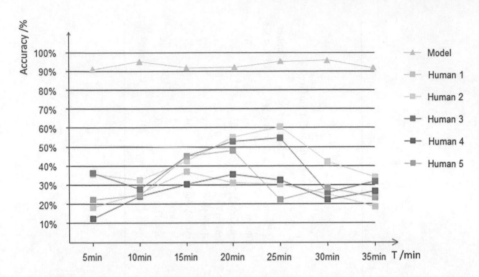

FIGURE 13.7   Model vs manual work.

101   HuaYu 22   HuaYu 25   JiHua 2   JiHua 4   JiHua 5   LaiNong 13   LuHua 9   LuHua 11   QingHua 6   TianFu 3   ZhongNong
                                                                                                                              108

FIGURE 13.8   Peanut images.

13.7   DISCUSSION

There is great significance to embryo and non-embryo surfaces in identifying the variety of maize. This work can be used to demarcate embryo area for maize gene mark. Experimental results show that the model can be used to identify embryo and non-embryo surfaces of maize with an accuracy as high as 99.58%. And it can be applied to practical work.

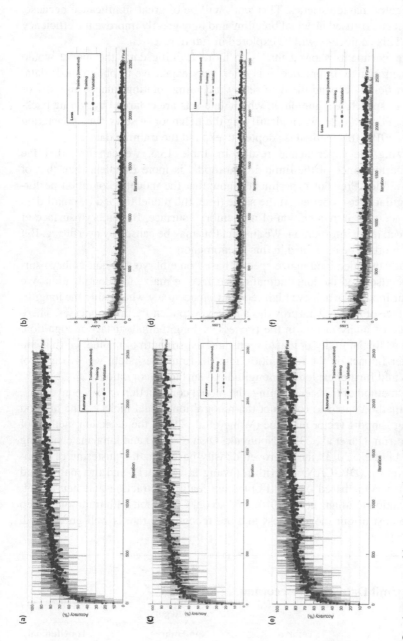

FIGURE 13.9 Accuracy and loss curve during training: (a) the accuracy curve of 12 kinds of peanut pods in the training process of right side surface identification, (b) the loss curve of 12 kinds of peanut pods in the training process of right side surface identification, (c) the accuracy curve of 12 kinds of peanut pods in the training process of left side surface identification, (d) the loss curve of 12 kinds of peanut pods in the training process of left side surface identification, (e) the accuracy curve of 12 kinds of peanut pods in the training process of front surface identification, and (f) the loss curve of 12 kinds of peanut pods in the training process of front surface identification.

We find that the grains of these seven maize varieties have a similar appearance. We make a guess that we can identify the appearance of maize traits to classify the genetic lineage of maize varieties. This work will be of great significance because, for example, it can be used in actual breeding and may greatly improve the efficiency of breeding. This conjecture will be explored in future research.

Previous experiments showed that with more training data, the model would get better performance. In practice, it is often very expensive to obtain more data, so we can artificially expand the data set by including rotating images and mirror images, – as in speech recognition, in which data sets are enlarged by adding background noise. For example, when identifying the behavior of sows during lactation (Zheng et al., 2018), this method is adopted to expand the training data set.

By analyzing the experimental results in Table 13.3, we can find that the recognition performance of the limited data model is more excellent than that of the big data model. Previous experiments show that the model recognition performance on big data sets is better. At the same time, the model trained by small data sets is more accurate in recognition of non-embryo surface. But embryo surface of maize has much more information. We suspect this may be caused by overfitting. But still, there is more to be explained in this phenomenon.

In this study, we identified maize species based on embryo and non-embryo surfaces, and we also used the longitudinal cut surface of maize for identification. We found that our model can achieve high recognition accuracy when using the longitudinal cut surface of maize. This provides a new way to identify maize species. There is interference of many factors in the process of longitudinal surface recognition. Factors such as the humidity of maize seeds and the strength of manual cutting can have a greater impact on the recognition rate. In future research, we will control these factors and further explore the longitudinal surface recognition.

Using the method of transfer learning, the model construction is very convenient. In the past, methods used to construct the recognition model included the method of identifying support vector machine (Wang et al., 2014), the detection method of infrared spectrum (Jia et al., 2015), Nobuyuki Otsu (OTSU) and k-means clustering algorithm (Ma et al., 2013), the improved density-based spatial clustering of applications with noise (DBSCAN) algorithm (Wang et al., 2011), and the maize seed variety identification based on multi-objective feature extraction and neural network optimization (Yuliang et al., 2010), and so on. Compared with traditional deep learning, the cost of our model is less and the recognition rate is well guaranteed.

TABLE 13.3

Model and Limit Data Model Accuracy

| | Category | | |
| | Embryo Surface (%) | Non-Embryo Surface (%) | Longitudinal Cut Surface (%) |
Items			
Model	91.61	91.25	87.25
Model (limited data)	92.59	95.24	81.05

TABLE 13.4
Number of Peanut Pod Samples and Model's Accuracy

	Category		
Items	Right Side Surface	Left Side Surface	Front Side Surface
Number of peanut pod samples	1,200	1,200	1,200
Accuracy	89.9%	86.3%	90.3%

Compared with the above-mentioned methods, our method is simpler. Deep learning methods could recognize features automatically, and more samples would allow more potential feature combination to be explored.

Although transfer learning allows the training of a highly accurate model with a relatively small training data set, its performance would be inferior to that of a model trained from a random initialization on an extremely large data set of images, since even the internal weights can be directly optimized for maize feature detection (Kermany et al., 2018). In this experiment, we adopted the method of local fine-tuning model weight and bias to rapidly train the model. We suspect that the global fine-tuning of model weight and bias method can further improve the performance of the model, and the time spent on training will increase accordingly. However, this method will consume less time than the traditional CNN model. The next step will be to study fine-tuning to prove our conjecture.

In our experiment, we used the model to identify the species of peanut pods, and the experimental results are shown in Table 13.4. Experimental results proved that our model had good robustness performance.

13.8 CONCLUSIONS AND FUTURE WORK

In this chapter, the study proves that the method of transfer learning can be used to identify maize species based on its surface. The method is further extended to the identification of peanut pods and achieved good results, which proves our model's stable performance and strong robustness.

Objectives of our future work will include to (1) further extend the application of the model (e.g. identification of weed and pest), (2) improve the anti-interference ability of the model, and (3) further study the identification of maize using longitudinal surface.

Finally, our goal is to integrate our methods into working systems that can be used to breeding.

REFERENCES

Boelt B, Shetty N, Gislum R, et al. Erratum: Classification of viable and non-viable spinach (Spinacia oleracea L.) seeds by single seed near infrared spectroscopy and extended canonical variates analysis. *Journal of Near Infrared Spectroscopy*, 2011, 19(3): 285–286.

Carneiro G, Nascimento J C, Bradley A P, et al. Unregistered multiview mammogram analysis with pre-trained deep learning models. *International Conference on Medical Image Computing and Computer Assisted Intervention*, Cham, Springer, 2015, pp. 652–660.

Donahue J, Jia Y, Vinyals O, et al. DeCAF: A deep convolutional activation feature for generic visual recognition. *International Conference on Machine Learning*, Beijing, China, June 21-26, 2014, pp. 647–655.

Duan L, Xu D, Tsang I W, et al. Learning with augmented features for heterogeneous domain adaptation. *Proceedings of the 29th International Conference on Machine Learning, ICML*, Edinburgh, Scotland, 2012, pp. 711–718.

Harel M, Mannor S. Learning from multiple outlooks. *Proceedings of the 28th International Conference on Machine Learning*, Bellevue, Washington, USA. Omnipress, 2011, pp. 401–408.

Jia S, An D, Liu Z, et al. Variety identification method of coated maize seeds based on near-infrared spectroscopy and chemometrics. *Journal of Cereal Science*, 2015, 63: 21–26.

Kamilaris A, Prenafetaboldu F X. Deep learning in agriculture: A survey. *Computers and Electronics in Agriculture*, 2018, 147: 70–90.

Kermany D, Goldbaum M H, Cai W, et al. Identifying medical diagnoses and treatable diseases by image-based deep learning. *Cell*, 2018, 172(5): 1122–1131.

Krizhevsky A, Sutskever I, Hinton G E, et al. ImageNet classification with deep convolutional neural networks. *Advances in Neural Information Processing Systems*, 2012: 1097–1105.

Kulis B, Saenko K, Darrell T, et al. What you saw is not what you get: Domain adaptation using asymmetric kernel transforms. *Computer Vision and Pattern Recognition*, 2011: 1785–1792.

Lee S H, Chan C S, Wilkin P, et al. Deep-plant: Plant identification with convolutional neural networks. *IEEE International Conference on Image Processing (ICIP)*, Quebec City, Canada. IEEE, 2015, pp. 452–456.

Liao K, Paulsen M R, Reid J F. Real-time detection of colour and surface defects of maize kernels using machine vision. *Journal of Agricultural Engineering Research*, 1994, 59(4): 263–271.

Liao W X, Wang X Y, An D, et al. Hyperspectral imaging technology and transfer learning utilized in identification haploid maize seeds identification. *International Conference on High Performance Big Data and Intelligent Systems (HPBD&IS)*, Shenzhen, China. IEEE, 2019, pp. 157–162.

Lu H, Fu X, Liu C, et al. Cultivated land information extraction in UAV imagery based on deep convolutional neural network and transfer learning. *Journal of Mountain Science*, 2017, 14(4): 731–741.

Ma D, Cheng H, Zhang W, et al. Maize embryo image acquisition and variety identification based on OTSU and K-means clustering algorithm. *International Conference on Information Science and Cloud Computing Companion*. IEEE, 2013, pp. 835–840.

Mahesh S, Jayas D S, Paliwal J, et al. Hyperspectral imaging to classify and monitor quality of agricultural materials. *Journal of Stored Products Research*, 2015, 61: 17–26.

Nam J, Fu W, Kim S, et al. Heterogeneous defect prediction. *IEEE Transactions on Software Engineering*, 2017, 44(9): 874–896.

Prettenhofer P, Stein B. Cross-language text classification using structural correspondence learning. *Proceedings of the 48th Meeting of the Association for Computational Linguistics*, Uppsala, Sweden, 2010: 1118–1127.

Ramcharan A, Baranowski K, Mccloskey P, et al. Using transfer learning for image-based cassava disease detection. *Frontiers in Plant Science*, 2017, 8–12.

Razavian A S, Azizpour H, Sullivan J, et al. CNN features off-the-shelf: An astounding baseline for recognition. *Computer Vision and Pattern Recognition*, Washington, DC, USA, 2014: 512–519.

Sánchez J J G, Goodman M M, Rawlings J O. Appropriate characters for racial classification in maize. *Economic Botany*, 1993, 47(1): 44–59.

Sladojevic S, Arsenovic M, Anderla A, et al. Deep neural networks based recognition of plant diseases by leaf image classification. *Computational Intelligence and Neuroscience*, 2016. Article ID 3289801, 11 pages.

Tan T, Li Z, Liu H, et al. Optimize transfer learning for lung diseases in bronchoscopy using a new concept: Sequential fine-tuning. *Computer Vision and Pattern Recognition*, 2018.

Wang P, Liu S, Liu M, et al. The improved DBSCAN algorithm study on maize purity identification. *International Conference on Computer and Computing Technologies in Agriculture*, Berlin, Heidelberg. Springer, 2011, pp. 648–656.

Wang Y, Xu L, Zhao X, et al. Maize seed embryo and position inspection based on image processing. *IFIP Advances in Information & Communication Technology*, 2014, 420: 1–9.

Yan X, Wang J, Liu S, et al. Purity identification of maize seed based on color characteristics. *Transactions of the Chinese Society of Agricultural Engineering*, 2010, 346: 620–628.

Yarnia M, Farajzadeh E, Tabrizi M. Effect of seed priming with different concentration of GA 3, IAA and Kinetin on Azarshahr onion germination and seedling growth. *Journal of Basic Sciences and Applied Research*, 2012, 2(3): 2657–2661.

Ye S, Wang Y, Huang D, et al. Genetic purity testing of F1 hybrid seed with molecular markers in cabbage (Brassica oleracea var. capitata). *Scientia Horticulturae*, 2013, 155(2): 92–96.

Yosinski J, Clune J, Bengio Y, et al. How transferable are features in deep neural networks. *Advances in Neural Information Processing Systems*, Montreal, Canada, 2014, pp. 3320–3328.

Yuliang W, Xianxi L, Qingtang S, et al. Maize seeds varieties identification based on multi-object feature extraction and optimized neural network. *Transactions of the Chinese Society of Agricultural Engineering*, 2010, 26(6): 199–204.

Zheng C, Zhu X, Yang X, et al. Automatic recognition of lactating sow postures from depth images by deep learning detector. *Computers and Electronics in Agriculture*, 2018, (147): 51–63.

Zhou J T, Tsang I W, Pan S J, et al. Heterogeneous domain adaptation for multiple classes. *International Conference on Artificial Intelligence and Statistics*, Reykjavik, Iceland, 2014, pp. 1095–1103.

Zhu Y, Chen Y, Lu Z, et al. Heterogeneous transfer learning for image classification. *Twenty-Fifth AAAI Conference on Artificial Intelligence*, San Francisco, CA 2011, pp. 1304–1309.

14 A Carrot Sorting System Using Machine Vision Technique

Carrot grading is a labor-intensive, time-consuming process and is usually performed manually. Manual inspection poses many problems in maintaining consistency and guaranteeing efficient detection. To improve the grading efficiency and achieve automatic detection, we developed an automated carrot sorting system using machine vision technology. The system consisted of an image processing system, an image acquisition system, a roller conveyor system, and a control system. It first picked out carrots with surface defects and then graded the qualified carrots by length. We proposed detection methods for three kinds of surface defects: irregular shape, fibrous root, and surface crack. Given the fact that carrots are regularly in convex shape, convex polygon was used to detect their shape. We proposed a concave point method to detect the fibrous root and adopted the Hough transform to detect surface crack. Experimental results showed that the proposed methods could not only achieve satisfying detection accuracy but also high efficiency. The accuracy rates of curvature, fibrous root, and surface crack were 95.5%, 98%, and 88.3%, respectively. The proposed methods and constructed sorting system could meet the demand of carrot grading and sorting.

14.1 INTRODUCTION

Appearance is a critical attribute of carrots and influences not only their market value but also consumers' preference and choice (Zhang et al., 2014). Products with good appearance and uniform shape are preferable to most consumers and have better sales appeal. The grading process can guarantee that the products meet defined grades and quality requirements for sellers and provide the expected level of quality for buyers (ElMasry et al., 2012).

The carrot grading and sorting process is repetitive, labor-intensive, time-consuming (Zhou et al., 1998), and is usually performed by workers who stand along the conveyor systems and remove all undesirable tubers based on their empirical knowledge. So, the carrot-packaging industry is facing many labor-related problems such as rising costs, low work productivity, and production waste owing to inconsistent grading and human errors (Al Ohali, 2011).

Computer vision provides a way to perform external quality inspection (Aleixos et al., 2002) with a high level of flexibility and repeatability and superior accuracy at relatively low cost (Sylla, 2002). Computer vision system has become a common scientific tool in industrial and agricultural productions due to its superior performance, low cost, and ease of use (Teena et al., 2013). In the past few decades, considerable research has been done in fruit and vegetable grading based on computer vision. But most of the work mainly focused on spherical or ellipsoidal products, such as apples (Ariana et al., 2006a), peaches (Esehaghbeygi et al., 2010), citrus (Aleixos et al., 2002; ElMasry et al., 2008), and potatoes (Barnes et al., 2010). Though some researches focused on long-shaped or thin products, they mainly concentrated on such products as cucumbers (Ariana et al., 2006b) and eggplants (Kondo et al., 2005; Kondo et al., 2007), which are very different from carrots in external quality attributes.

In the past few years, several efforts have been made on carrot inspection and grading using image processing. Howarth et al. (1992) designed a Bayes decision function to classify carrot tips into five classes using a curvature profile. The average misclassification rate on 250 carrots was 14%. Howarth and Searcy (1992) used Bayesian classifiers to classify fresh market carrots based on forking, surface defects, curvature, and brokenness. Overall, Type I and II errors were 11.1% and 19.1%, respectively, and the system could classify approximately 2.5 carrots per second. Hahn and Sanchez (2000) developed a prototype toward a fruit volume detector by rotating a CCD camera around the produce. A regression coefficient between read and predicted volume was achieved with carrots. Han et al. (2013) proposed extraction and detection methods for identifying carrots with fibrous roots, green shoulder, and cracks. The accuracy rates of detecting green shoulder, fibrous roots, and surface-cracking are 97.5%, 81.8%, and 92.3%, respectively.

Several carrot sorting systems are currently available from a number of companies, such as TOMRA, Raytec Vision, and Visar. TOMRA's carrot sorting machine is an all-in-one sorting machine; it is mainly used for separating good carrots, misshapen carrots and black spots carrots. The sorting machine of Raytec Vision could detect the presence of organic or inorganic foreign bodies, rotten products, and color defects.

Though some work has been done on carrot grading, most of them adopted one index, either the size or the shape or the color. Meanwhile, the detection accuracy and processing speed could not meet the requirement of real applications. Most commercial carrot sorting systems were all-in-one machine and mainly detected some common defects such as rotten products or color defects. So, it is necessary to develop new detection methods using computer vision system specifically for sorting and grading carrots with improved performance, which may increase the profits in carrot business.

The main goal of this chapter was to develop an automated carrot grading system by designing special mechanical equipment and develop new algorithms to detect the majority of defects and sort the carrots by length. The system could automatically pick out defective carrots and grade acceptable carrots into three classes at a speed of 12–15 carrots per second.

14.2   MATERIALS AND METHODS

14.2.1   Carrot Samples

Carrot samples were collected from a vegetable packinghouse (Qingdao, Shandong, China), which carried out carrot grading and sorting manually. In the packinghouse, carrots were first mechanically cleaned and washed to remove all the dirt, and then were transferred through a conveyor, from which the workers pick out irregular, damaged, and all undesirable carrots. Instead of sampling carrots randomly, we intentionally selected tubers of various sizes and shapes to cover regular and defective tubers. We selected three batches of carrot samples (600 carrots per batch) for evaluating three types of defects. The carrots were first manually classified by a professional inspector into regular and defective ones.

14.2.2 GRADING SYSTEM

The carrot grading system (Figure 14.1) consisted of an image processing system (a), an image acquisition system (b), a roller conveyor system (c), and a control system. The image acquisition system (Figure 14.2) consisted of a lighting chamber (a), three CCD cameras (b) (MV-VDM033SM/SC, Microvision, China), and some LED strips (c). CCD cameras were mounted right above three adjacent trays in a horizontal line and parallel to the trays. The cameras were about 30 cm away from the trays, and the images acquired by CCDs had a resolution of 320×240 pixels. LED strips were fixed on all sides of the lighting chamber to provide a uniform illumination on the carrots when they passed below the cameras. The image processing system handled all routines of carrot grading process such as image acquisition, image preprocessing, image segmentation, and detection algorithms which were programed in C++ language to ensure real-time control and response of the operations. The conveyor system consisted of a long platform connected with a rotating equipment which transferred the carrots one by one; the conveyor was composed of many rollers, and every two adjacent rollers were grouped into a moveable tray. The conveyor was capable of rotating and moving carrots simultaneously, making carrots be captured by the cameras in all angles and sides as much as possible. The control system included a single chip microcomputer (SCM), two photoelectrical sensors, four air rifles, and switch devices. The photoelectrical sensors were used to detect whether a tray was passing by; the air rifle was used to knock over the trays and let the carrots roll to the corresponding grading field. The SCM received signals from the photoelectrical

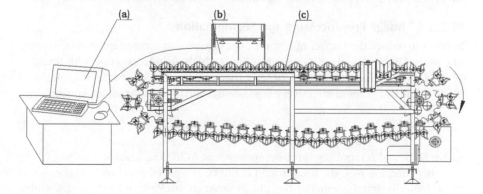

FIGURE 14.1 Schematic diagram of carrot grading system: (a) image processing system, (b) image acquisition system, and (c) roller conveyor system.

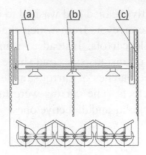

FIGURE 14.2 Image acquisition system: (a) lighting chamber, (b) CCD camera, and (c) LED strip.

sensor and compare them with the data in the computer, and then sent the signal to the air rifles for grading the carrots, thus controlling the grading process.

The workflow of carrot grading was as shown below:

1. A carrot was placed on a tray and moved forward in the conveyer belt.
2. When the carrots passed through, the photoelectrical sensors detected them and created a signal which would be sent to the control system; the control system induced the cameras to capture the images; and then, the image processing system would detect whether the carrot is defective – if defective, it graded the carrots as 0, and if not, it determined the grade (1–3) of the carrots based on its length.
3. As the carrots moved on, the control system operated the grading mechanism according to the results from image processing.
4. Steps 1–3 were repeated until all carrots were graded.

14.2.3 IMAGE PROCESSING AND DETECTION ALGORITHMS

All the image processing and detection algorithms were programed in C++ language on Visual Studio 2010 to meet the demand of real-time detection and control.

14.2.3.1 Image Preprocessing and Segmentation

In order to reduce the impact of complex background on image segmentation, referring to the Ohta color space (Ohta et al., 1980) which included a set of color features: $(R+G+B)/3$, $R - G$ and $(2G - R - B)/2$, we found that a gray-value image – calculated by equation 14.1 – could effectively separate the carrots from complex background.

$$I = 2.5R - 2G - 0.5B \tag{14.1}$$

where R, G, and B represent three components of RGB image, respectively.

Using equation 14.1, the background could be converted into black; hence, it was possible to accurately segment the object from its background by using a global threshold (the threshold was 40), and it would result in a binary image where the object (the carrot) was represented in white (ones) and its background in black (zeros).

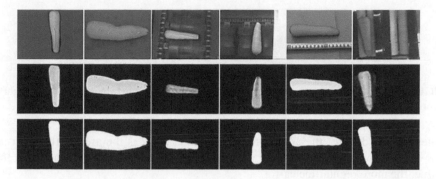

FIGURE 14.3   Illustration of image segmentation.

Figure 14.3 illustrates the segmentation results under different backgrounds. The sixth column shows that the proposed method is effective not only for image segmentation but also for image de-noising.

The first row shows original carrot images under different backgrounds, the second row is the gray-value images calculated by equation 14.1, and the last row is the binary image segmented by a global threshold.

For the proposed image segmentation method, the gray image was the weighted sum of the three components of RGB image. This method can be applied to the cases that the object is single-colored and the background is very complex. When the method is used to other objects, the weights in equation 14.1 ought to be tuned. We have also used the method to image segmentation of corn leaves (Deng et al., 2013) and potatoes, and achieved good results. We believe that there must be some relationship between the weights and the color of objects to be detected. But the relationship is not clear and needs further research. If we can determine the three weights by the color model of the objects to be detected, it will provide a new approach for image segmentation and object detection.

14.2.3.2   Shape Detection Algorithm

According to grading requirements of the carrot trade (MOFCOM, 2008), a carrot should be in a natural and uniform shape without any obvious twisting or bending. Typically, carrots without a mutational edge are convex-shaped. As to misshapen (crooked or bent) carrots, there would be an obvious concave region on the contour. Based on this idea, we proposed a new method, called "convex polygon method", to detect the degree of curvature of a carrot. First an external convex polygon was created to fit the contour of the carrot, as shown in Figure 14.4.

Figure 14.4 demonstrates that the contour of a typical carrot is very close to its external convex polygon. But for bent or twisted carrots, the external convex polygon does not so closely approximate to the carrot contour. We used the ratio of the area of carrot to its external convex polygon to describe the degree of curvature of the carrot, which is defined as follows:

$$D = \frac{\text{Area1}}{\text{Area2}} \tag{14.2}$$

FIGURE 14.4 External convex polygons of different shapes: (a) typical carrot, (b) slightly bent carrot, (c) significantly bent carrot, and (d) twisted carrot.

where Area1 denotes the area of the carrot and Area2 is the area of the external convex polygon. D measures the regularity degree of the carrot shape. The larger the value, the more regular the carrot.

14.2.3.3 Fibrous Root Detection

Fibrous root greatly influences carrots' quality, and any carrot with fibrous root should be rejected. For a carrot with a fibrous root, there will be an obvious depression area. Recent research suggests that if the lines between two arbitrary points on an object edge all fall inside the object, there must be a convex area on the object; otherwise, if all fall outside of the object, there will be a depression area (Liu and Wang, 2001). Based on this research, we proposed a simple but effective fibrous detection algorithm, which is described in Figure 14.5.

First, we extracted the carrot edge from the binary image and then judged whether an edge point was concave point by point. We used the extended border following algorithm (Suzuki, 1985) to extract the outer border of the binary image. Sample results of edge extraction are illustrated in Figure 14.6 in which the borders are outlined in light color. The edge could be accurately extracted from the binary image.

Suppose the current point was A, the points equidistant from A were B and C, the distance between A and B was d, and the midpoint of B and C was M, as shown in Figure 14.7. Selecting an appropriate value for parameter d was of great importance. We randomly selected 50 samples and increased d from 5 to 100 pixels with a step size of 2. Next, we computed the detection accuracy for all the values of d. Then, we selected the value (20 pixels) corresponding to the highest accuracy as the value of parameter d.

Then, we determined whether the point M was inside the carrot. If so, point A might be a convex point; otherwise, it was not.

Due to the diversity of carrots' shape, their edges were not always regular. So, the obtained concave points should be judged further. First, we calculated the angle between lines AB and AC by:

$$\theta = \arccos\frac{b^2 + c^2 - a^2}{2bc} \tag{14.3}$$

where a, b, and c are the lengths of lines BC, AC, and AB, respectively.

Then, we could judge whether there was a fibrous root by setting a threshold to θ. If θ was smaller than a preset threshold T (2.6 rad), which was determined in the same way as parameter d, we considered point A as the fibrous root's location. However, it was often the case that some successive points were detected around a

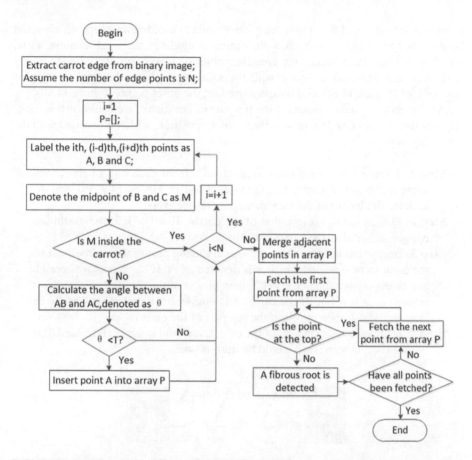

FIGURE 14.5 Flowchart of fibrous detection.

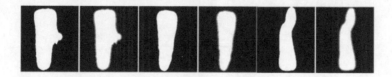

FIGURE 14.6 Sample results of edge extraction.

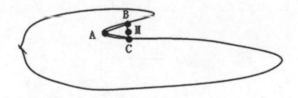

FIGURE 14.7 Schematic diagram of concave point detection.

fibrous root (Figure 14.8a). These successive points should be merged, which could be achieved by selecting a point with minimum angle as the final location. First, we searched for these successive concave points by traversing all the edge points (Figure 14.8b), selected the point with the smallest angle as the final location of fibrous root (Figure 14.8c), and then removed all the other points (Figure 14.8d).

As fibrous roots never appear at the top part of the carrot, we could further identify whether the obtained point was fibrous by its position, which proceeded with the following steps:

Step 1: Intercept two diameters, respectively, from each end of the carrot, about one-tenth of carrot length, as depicted in Figure 14.9. $W1$ and $W2$ indicate the length of the two diameters, respectively.

Step 2: Determine the orientation of the carrot. If $W1 > W2$, the carrot was upright; otherwise, the carrot was upside down.

Step 3: Determine whether the point was at the top part of the carrot. Assume the point to be detected which was denoted as pt. If the carrot was upright and pt was above diameter $V1V2$, then pt was at the top part of the carrot; otherwise, it was not. If the carrot was upside down and pt was below the diameter $V3V4$, then pt was at the top part of the carrot.

Step 4: If pt was at the top part of the carrot, it would not be determined to be fibrous root; otherwise, it would be fibrous one.

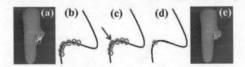

FIGURE 14.8 Process of point merging: (a) before merging, (b) successive concave points, (c) point with the smallest angle, (d) removing other points, and (e) after merging.

FIGURE 14.9 Illustration of diameter interception.

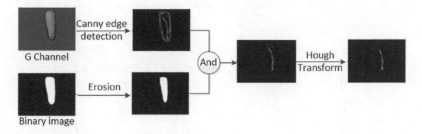

FIGURE 14.10 Process of crack detection.

14.2.3.4 Surface Crack Detection

The Hough transform is an effective method for line detection in the field of pattern recognition (Zeng et al., 2005). The traditional application of the Hough transform is to detect line, and it is also used to detect other shapes such as curve, circle, and oval. We use the Hough transform to detect the surface crack as it is a long and narrow line or a curve, which can be considered as a number of lines, in appearance.

Figure 14.10 depicts the process of crack detection. As the crack is relatively obvious on the G channel image, we first performed canny operator on G channel to detect edges. Next, an erosion operation was made on the binary image to remove the border. Then, an "And" operation was made between the canny edge image and binary image to remove the contour and only keep the crack. Finally, linear Hough transform was used to detect the curve. The Hough transform received edge image as input and outputted a sequence of line segments. Connecting the adjacent lines would form the crack curve. In the real grading process, carrots were picked out as defective once a crack was found on their surface. So, detecting whether there was crack or not could meet the requirement of carrot grading.

14.2.4 Bayes Classifier

We used Bayes classifier to determine the curvature threshold, and the Bayes classifier function is defined by

$$g(x) = w^T x - w_0 \qquad (14.4)$$

Regardless of different prior probabilities, the threshold can be computed by

$$w_0 = \frac{1}{2}(\tilde{m}_1 + \tilde{m}_2) \qquad (14.5)$$

where $\tilde{m}_1$ and $\tilde{m}_2$ are the mean of the two groups of samples, respectively.

14.3 EXPERIMENTAL RESULTS

14.3.1 Detection

We selected 160 carrots randomly as training samples to determine the threshold for judging whether a carrot was regular, and the remaining 440 carrots were used for

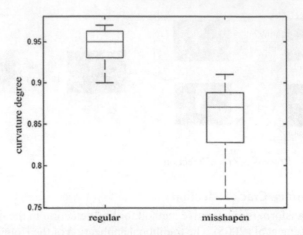

FIGURE 14.11   Boxplot of curvature degree.

TABLE 14.1
Shape Detection Results on Testing Samples

Items	Correct	Error	Total	Accuracy (%)
Regular	314	14	328	95.7
Misshapen	106	6	112	94.6
Total	420	20	440	95.5

testing. The training samples were classified into regular and misshapen (crooked or bent) by a professional inspector. We calculated the curvature degree D using equation 14.2 for all training samples and created the boxplot (see Figure 14.11).

We used equation 14.5 to determine the threshold value (0.9), and then evaluated the detection performance on the testing samples, and the results are presented in Table 14.1.

Results show that the curvature degree represented by equation 14.2 was able to predict carrot shape with an overall success rate of 95.5%. Regular tubers were predicted with an accuracy rate of 95.7%, and the accuracy rate of predicting misshapen tubers was 94.6%. The proposed method demonstrated an outstanding performance in shape detection.

14.3.2   FIBROUS ROOT DETECTION

To evaluate the performance of the proposed method, we conducted experiments on 600 samples, and the experimental results are presented in Table 14.2.

Table 14.2 shows that the proposed detection algorithm was able to predict fibrous root with an overall accuracy rate of 98%. Carrots with fibrous root were predicted with a success rate of 97.9%, and the accuracy rate of predicting normal ones was 98%. There were two conditions that led to misidentification of normal carrots as

TABLE 14.2
Fibrous Root Detection Results

Items	Correct	Error	Total	Accuracy (%)
Normal	491	10	501	98.0
Fibrous	97	2	99	97.9
Total	588	12	600	98.0

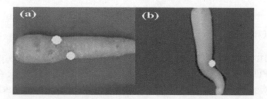

FIGURE 14.12 Two misjudgment conditions: (a) decayed carrot and (b) twisted carrot.

fibrous ones. One was rot on the surface leading to inaccurate edge extraction, as presented in Figure 14.12a. The other was the twisted carrot in which there was obvious depression, as depicted in Figure 14.12b.

Though the above two conditions influenced the detection performance of fibrous root, these carrots were also considered defective and discarded; so, these would not affect the overall performance. Meanwhile, occurrences of the two conditions were very low and less than 1%. To balance the detection performance with the processing speed, we did not take any measure to deal with these two cases.

For the image of 320 × 240 pixels, the average width of root fibers was 24 pixels. Our proposed method was able to detect those root fibers whose width was no less than 4 pixels.

14.3.3 CRACK DETECTION

We tested the crack detection algorithm on 600 carrot samples, and the results are listed in Table 14.3.

Experimental results show that the proposed crack detection method had a good performance on the crack carrots with an accuracy rate of 98.3%. But for normal

TABLE 14.3
Crack Detection Accuracy

Items	Correct	Error	Total	Accuracy (%)
Normal	408	68	476	85.7
Crack	122	2	124	98.3
Total	530	70	600	88.3

TABLE 14.4
Average Computational Time of Detection Algorithms

Item	Curvature	Fibrous	Crack	Total
Time (ms)	2.9	4.6	10.3	17.8

carrots, there were lots of normal carrots detected as cracked ones, and the accuracy rate was only 85.7%.

14.3.4 TIME EFFICIENCY

To test the time efficiency of the proposed algorithms, we run all algorithms on a laptop computer (CPU: Intel®Core™ i5-4200M @ 2.50 GHz, RAM: 4 GB; OS: Win7). Each algorithm was run ten times, and the results were averaged, as presented in Table 14.4.

As each carrot had three images during the process of grading, the total time of each carrot was about 54 ms. Taking the additional time of image preprocessing into consideration, about 12–15 carrots can be processed per second, which can meet real-time requirements well.

14.4 DISCUSSIONS

14.4.1 SHAPE DETECTION

In this chapter, we have presented an effective method, denoted as "convex polygon method", to detect carrot shape. Meanwhile, the algorithm was independent of image size or carrot orientation. There were some factors affecting the detection performance, such as stem attached to carrot and part of carrot extending beyond image region. The stem attached to carrot would lead to irregular shape and influence the detection performance to some degree. As in manual process, stems should be removed before grading, so the probability of this case to occur is very low. In the case of part of carrot extending beyond imaging region, carrots would be detected as defective ones due to incomplete recognition. To avoid this, we adjust the height of cameras to ensure that the longest carrot is fully inside camera's view field.

Comparatively, our method is not very sensitive to noise and has the advantages of both detection accuracy and time efficiency. In this chapter, we only classified the carrots into two grades: regular and misshapen, which can meet the demand of the carrot grading in real carrot-packaging industry. The proposed method can also classify carrots into more grades via the curvature degree for more accurate demand.

14.4.2 FIBROUS ROOT DETECTION

In the fibrous root detection method, parameter d was tuned for image with a resolution of 320 × 240 pixels and its value depended on the size of images. In order to make our method adapt to different image sizes, there are two solutions:

Resizing the image to 320×240 pixels.

Establishing the relationship between the parameters d and the carrot length L using least-squares fitting algorithm.

$$d = \frac{L}{12} \tag{14.6}$$

The proposed method provided an effective way to detect fibrous root. It has demonstrated not only outstanding accuracy but also high efficiency. In practical application, the computational complexity is an essential criterion. The computational complexity of our method is proportional to the number of edge points N, so the time complexity of the algorithm is $O(N)$, which indicates there is a linear relationship between the times of statement execution and N. Han et al. (2013) proposed a fibrous root detection method which detected the fibrous root by extracting the skeleton of the carrot image. However, the accuracy rate of detection was only 81.8%. Moreover, the skeleton extraction algorithm is more time-consuming than our method. So, our method has a comparative advantage in detection performance, which can meet the demand of an automatic system working in real-time operation. We have proposed two solutions to make our method be flexibly used in carrot fibrous root detection of different resolutions. As long as the carrot edge can be accurately extracted, this method will have first-rate detection performance.

14.4.3 CRACK DETECTION

In the crack detection method, lots of normal carrots were judged as defective, which mainly arose from the properties of canny operators. As the canny algorithm is very sensitive to the change of gray values, nonuniform illumination would influence the detection result to some degree. This could be improved in two ways. One is to improve the lighting environment and provide uniform illumination as far as possible. The other is to perform enhancement operations such as homomorphic filtering and Retinex enhancement.

14.4.4 FUTURE WORK

In respect of detection method, about 12–15 carrots could be processed per second. But concerning the whole grading system, the detection speed was limited by the mechanical equipment with roller mechanism which could not run fast enough and the CCD cameras which were unable to capture clear pictures when the carrots moved in high speed. To alleviate these limits, the roller conveyor system needs to be improved; perhaps tape transport system would be a good choice. In the case of camera, high-sensitivity line scan CCD camera would meet the requirement of high-speed applications much better.

As the overall accuracy rate of crack detection is only 88.3%, future work should be done to improve the crack detection performance by referring to pavement crack detection methods such as Gabor filter (Salman et al., 2013), subregion, and multiscale analysis (Lu et al., 2014), learning from samples' paradigm (Oliveira and Correia, 2013).

14.5 CONCLUSIONS

In order to achieve automatic carrot grading, we have constructed a machine vision system and proposed external quality inspection algorithms including curvature, fibrous root, and crack detection. Experimental results showed that the proposed detection methods demonstrated good detection performance and the accuracy rates of detection were 95.5%, 98%, and 88.3%, respectively. Meanwhile, 12–15 carrots could be processed per second, which could meet the requirements of automatic carrot grading and sorting.

REFERENCES

Al Ohali Y Computer vision based date fruit grading system: Design and implementation. *Journal of King Saud University. Computer and Information Science*, 2011, 23(1): 29–36. http://dx.doi.org/10.1016/j.jksuci.2010.03.003

Aleixos N, Blasco J, Navarron F, Moltó E. Multispectral inspection of citrus in real-time using machine vision and digital signal processors. *Computers and Electronics in Agriculture*, 2002, 33(2): 121–137. http://dx.doi.org/10.1016/S0168-1699(02)00002-9

Ariana D, Guyer D E, Shrestha B. Integrating multispectral reflectance and fluorescence imaging for defect detection on apples. *Computers and Electronics in Agriculture*, 2006a, 50(2): 148–161. http://dx.doi.org/10.1016/j.compag.2005.10.002

Ariana D P, Lu, R, Guyer, D E. Near-infrared hyperspectral reflectance imaging for detection of bruises on pickling cucumbers. *Computers and Electronics in Agriculture*, 2006b, 3(1): 60–70. http://dx.doi.org/10.1016/j.compag.2006.04.001

Barnes M, Duckett T, Cielniak G, Stroud G, Harper G. Visual detection of blemishes in potatoes using minimalist boosted classifiers. *Journal of Food Engineering*, 2010, 98(3): 339–346. http://dx.doi.org/10.1016/j.jfoodeng.2010.01.010

Deng L, Chen H, Ma W. Study on identification of waxy corn leaf by computer vision based on reflection and transmission image. *Science and Technology of Cereals, Oils and Foods*, 2013, 21(4): 80–83.

ElMasry G, Cubero S, Moltó E, Blasco J. In-line sorting of irregular potatoes by using automated computer-based machine vision system. *Journal of Food Engineering*, 2012, 112(1): 60–68. http://dx.doi.org/10.1016/j.jfoodeng.2012.03.027

ElMasry G, Wang N, Vigneault C, Qiao J, ElSayed A. Early detection of apple bruises on different background colors using hyperspectral imaging. *LWT - Food Science and Technology*, 2008, 41(2): 337–345. http://dx.doi.org/10.1016/j.lwt.2007.02.022

Esehaghbeygi A, Ardforoushan M, Monajemi S A H, Masoumi A A. Digital image processing for quality ranking of saffron peach. *International Agrophysics*, 2010, 24(2): 115–120.

Hahn F, Sanchez S. Carrot volume evaluation using imaging algorithms. *Journal of Agricultural Engineering Research*, 2000, 75(3): 243–249. http://dx.doi.org/10.1006/jaer.1999.0466

Han Z, Deng L, Xu Y, Feng Y, Geng Q, Xiong K. Image processing method for detection of carrot green-shoulder, fibrous roots and surface cracks. *Transations of the Chinese Society of Agricultural Engineering*, 2013, 29(9): 156–161.

Howarth M S, Brandon J R, Searcy S W, Kehtarnavaz N. Estimation of tip shape for carrot classification by machine vision. *Journal of Agricultural Engineering Research*, 1992, 53: 123–139. http://dx.doi.org/10.1016/0021-8634(92)80078-7

Howarth M S, Searcy, S W. Inspection of fresh carrots by machine vision. *Proceedings of Food Processing Automation II Conference*. ASAE, St. Joseph, MI, 1992.

Kondo N, Chong, V K, Ninomiya K, Nishi T, Monta M. Application of NIR-color CCD camera to eggplant grading machine. ASAE Paper No. 056073. ASAE, St. Joseph, MI, 2005.

Kondo N, Ninomiya K, Kamata J, Chong V K, Monta M, Ting K C. Eggplant grading system including rotary tray assisted machine vision whole fruit inspection. *Japanese Society of Agricultural Machinery*, 2007, 69(1): 68–77.

Liu X, Wang L. A study on computing the concavity of object. *Computer and Modernization*, 2001, (6): 1–4.

Lu Z, Wu C, Chen D, Shang, S. Pavement crack detection algorithm based on sub-region and multi-scale analysis. *Journal of Northeastern University*, 2014, 35(5): 622–625.

MOFCOM. SB/T 10450-2007: Grading of purchase and sale of carrot. Ministry of Commerce of the People's Republic of China, 2008.

Ohta Y, Kanade T, Sakai T. Color information for region segmentation. *Computer Graphics and Image Processing*, 1980, 13(3): 222–241. http://dx.doi.org/10.1016/0146-664X(80)90047-7

Oliveira H, Correia P L. Automatic road crack detection and characterization. *IEEE Intelligent Transportation Systems Society*, 2013, 14(1): 155–168. http://dx.doi.org/10.1109/TITS.2012.2208630

Salman M, Mathavan S, Kamal K, Rahman M. Pavement crack detection using the Gabor filter. *16th International IEEE Conference on Intelligent Transportation Systems (ITSC 2013)*, 2013, pp. 2039–2044. http://dx.doi.org/10.1109/ITSC.2013.6728529

Suzuki, S. Topological structural analysis of digitized binary images by border following. *Computer Vision, Graphics, and Image Processing*, 1985, 30(1): 32–46. http://dx.doi.org/10.1016/0734-189X(85)90016-7

Sylla, C. Experimental investigation of human and machine-vision arrangements in inspection tasks. *Control Engineering Practice*, 2002, 10(3): 347–361. http://dx.doi.org/10.1016/S0967-0661(01)00151-4

Teena M, Manickavasagan A, Mothershaw A, El Hadi S, Jayas D S. Potential of machine vision techniques for detecting fecal and microbial contamination of food products: A review. *Food and Bioprocess Technology*, 2013, 6(7): 1621–1634. http://dx.doi.org/10.1007/s11947-013-1079-7

Zeng J X, Zhang G M, Chu J, Lu Y M. The application of hough transform in the detection of exponent function curve. *Journal of Image and Graphics*, 2005, 10(2): 236–240.

Zhang B, Huang W, Li J, Zhao C, Fan S, Wu J, Liu C. Principles, developments and applications of computer vision for external quality inspection of fruits and vegetables: A review. *Food Research International*, 2014, 62: 326–343. http://dx.doi.org/10.1016/j.foodres.2014.03.012

Zhou L, Chalana V, Kim Y. PC-based machine vision system for real-time computer-aided potato inspection. *International Journal of Imaging Systems and Technology*, 1998, 9(6): 423–433. http://dx.doi.org/10.1002/(SICI)1098-1098(1998)9:6<423::AID-IMA4>3.0.CO;2-C

15 A New Automatic Carrot Grading System Based on Computer Vision

To speed up the process of carrot grading and realize automatic carrot grading, we have designed and implemented an automatic system based on computer vision, which consists of a feeding machine, a conveyor system, an image processing system, an image acquisition system, and a grading control system. Carrots were transferred based on three-level accelerated relay transmission techniques, and the ultimate speed is about $4\,\text{m s}^{-1}$. To obtain a high-quality image of carrot moving at high speed, frame-trigger linear push-broom imaging technology was used to acquire the carrot image. The system was controlled by the single-chip microcomputer (SCM) and a computer. Pneumatic control and time-delay techniques were applied to control the whole process of carrot grading. Detection methods were proposed to detect surface defects like cracks, fracture, abnormity, and so on. Experimental results show that the proposed defect detection method can achieve an overall accuracy rate of 90% and about 15 carrots could be graded per second. The constructed grading system could meet the demand of carrot grading and sorting.

15.1 INTRODUCTION

In post-harvesting operations, the carrot grading and sorting process poses great problems for growers and traders. The reason is that it is a repetitive, labor-intensive, and time-consuming process, and it is mainly carried out by humans manually through visual inspection (Al Ohali, 2011). Meanwhile, the manual processing poses additional problems of maintaining the consistency and uniformity of the carrot grading process.

Over the past few decades, with the rapid development of information science, image processing, and pattern recognition technology, computer vision has been developed as a scientific inspection tool for quality and safety of a variety of food and agricultural products (Zhang et al., 2014). The computer vision grading technology is real-time, objective, nondestructive, and can detect multiple indices simultaneously, such as size, defect, color, shape, and maturity (Liming and Yanchao, 2010). The methods extend the manual-machine grading, in which the features are determined manually under laboratory conditions (Mustafa et al., 2009).

In the recent several decades, several efforts have been made on carrot inspection and grading based on computer vision. Howarth et al. (1992) developed a machine vision system to inspect fresh market carrots; however, it only classified approximately 2.5 carrots per second. Brown and Gracie (2000) identified the major factors influencing variability in root sizes of carrots and developed recommendations for minimizing variability in carrot sizes. Hahn and Sanchez (2000) developed a fruit volume detector, and a regression coefficient of 0.98 was achieved between real and predicted volumes of carrots. Han et al. (2013) proposed external feature extraction methods for carrots, and the accuracy rates of detection for green shoulder, fibrous root, and carrot surface crack were, respectively, 97.5%, 81.8%, and 92.3%.

Though lots of achievements have been made on carrot quality detection, some of them only adopted one index (Brown and Gracie, 2000; Hahn and Sanchez, 2000; Howarth et al., 1992), either size or shape or color. Some others could detect multiple indices simultaneously, and they only focused on the detecting algorithms (Han et al., 2013). Some systems were designed and constructed for carrot grading; however, the accuracy rate of detection or processing speed could not meet requirements of practical applications (Howarth et al., 1992; Koszela et al., 2013; Du et al., 2015). Several carrot sorting systems are currently offered by some companies, such as TOMRA, Raytec Vision, and Visar. But most commercial sorting systems were all-in-one machine and mainly focused on some common defects such as rots or color defects. So, developing special mechanical machinery using computer vision is necessary to improve the performance of carrot grading systems and increase the profits from carrot business.

In this chapter, we investigate the requirements for designing a device-mediated carrot quality assessment and grading system, develop a machinery that can effectively meet these requirements, and test its effectiveness on actual production data.

15.2 MATERIALS AND METHODS

15.2.1 MATERIALS

Carrot samples were obtained from a vegetable processing plant (Qingdao, Shandong, China) which carried out carrot grading and sorting manually. The carrots were first mechanically cleaned and washed to remove all the dirt, and then were transported through a conveyor where the workers pick out irregular, damaged, and all undesirable carrots. Instead of sampling carrots randomly, we intentionally selected tubers of various sizes and shapes to cover regular and defective tubers. Several batches of carrot samples (about 600 carrots per batch) were selected for evaluating the performance of the proposed methods.

15.2.2 DESIGN OF CARROT GRADING SYSTEM

The automatic carrot grading system (Figure 15.1) comprises a feeding machine 1, a conveyor system 2, an image processing system 3, an image acquisition system 4, and a grading control system 5.

The feeding machine 1 consists of a vibration hopper 1.1, a jack device 1.2, two limit inclined chutes 1.3, some step-like pushing plates 1.4, connecting rods 1.5, eccentric wheel 1.6, electromotor 1.7, and a fixed plate 1.8. The washed carrots are dropped into the vibration hopper 1.1 and shifted up by the jack device 1.2. And then they are driven upward by the step-like pushing plate group 1.4, which was driven

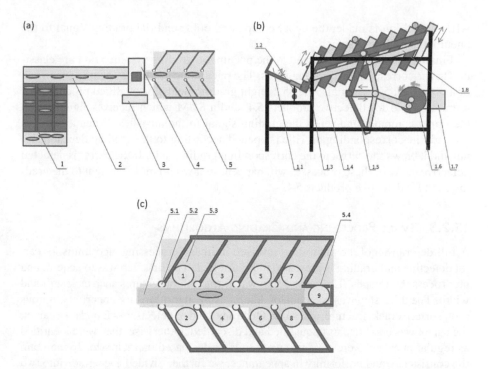

FIGURE 15.1 Schematic diagram of the automatic carrot grading system: (a) 1, feeding machine; 2, conveyor system; 3, image processing system; 4, image acquisition system; and 5, pneumatic grading system. (b) 1.1, vibration hopper; 1.2, jack device; 1.3, limit inclined chute; 1.4, step-like pushing plate; 1.5, connecting rod; 1.6, eccentric wheel; 1.7, electromotor; and 1.8, fixed plate. (c) 5.1, plane belt; 5.2, air rifle #2; 5.3, grading field #1; and 5.4, grading field for defective products.

by the connecting rods 1.5 and the electromotor 1.7. During the process, the step-like pushing plate groups alternatively raise and drop with the action of limit inclined chutes 1.3. Under the action of the thrust and gravity, carrots move to the very top and fall onto the conveyor system 2.

The conveyor system 2 consists of two sections of v-type conveyors and a plane belt, and transfers carrots one by one in a line. The conveyor system was accelerated gradually. The first conveyor ran with a speed of about 2–3 m s^{-1}, and the second and third ran at a speed of 4 m s^{-1}. The carrots are transferred forward to the image acquisition system 4. The image acquisition system consists of a lighting chamber, a linear camera mounted on top of the lighting chamber, two mirrors at an angle about 135°, and two ring LED strips. Each mirror was about 5 cm away from the carrot so that the image acquisition system can obtain three images of a carrot and get complete information of the carrots' features.

After the images are acquired, they are processed by the image processing system 3. The image processing system 3 (upper computer) deals with all routines of grading process such as image acquisition, image preprocessing, and detection algorithms which are programed in C++ language to ensure real-time control and response.

After a decision is made, the upper computer would send the grading signal to the pneumatic grading system 5.

Finally, carrots are transferred to the pneumatic grading system 5 and are classified to the corresponding grading fields. The pneumatic grading system is composed of a plane belt 5.1, eight air rifles 5.2, eight grading fields for qualified products 5.3, a grading field for defective products 5.4, and a SCM which receives signals from the upper computer and sends the grading signal to the air rifles. When a carrot is arrived, the corresponding air rifle is opened according to the grading decision signal and it blows the carrot to the corresponding grading field. If the carrot is detected defective, none of these processes will happen, so it would move straight to the grading field for defective products 5.4.

15.2.3 IMAGE PROCESSING AND GRADING ALGORITHMS

A full description of the main steps involved in image processing algorithms for carrot detecting and grading is presented in Figure 15.2. The first step was to acquire and preprocess the images. The second step was extraction of features such as length and width. The third step was detection of defects using algorithms for abnormity, fibrous root, surface crack, fracture, and so on. Once any kind of defects was found on a carrot, the carrot was classified as unqualified and discarded. Otherwise, they were identified as regular ones and were classified up to four grades based on their size. To maintain the consistency and uniformity in appearance, we further divided each grade into two levels: slenderness and tubbiness. So eventually, there would be eight grades.

15.2.3.1 Image Acquisition

As the carrots moved forward at a high speed about 4 m s^{-1}, a color bilinear camera (DALSASpyder3 SG-34 GEV, 18 KHz line rates, 2,048 pixels) (Teledyne DALSA, 2012) was used to obtain clear images of each carrot. Frame-trigger linear push-broom

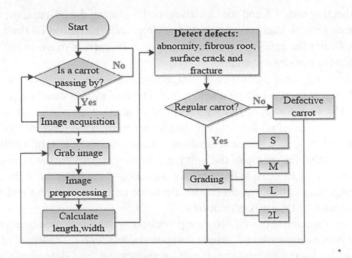

FIGURE 15.2 Flowchart of image processing and carrot grading.

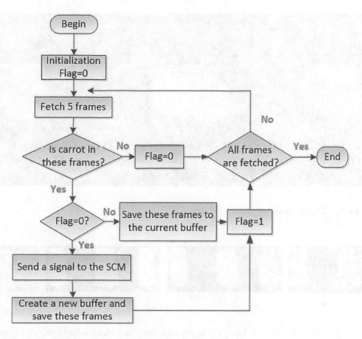

FIGURE 15.3   Flowchart of frame processing.

imaging technology was used to acquire the images of carrots. The camera continuously acquired images frame by frame. At the same time, the upper computer concurrently fetched the frames from the camera and only kept the images of carrots, which is detailed in Figure 15.3. Each time, five frames (5 × 2,048, about 5 mm width) were fetched from the buffer, and then checked for the presence of carrots. If the presence of carrot was detected in these frames, which was judged by a flag variable, the process of creating carrot image began. A new buffer was created and the frames were saved in it. Otherwise, the frames are saved in the current buffer. If there is no carrot on these frames, they were discarded. Repeat the process again and again until all frames are fetched. As a result, all carrot images would be saved in the buffers.

In order to accelerate the system and perform all processes in real time, acquisition and processing of the images were conducted simultaneously by parallel processing technology, making it possible to perform the classification in an online mode. Views from three sides were acquired for each carrot, as shown in Figure 15.4. The middle one is the real image of a carrot, and the other two are its mirror images. The middle image was used to determine the length and detect surface defects such as green shoulder, fibrous root, crack, and fracture. The other two images were used to detect surface crack and erosion.

15.2.3.2   Image Preprocessing

First, the input image I was converted to grayscale image by the following equation:

$$I_G = 2.5I_r - 2I_g - 0.5I_b \qquad (15.1)$$

FIGURE 15.4 Three sides of carrot image.

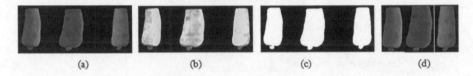

(a)   (b)   (c)   (d)

FIGURE 15.5 Illustration of image preprocessing: (a) source image, (b) gray image, (c) binary image, and (d) individual image.

where I_r, I_g, and I_b, respectively, represent three components of color image I.

Thus, the background of image I_G (Figure 15.5a) could be converted into black (Figure 15.5b), which would result in a binary image where the object (the carrot) was represented in white (ones) and the background in black (zeros), as shown in Figure 15.5c; hence, it was possible to segment the object accurately from its background by using a global threshold (the threshold was 40). Then, three object (carrot) areas could be extracted by labeling the connected regions on the binary image, which would result in three individual images (Figure 15.5d).

15.2.3.3 Defect Detection Algorithms

After images of the carrot was separated from its background, detection algorithms were used to detect the surface defects such as cracks, fractures, and green shoulders.

1. Surface crack detection

    As the cracks were obvious on the G component, we operated on G image to detect surface crack. First, histogram equalization was used to adjust luminance and contrast. Then, canny operator was applied to extract and obtain the edges. Then, an "And" operation was made between the canny edge image and binary image to remove the contour and to keep only the crack.

2. Fracture detection

    For carrots, there are two types of fractures. One is that the carrot was broken from the top (see Figure 15.6a) and the other is from the bottom

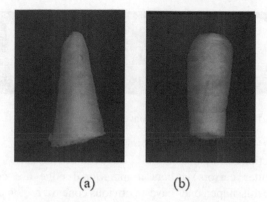

(a)   (b)

FIGURE 15.6   Samples of broken carrots: (a) top broken and (b) bottom broken.

(see Figure 15.6b). These two conditions should be considered separately. In the process of grading, carrots with either fractures would be rejected.

Generally speaking, the broken carrots are shorter than the normal ones, while the width remains unchanged. So, we can make a preliminary judgment by the ratio between width and length. Then, we presented two kinds of algorithms for the above two cases. The flowchart of fracture detection is depicted in Figure 15.7.

Normal carrots are cone-shaped and their bottom is pointy, whereas broken carrots are approximately trapezoid-shaped and their bottom is relatively flat. Du et al. (2015) proposed a detection algorithm based on transverse diameter, which failed to detect the fracture if it occurred very close to the bottom. So, we extended their method and added another criterion to detect fracture.

Though the carrot broken on top is still cone-shaped on the whole, the top part is relatively flat (Figure 15.7), which is arc-shaped in normal carrots. So, we can judge by the shape of the top region. First, we sought for two boundary points (as shown in Figure 15.7) using local maximum method and then obtained a sequence of points between the two points. Then, we calculated the shape of the top region by accumulating the distance from each point to line *AB*.

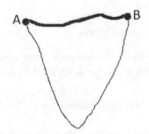

FIGURE 15.7   Schematic diagram of top-broken detection.

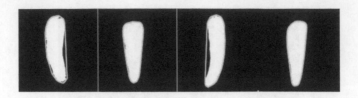

FIGURE 15.8   Carrots and their external convex polygons.

3. Shape detection

The regular carrots without a mutational edge are convex-shaped, whereas the misshapen ones have an obvious concave region on the contour. So, an external convex polygon was created to fit the contour of the carrot (as shown in Figure 15.8). The ratio of the area of carrot to its external convex polygon was used to describe the degree of curvature of the carrot.

Some other defect detection algorithms have been involved in our previous studies (Du et al., 2015; Deng et al., 2017), so that won't be discussed here again.

15.2.3.4   Grading Regular Carrots by Size

Once defect was detected from carrots, they would be isolated in a separate class; meanwhile, the regular carrots were classified into different grades based on their size. To obtain the average length of carrots, we first measured the length of 100 carrots by hand. Then we took pictures of these carrots and calculated the length (in pixels) of these carrots. Next, we built the relationships between the actual length L_a and the pixel length L_p by linear fitting (Tomé and Miranda, 2004):

$$L_a = 0.1115L_p - 0.3241$$

The fitness error of carrot length between original carrot and shot carrot was less than 0.5 mm.

Finally, three thresholds were applied to the algorithm to segregate the carrots into four different grades by length: S, M, L, and $2L$ (i.e., <15, 15–20, 20–25, and >25 cm, respectively). Furthermore, each grade was divided into two levels (slenderness and tubbiness) to maintain the consistency and uniformity in appearance, which could be achieved by calculating the ratio between carrot width and length.

15.2.4   CONTROL OF GRADING SYSTEM

Pneumatic control and time-delay techniques were employed to control the whole process of carrot grading, which can be described as follows:

1. The SCM receives the starting signal of a carrot from the upper computer and starts a timer.
2. When the SCM receives carrot grading signal from the upper computer, it would insert the signal into a queue. When the corresponding timer is up

to preset time (the time carrot moving from the camera to the air rifle), the SCM picks up the grading signal for the queue head and sends the grading signal to the corresponding air rifle. Then, the air rifle would blow the carrot to the corresponding grading field.

3. Steps 1 and 2 are repeated until all carrots are graded.

15.3 RESULTS

15.3.1 DEFECT DETECTION

To evaluate the performance of our detection methods, we tested each detection algorithm individually on different batches of samples. The detection results of all kinds of defects are presented in Table 15.1. In general, detection of all kinds of defects achieved accuracy rate above 89.6%, and the mean accuracy rate was 93.2%.

15.3.2 REGULAR CARROT GRADING

Regular carrots were classified into four grades by length. To evaluate the performance, we selected 1,158 carrots as testing samples covering the four grades. First, we measured their actual length by a ruler and classified them into four grades (S, M, L, and 2L). Then, we classified these carrots both by a professional worker and by running our grading algorithm. The results are, respectively, presented in Tables 15.2 and 15.3. Experimental results show that our grading method was more precise and greatly outperformed the manual way. Our method has shown an improvement in grading accuracy of approximately 9% over human grading.

15.3.3 TIME EFFICIENCY

To evaluate the time efficiency of the proposed detection algorithms, we ran all algorithms on a computer (CPU: Intel Core TM i5-4590M, RAM: 8 GB; OS: Win7). Each algorithm was run for ten times and the results were averaged, as presented in Table 15.4. Though the processing time of a carrot is about 118 ms, as the process was performed in parallel, the processing speed would not affect the overall efficiency, which is mainly affected by the transfer speeds.

TABLE 15.1

Detection Results of Defective Carrots

Item	Total	Correct	Accuracy (%)
Top broken	600	574	95.6
Bottom broken	600	543	92.4
Abnormity	560	546	97.5
Crack	580	520	89.6
Fibrous root	560	542	96.7
Rotten	600	538	89.6
Total	3500	3263	93.2

TABLE 15.2
Grading Results by a Professional Worker

From/To	S	M	L	2L	Total	Accuracy (%)
S	189	43	9		240	78.8
M	13	172	61		246	69.9
L		2	196	18	216	90.7
2L			9	207	216	95.8
Total	202	217	275	225	918	83.2

TABLE 15.3
Grading Results of the Proposed Method

From/To	S	M	L	2L	Total	Accuracy (%)
S	220	20			240	91.7
M	24	210	12		246	85.4
L		6	204	6	216	94.4
2L			8	208	216	96.3
Total	244	236	224	214	918	91.7

TABLE 15.4
Processing Time (ms) of Detection Algorithms

Times	Top Broken	Bottom Broken	Crack	Abnormity	Fibrous Root	Rotten	Total
1	27	25	109	24	58	32	125
2	23	23	109	23	22	28	125
3	22	22	109	24	24	30	109
4	22	25	94	24	21	28	109
5	22	23	109	22	21	29	110
6	22	27	110	22	22	28	124
7	24	24	109	22	22	27	125
8	23	26	109	22	26	29	109
9	21	21	109	23	22	30	124
10	22	24	109	22	22	28	124
Average	22.8	24	107.6	22.8	26	28.9	118.4

15.3.4 Performance Parameter

The performance parameters of the grading system are listed in Table 15.5, which can meet the requirements of national standards for carrot grading.

TABLE 15.5
Performance Parameters of the Grading System

Item	Value
Accuracy	≥93%
Damaged rate	≤5%
Bearing temperature rise	≤20°C
Noise	≤85 dB

15.4 DISCUSSION

On the aspect of mechanical equipment, we have designed feeding machine based on pushing plate device. The push plate device loads the transfer system with carrots one by one. At the same time, three feeding channels can load carrots simultaneously, which ensure the feeding efficiency. The material of conveyor belt would have an effect on the transfer process.

In the pneumatic grading system, we used a plane belt, which has some advantages compared to the V-shaped belt. First, because of their simple mechanical structure, they are convenient for manufacture, installation, and maintenance. Second, it can reduce noise, especially the friction noise. Third, pneumatic gun only needs to drop the carrot down to the plane conveyer, which would reduce the damage rate of carrots.

In terms of the image acquisition system, mounting position of the camera would have some impacts on the image acquisition. In our system, the camera is mounted on top of the lighting chamber and two mirrors are installed right below the camera. As the carrots are to be graded immediately after being washed, some water would be present on their surface. When they passed by the lighting chamber, water would drip onto the mirrors, which would spoil the mirror and affect the image quality. We have attempted to mount the camera on the side of the lighting chamber. However, focal length of the camera is not long enough. Maybe placing the camera at the bottom would be a good choice, which needs to be studied further.

The images were acquired in frame-trigger mode, rather than photoelectric trigger, of the camera. Under the photoelectric trigger mode, two sets of photoelectric sensors were, respectively, mounted before the camera and pneumatic gun. When the photoelectric sensor detects a carrot is passing by, it would send a signal to the camera, and the camera would begin to acquire the image. When a carrot is coming to the pneumatic gun, the SCM would fetch a grading signal from the signal queue and send it to the pneumatic gun. However, the photoelectric sensor is not always reliable, and there may be the case that the two sets of photoelectric sensors were not inconsistent, which would result in disorder of the signal and greatly affect the grading result. However, in the mode of frame trigger, the camera continuously acquires images so that none of the images would be missed. Hence, it is more reliable than the photoelectric trigger mode.

Compared with the previous machine (Deng et al., 2017), some improvements have been made in the proposed system:

1. The mechanical structure has been changed completely, which has greatly enhanced the grading efficiency and made it more suitable for carrots than the traditional grading machine. We used belt transport system instead of the roller conveyor system, which greatly improved the speed of transportation. As mentioned above, a feeding machine specially designed for carrots was used.
2. A color bilinear camera was used to obtain high-quality images of carrots at a high speed instead of the common CCD camera. Meanwhile, linear push-broom imaging technology with frame trigger was used, which was usually used for high-resolution satellite imaging (Weser et al., 2008).
3. To realize precision control and high-efficiency grading, we have employed some prevailing techniques used in the aerospace field, such as accelerated relay transmission and high-speed pneumatic intercept (Koren et al., 2008). Three-level accelerated relay transmission technology was used in the process of transportation, and the conveyor system consisting of three conveyors was accelerated gradually. Meanwhile, high-speed pneumatic intercept technology was applied to the process of grading control.

15.5 CONCLUSION

In order to increase the performance and speed of carrot grading and sorting, an automated system has been set up using machine vision technology. All design aspects including electronic hardware, software, lighting, mechanism, and control are discussed in this chapter. The mechanical methods and image processing techniques for real-time quality evaluation were introduced in detail. System specifications, such as accuracy and time efficiency, and production test results are also discussed in detail. The experiment has shown an improvement in grading accuracy of approximately 9% over human grading, while reducing labor costs by almost 75% and shortening the processing time.

REFERENCES

Al Ohali Y. Computer vision based date fruit grading system: Design and implementation. *Journal of King Saud University-Computer and Information Sciences*, 2011, 23(1): 29–36. http://dx.doi.org/10.1016/j.jksuci.2010.03.003

Brown P H, Gracie A J. Managing carrot root size. *Organising Committee*, 2000: 18–19.

Deng L, Du H, Han Z. A carrot sorting system using machine vision technique. *Applied Engineering in Agriculture*, 2017, 33(2). http://dx.doi.org/10.13031/aea.11549

Du H, Deng L, Xiong K, Han Z. Quality grading system and equipment of carrots based on computer vision. *Journal of Agricultural Mechanization Research*, 2015, 37(1): 196–200.

Hahn F, Sanchez S. Carrot volume evaluation using imaging algorithms. *Journal of Agricultural Engineering Research*, 2000, 75(3): 243–249. https://doi.org/10.1006/jaer.1999.0466

Han Z, Deng L, Xu Y, Feng Y, Geng Q, Xiong K. Image processing method for detection of carrot green-shoulder, fibrous roots and surface cracks. *Transactions from the Chinese Society of Agricultural Engineering*, 2013, 29(9): 156–161.

Howarth M S, Brandon J R, Searcy S W, Kehtarnavaz, N. Estimation of tip shape for carrot classification by machine vision. *Journal of Agricultural Engineering Research*, 1992, 53: 123–139.

Koren A, Idan M, Golan O M. Integrated sliding mode guidance and control for missile with on-off actuators. *Journal of Guidance, Control, and Dynamics*, 2008, 31(1): 204–214. http://dx.doi.org/10.2514/1.31328

Koszela K, Weres J, Boniecki P, Zaborowicz M, Przybył J, Dach J, Janczak D. Computer image analysis in the quality procedure for selected carrot varieties. *Proceedings Volume 8878, Fifth International Conference on Digital Image Processing (ICDIP 2013), 8878*, 2013, pp. 887811. http://dx.doi.org/10.1117/12.2030701

Liming X, Yanchao Z. Automated strawberry grading system based on image processing. *Computers and Electronics in Agriculture*, 2010, 71: S32–S39. http://dx.doi.org/10.1016/j.compag.2009.09.013.

Mustafa N B A, Ahmed S K, Ali Z, Yit W B. Agricultural produce sorting and grading using support vector machines and fuzzy logic. *IEEE International Conference on Signal and Image Processing Applications*, 2009, pp. 391–396. https://doi.org/10.1109/ICSIPA.2009.5478684

Teledyne DALSA. User's manual of Spyder3 SC-34. March 2012.

Tomé A R, Miranda P M A. Piecewise linear fitting and trend changing points of climate parameters. *Geophysical Research Letters*, 2004, 31(2). http://dx.doi.org/10.1029/2003GL019100

Weser T, Rottensteiner F, Willneff J, Poon J, Fraser C S. Development and testing of a generic sensor model for pushbroom satellite imagery. *Photogrammetric Record*, 2008, 23(123): 255–274. http://dx.doi.org/10.1111/j.1477-9730.2008.00489.x

Zhang B, Huang W, Li J, Zhao C, Fan S, Wu J, Liu C. Principles, developments and applications of computer vision for external quality inspection of fruits and vegetables: A review. *Food Research International*, 2014, 62: 326–343.

16 Identifying Carrot Appearance Quality by Transfer Learning

It is extremely important to correctly identify the carrot appearance quality in design and manufacture of carrot sorter. In this chapter, we have established a carrot appearance quality control system based on deep learning framework. The information of carrot is collected using the image, and thereafter, the recognition model is erect on AlexNet network, which is pretrained by a large-scale computer vision database (Image-Net). Our framework uses transfer learning, which trains neural networks with small amounts of data compared to the traditional convolutional neural network (CNN). Applying this approach to the dataset of carrot images, we demonstrate the performance of the proposed model by comparing it with the performance of human experts. The different grades can be recognized with great accuracy from a large amount of carrots under different surface conditions. Further, we demonstrate the general applicability of our system in potato quality recognition. At the same time, when the number of training samples is small, high recognition accuracy can still be achieved through transfer learning. The model can not only meet the requirements of classification recognition but can also greatly reduce the amount of cost spent in sample collection.

16.1 INTRODUCTION

Global carrot yield area is 1,196,000 hm², its total annual output is 36,917,000 tons, and the average yield per unit is 30.9 ton hm⁻². The yield of carrot is miraculous, and carrot is one of the most competitive agricultural products. At present, most production and processing enterprises of carrots mainly use manual methods for sorting, which has low efficiency, strong subjectivity, loose standards, and defects. Therefore, it is not suitable for large-scale production and processing.

In the field of carrot detection classification, a traditional method based on image processing (Han et al., 2013) was proposed. The traditional image classification algorithm is first used to segment the image manually, and then identify each segmented object using statistical classifier or shallow neural computing machine learning classifier specially designed for each object class, and finally classify the image (Goldbaum et al., 1996).

The development of CNN has greatly improved the ability of image classification and object detection (Zeiler and Fergus, 2014; Krizhevsky et al., 2012). These are multiple layers of processing that apply image analysis filters (or convolution). Abstract representation of each layer of the image is constructed by systematically convolving multiple filters on the image, generating a feature map, and taking it as input of the next layer. This architecture enables image processing in the form of pixels and provides the required classification as output. The image to classification method in a classifier replaces many steps of previous methods with image analysis.

In many real-world applications, regathering the required training data and rebuilding the model (Oquab et al., 2014) are expensive or even impossible. It is good to reduce the need and workload of regathering training data. In this case, knowledge transfer or transfer learning is required in task domains (Pan and Yang, 2010). One way to address the lack of data in a given domain is to use data from a similar domain, a technique known as transfer learning. Transfer learning has been successfully applied to text classification (Wang and Mahadevan, 2011), image classification (Duan et al., 2012; Kulis et al., 2011; Zhu et al., 2011), human activity classification (Harel and Mannor, 2011), software defect classification (Nam et al., 2017), and multilingual text classification (Carneiro et al., 2015; Prettenhofer and Stein, 2010; Zhou et al., 2014ab).

Transfer learning has proven to be a very effective technology, especially when data are limited (Razavian et al., 2014; Donahue et al., 2014; Yosinski et al., 2014). This chapter describes the schematic diagram of training CNN on 1,000 ImageNet datasets, which can significantly improve the accuracy of training neural network on new datasets of carrot image and reduce the training time. The layers of local connection (convolutional layers) are frozen and moved to a new network, while the final layer of full connection is recreated and trained, starting with random initialization above the transport layer (Figure 16.1).

We combined computer vision and transfer learning, and the model performed better than image processing method. The main contributions of this chapter are as follows: we try to develop an effective transfer learning algorithm to process carrot images and provide an accurate and real-time identification of key features of each image. The main demonstration of the technology was focused on carrot images, but the algorithm was also performed on potato images to verify the scalability of the technology.

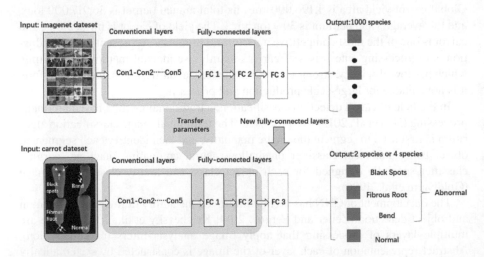

FIGURE 16.1   Schematic of a CNN.

16.2   MATERIALS AND METHOD

16.2.1   IMAGE CHARACTERISTICS

In this experiment, we collected images of normal carrots and carrots with defects such as black spots, fibrous roots, and bend, as shown in Figure 16.2. The specific characteristics of these carrots are as follows: The left is the image of a carrot with black spot which is pointed by white arrows. The second one from the left is the image of carrot with a fibrous root pointed by white arrows. The third one from the left is a bend carrot image (the white arrow points to the position). The right image is a normal carrot without any damage.

16.2.2   WORKFLOW DIAGRAM

The overall experimental design flowchart of carrot grading process is described in Figure 16.3, and the model of transfer learning is established for training and subsequent testing. The training dataset contains enough quality and standard images.

To identify two categories, we obtained 5,376 images (3,667 abnormal, 1,709 normal) to train our model. The model was tested on 2,304 images (733 from normal category and 1,571 from abnormal category). To identify the four categories, we obtained 2,286 images (206 with fibrous roots, 411 with black spots, 969 with bent shape, and 700 with normal appearance) to train the model. The model was tested on 980 images (300 from normal category, 88 from fibrous roots category, 176 from black spots category, and 415 from bent category) (Figure 16.3).

16.2.3   CONVOLUTIONAL NEURAL NETWORK

Since its introduction in the early 1990s, CNN has performed successfully in such tasks as handwriting classification and face detection (Zeiler and Fergus, 2014). At the same time, CNN has achieved good results in human posture evaluation (Ji et al., 2013), document recognition (Lecun et al., 1998), speech recognition (Abdelhamid et al., 2014), and medical image diagnosis (Litjens et al., 2017). Therefore, we carried out transfer learning on the basis of CNN to obtain excellent performance.

In the experiment, a CNN AlexNet (Krizhevsky et al., 2012) is utilized to extract the features of the carrot image and identify carrot quality. The AlexNet network was proposed by Hinton and his student Alex Krizhevsky in the 2012 ImageNet competition.

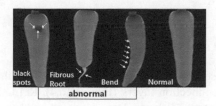

FIGURE 16.2   Carrot images.

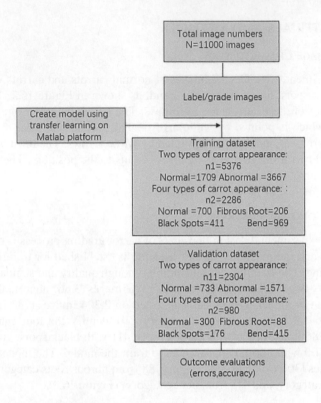

FIGURE 16.3   Workflow diagram and data amount.

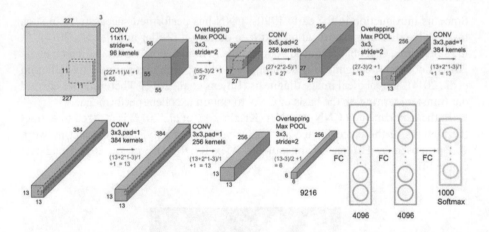

FIGURE 16.4   Detailed structure diagram.

AlexNet is comprised of eight layers of network, including five convolutional layers and three full connection layers (Figure 16.4). ReLU is taken as activation function in each convolutional layer. Dropout is adopted to prevent overfitting in full connection layers 1 and 2. Finally, softmax is adopted as loss function. The first seven layers are used for feature extraction, and the last layer is used for classification. The network uses a convolution kernel of 11×11 in the first convolutional layer, and to further extract image features, the network uses convolution kernel at 5×5 in the second convolutional layer. In addition, it uses a convolution kernel of 3×3 in the last convolutional layer. The network adopts small-size convolutional layer with three to five convolutional layers, because the small-size convolutional layer is more nonlinear than a large-size convolutional layer, making the function more accurate and reducing the number of parameters. Because the depth of the network and the small size of the convolution kernel help to achieve implicit regularization, the network starts to converge after a few iterations. Therefore, we have selected AlexNet network for the identification task of carrots.

16.2.4 TRANSFER LEARNING

The AlexNet pretrains the ImageNet dataset sufficiently, and the fully connection layer at the end of the network is 1,000 tensors (Krizhevsky et al., 2012). Although the neural network cannot recognize the carrot before training, it provides a good initial value for the carrot recognition network. Good starting values are critical to network training. Transfer the last layer to the new classification task by replacing them with a layer of full connection, a softmax layer, and a classification output layer. Based on the new data, renew the full connection layer. Set the total layer of connection to the same size as the number of classes in the new data. In order to learn faster in the new layer than in the transfer layer, the weight learning rate factor and the bias learning rate factor value of the fully connected layer were increased. For binary classification problem, the output was set to normal and abnormal, and for multiple classification problems, the output was set to four categories, namely normal, black spots, fibrous roots, and bending. The network is retrained using the carrot images, and after the weight of network training is fine-tuned, the network is applied to the carrot recognition. In the training process, random-gradient descent parameter optimization was adopted, and the overall learning rate was 0.0001.

16.3   PERFORMANCE OF THE MODEL

The recognition accuracy and loss during training are illustrated in Figure 16.5. Figure 16.5a and c shows the accuracy of training iterations, and Figure 16.5b and d shows the loss during training. The training iterations of the binary classifiers were 1,400, and the training iterations of the multi-classifier were 1,000. The graph was normalized, and the smoothing coefficient was 0.6 to clearly show the trend. The accuracy and loss of verification show better performance, because the training set also contains the image with more noise and lower quality, so as to reduce overfitting and contribute to the generalization of classifier. In the plot of accuracy curve, the training dataset is the solid line, and the validation dataset is the dashed line. In the plot of cross entropy loss, the training dataset is the solid line, and the validation dataset is the dashed line.

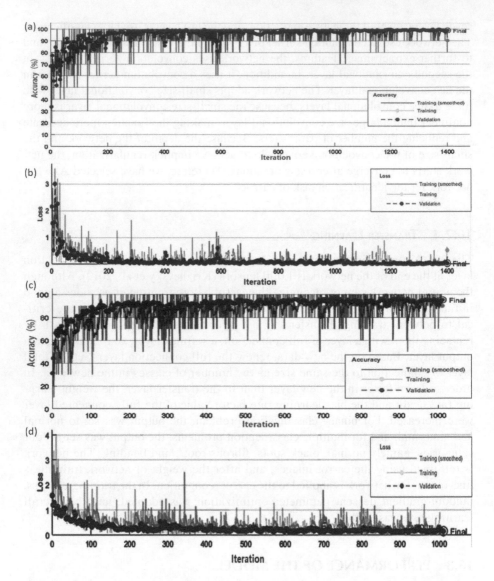

FIGURE 16.5 Accuracy and loss curve during training: (a) the accuracy curve during the training process of carrot identification of two kinds of carrots, (b) the curve of loss during the training process of two kinds of carrots, (c) the curve of accuracy during the training process of carrot identification of four kinds of carrots, and (d) the curve of loss during the training process of carrot identification of four kinds of carrots.

We assessed our system for identifying the appearance quality of carrots. For binary classification recognition, the accuracy rate was 98.7%, the sensitivity was 98.3%, and the specificity was 99%. Among the multiple classification recognition of black spots, fibrous roots, curvature, and normal values, our accuracy reached 95.30%, with a sensitivity of 96.8% and a specificity of 95.4%.

TABLE 16.1
Performance of the Two-Class Model

Items	Number of Total Images	Number of Trained Images	Number of Verified Images	Number of System Identification	Accuracy Rate (%)
Abnormal	5238	3667	1571	1542	98.15
Normal	2442	1709	733	712	97.13
Total	7680	5376	2304	2254	98.70

TABLE 16.2
Performance of the Four-Class Model

Items	Number of Total Images	Number of Trained Images	Number of Verified Images	Number of System Identification	Accuracy Rate (%)
Fibrous roots	294	206	88	83	94.31
Black spots	587	411	176	165	93.75
Bend	1384	969	415	399	96.15
Normal	1000	700	300	286	95.33
Total	3265	2286	979	933	95.33

Recognition accuracy of the proposed model for each category is shown in Tables 16.1 and 16.2.

In order to further evaluate the recognition performance of the system, we use less data to train the network. We trained a "limited data model" to distinguish the two categories and used only 300 images randomly selected from each category to train the neural network and to compare the transfer learning performance of results using limited data with that using big datasets. The statistical model was 97.3% accurate (Figure 16.6a), 98% sensitive, and 96.6% specific.

We also trained a "limited data model", similarly for four categories, randomly selected 75 images from each category to train the neural network, and compared the conversion learning performance of the results of limited data with that of big datasets. Using 75 images of each type to test our model, the accuracy rate of the model was 94.3% (Figure 16.6c), the sensitivity was 90.6%, and the specificity was 95.5%.

The loss of the "limited data model" for the training step is plotted in Figure 16.6b and d. The recognition accuracy of the "limited data model" for each category is shown in Tables 16.3 and 16.4.

16.4 COMPARISON WITH MANUAL WORK

We used a separate set of 600 images to compare the recognition performance of the system with that of human experts (Figure 16.7). Six experts with significant experience of classification in the laboratory were instructed to make classification

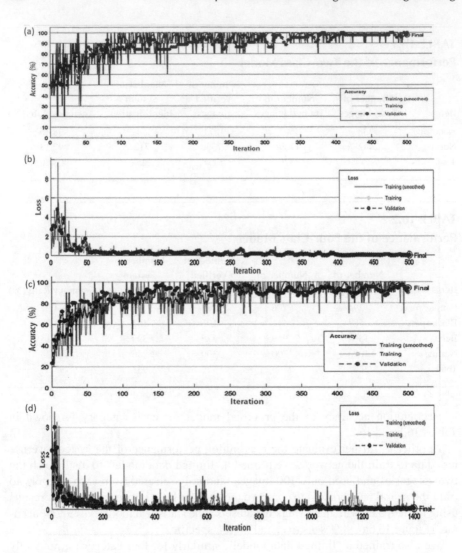

FIGURE 16.6 "Limited data model" classifying accuracy and loss: (a) the curve of accuracy in the process of identification training of two kinds of carrots, (b) the curve of loss in the process of identification training of two kinds of carrots, (c) the curve of accuracy in the process of identification training of four kinds of carrots, and (d) the curve of loss in the process of identification training of four kinds of carrots.

decisions for each tested carrot using just carrot images. For the binary recognition problems, the receiver operating characteristic (ROC) curve is plotted according to the classification performance of normal and abnormal carrots. It is obvious that human experts and our system are comparable in identifying carrots. Meanwhile, the sensitivity and specificity of the experts are plotted on the ROC curve of the training model (Figure 16.8).

TABLE 16.3

Performance of the Two-Class Model Using "Limited Data Model"

Items	Number of Total Images	Number of Trained Images	Number of Verified Images	Number of System Identification	Accuracy Rate (%)
Abnormal	600	300	300	294	98
Normal	600	300	300	290	96.6
Total	1200	600	600	584	97.3

TABLE 16.4

Performance of the Four-Class Model Using "Limited Data Model"

Items	Number of Total Images	Number of Trained Images	Number of Verified Images	Number of System Identification	Accuracy Rate (%)
Black spots	150	75	75	71	94.6
Fibrous roots	150	75	75	75	100
Bend	150	75	75	69	92
Normal	150	75	75	68	90.6
Total	600	300	300	283	94.3

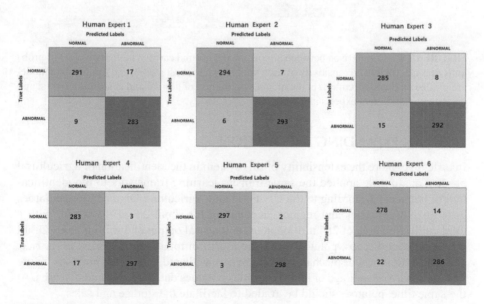

FIGURE 16.7 Situations of human experts.

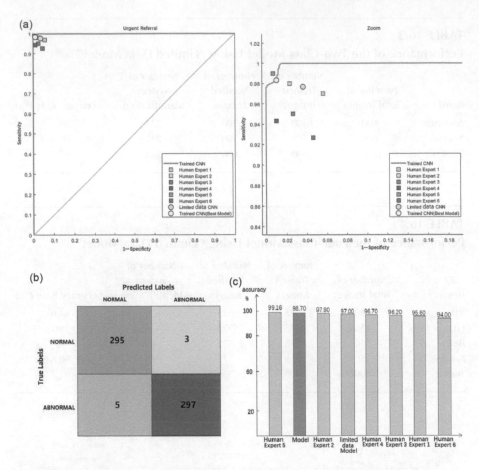

FIGURE 16.8 Comparison between abnormal and normal carrots: (a) for two class of problems, the specificity and sensitivity of each human expert, best model, and limited data model are plotted on the ROC curve; (b) the model's prediction of normal and abnormal carrots; and (c) accuracy of human experts and models.

16.5 AN EXPANDING APPLICATION

In order to explore the extensibility of this system in the identification of agricultural product quality, we applied the same transfer learning framework to the identification of potatoes. According to the UN Food and Agriculture Organization, potatoes grow in 148 countries and regions, accounting for about 2/3 of the world, with a total cultivation area of 285 million acres and a total output of 3 million tons. Of all the world's food crops, potatoes rank the fourth in total output, behind corn, rice, and wheat. Removal of mechanically damaged, withered, and rotten potatoes before storage can avoid spoilage of large quantities potatoes due to rotting and diseases. At the same time, potatoes should be graded to facilitate the storage and sales.

A total of 200 potato images are collected and labeled manually from the sorting line. The images were divided into five categories, each containing 40 images of

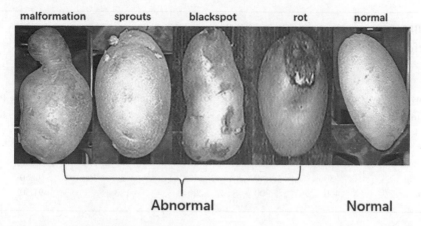

FIGURE 16.9 Representative potato images.

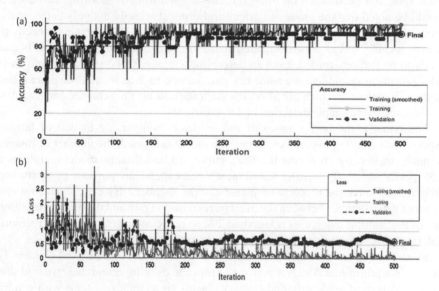

FIGURE 16.10 The accuracy and loss of potato classification model: (a) the curve of accuracy in the process of identification training of five kinds of potatoes and (b) the curve of loss in the process of identification training of five kinds of potatoes.

normal and abnormal potatoes. Abnormal images include malformation, black spots, rots, and sprouts (Figure 16.9). After 500 iterations, the training was stopped because the loss and precision were not further improved (Figure 16.10a and b). Recognition accuracy of the model for each category is shown in Table 16.5.

16.6 DISCUSSION

This chapter discusses the process of detecting surface key features of carrot using computer vision and deep learning, and proposes detection algorithm of fibrous root,

TABLE 16.5

Performance of the Potato Identification Model

Items	Number of Total Images	Number of Trained Images	Number of Verified Images	Number of System Identification	Accuracy Rate (%)
Malformation	80	40	40	38	95.00
Black spots	80	40	40	37	92.50
Rots	80	40	40	35	87.5
Sprouts	80	40	40	36	90.00
Normal	80	40	40	37	92.50
Total	400	200	200	183	91.50

black spot, and bend based on image processing and transfer learning. However, it should be noted that the grade of agricultural products should not only consider the appearance but also consider internal quality (Hong et al., 2007a), which needs to be processed by means of near-infrared spectrum detection (Hong et al., 2007b) and needs to be further studied. Even in determination of carrots' grade through their external appearance, there are some key parameters, such as length, roughness, and surface smoothness, which are also important indicators. This chapter does not discuss the relatively simple detection of some features.

Additionally, the performance of the model is comparable to that of human experts with working experience. When the model is trained, the number of images is much smaller. By observing the ROC curve, we find that the model maintains a high performance. By transfer learning, we can expand the purpose of pretrained network and solve the problem of sparse sample numbers. By experiment, we can find that the difference between the average recognition rate and the high recognition rate of the multiple categories is less than 2%, so reducing the training data, recognition rate of the model does not decline significantly.

Experimental results suggest that the new CNN model can be used to identify carrots in a pretrained AlexNet network. From the existing researches, we find that most models used to identify and classify objects are complex systems with combinations of multiple models (Cen et al., 2014). Compared with the previous sorting model, the model used in this chapter does not need additional complex algorithms for feature extraction. CNN can extract and understand the characteristics of data and classify them.

Compared with the big dataset model, the model training with limited dataset takes the risk of overfitting, so the model with limited dataset must face the additional problem of image selection. In previous experiments, important features were selected from images to improve the accuracy and speed of recognition and sequencing. In our experiment, we made the background of the image identical, so as to avoid the impact caused by background differences.

Although transfer learning allows highly accurate model training for relatively small training data, its performance is lower than the model training of randomly initializing model training from a very large image dataset, because the internal weight can be directly optimized for carrot feature detection. In practice, however, a new CNN starts with random initialization. Even with unlimited training data, it can take several weeks to achieve good accuracy, while the model of transfer learning completes training and testing on different data within 2 h (Kermany et al., 2018). In this experiment, our model only fine-tuned the posterior three-layer weight and achieved good performance. Next, we will fine-tune each layer of the network; we guess that the accuracy of the model will be improved further; and the training time of the global fine-tuning model will be less than that of the traditional convolution network, but more than the local finc-tuning model.

The performance of our model depends largely on the weight of the pretrained model. Therefore, the performance of the model may be enhanced when testing on a larger ImageNet dataset using more advanced deep learning techniques and architectures. In addition, the rapid development of CNNs in areas other than the quality identification of agricultural products will also improve the performance of our method.

In addition, our network provides a common platform that can potentially be applied to a wide range of images (such as potato images) to quickly make decisions. We proved this by training our network on the potato image dataset. The resulting high-accuracy model shows that the system can effectively use relatively small dataset to learn from increasingly complex images and has a high degree of generalization. By presenting effects with multiple types of images and various agricultural products, this framework of transfer learning provides a compelling system for further exploration and analysis in the image of agricultural products and more general application of automatic sorting artificial intelligence system of common agricultural products. From the accuracy curve, we found that there were abnormalities in the training process. Therefore, we suspect that overfitting phenomenon occurred due to the small amount of potato data. In the future, we will study and discuss this problem.

In the end, as mentioned earlier, we use the carrot image as a conventional method of interpretation and subsequent judgment of the produce. Our framework effectively identifies favorable conditions for making decisions as good as (sometimes better than) human experts. We are trying to build more efficient classification systems for agricultural products to reduce heavy workloads, especially in areas with severe labor shortages.

16.7 CONCLUSIONS AND FUTURE WORK

In this chapter, the classification model of carrot recognition is established based on the technology of transfer learning, and the AlexNet network is pretrained using ImageNet. The abnormal carrots were identified from the carrot samples with high accuracy.

Our future work will (1) further extend the application of models (such as maize seed recognition), (2) improve the robustness of model identification, (3) study the difference between local and global fine-tuning weights of the model, and finally integrate our methods into working systems that can be used to automate field operations.

REFERENCES

Abdelhamid O, Mohamed A, Jiang H, et al. Convolutional neural networks for speech recognition. *IEEE Transactions on Audio, Speech, and Language Processing*, 2014, 22(10): 1533–1545.

Carneiro G, Nascimento J C, Bradley A P, et al. Unregistered multiview mammogram analysis with pre-trained deep learning models. *Medical Image Computing and Computer Assisted Intervention*, 2015: 652–660.

Cen H, Lu R, Ariana D P, et al. Hyperspectral imaging-based classification and wavebands selection for internal defect detection of pickling cucumbers. *Food & Bioprocess Technology*, 2014, 7(6): 1689–1700.

Donahue J, Jia Y, Vinyals O, et al. DeCAF: A deep convolutional activation feature for generic visual recognition. *International Conference on Machine Learning*, Beijing, China, June 21-26, 2014, pp. 647–655.

Duan L, Xu D, Tsang I W, et al. Learning with augmented features for heterogeneous domain adaptation. *International Conference on Machine Learning*, Edinburgh, Scotland, 2012, pp. 667–674.

Goldbaum M H, Moezzi S, Taylor A L, et al. Automated diagnosis and image understanding with object extraction, object classification, and inferencing in retinal images. *International Conference on Image Processing*, 1996, pp. 695–698.

Han Z, Deng L, Xu Y, et al. Image processing method for detection of carrot green-shoulder, fibrous roots and surface cracks. *Transactions of the Chinese Society of Agricultural Engineering (Transactions of the CSAE)*, 2013, 29(9): 156–161.

Harel M, Mannor S. Learning from multiple outlooks. *International Conference on Machine Learning*, Bellevue, Washington, USA, 2011, pp. 401–408.

Hong T, Li Z, Wu C, et al. Review of hyperspectral image technology for no n-destructive inspection of fruit quality. *Transactions of the Chinese Society of Agricutural Engineering (Transactions of the CSAE)*, 2007a, 23(11): 280–285.

Hong T, Qiao J, Ning W, et al. Non-destructive inspection of Chinese pear quality based on hyper spectral imaging technique. *Transactions of the Chinese Society of Agricultural Engineering (Transactions of the CSAE)*, 2007b, 23(2): 15F 155.

Ji S, Xu W, Yang M, et al. 3D convolutional neural networks for human action recognition. *IEEE Transactions on Pattern Analysis and Machine Intelligence*, 2013, 35(1): 221–231.

Kermany D, Goldbaum M H, Cai W, et al. Identifying medical diagnoses and treatable diseases by image-based deep learning. *Cell*, 2018, 172(5): 1122–1131.

Krizhevsky A, Sutskever I, Hinton G E, et al. ImageNet classification with deep convolutional neural networks. *Neural Information Processing Systems*, 2012: 1097–1105.

Kulis B, Saenko K, Darrell T, et al. What you saw is not what you get: Domain adaptation using asymmetric kernel transforms. *Computer Vision and Pattern Recognition*, 2011: 1785–1792.

Lecun Y, Bottou L, Bengio Y, et al. Gradient-based learning applied to document recognition. *Proceedings of the IEEE*, 1998, 86(11): 2278–2324.

Litjens G J, Kooi T, Bejnordi B E, et al. A survey on deep learning in medical image analysis. *Medical Image Analysis*, 2017, 42: 60–88.

Nam J, Fu W, Kim S, et al. Heterogeneous defect prediction. *IEEE Transactions on Software Engineering*, 2017, 44(9): 874–896.

Oquab M, Bottou L, Laptev I, et al. Learning and transferring mid-level image representations using convolutional neural networks. *Computer Vision and Pattern Recognition*, 2014: 1717–1724.

Pan S J, Yang Q. A survey on transfer learning. *IEEE Transactions on Knowledge and Data Engineering*, 2010, 22(10): 1345–1359.

Prettenhofer P, Stein B. Cross-language text classification using structural correspondence learning. *Meeting of the Association for Computational Linguistics*, Uppsala, Sweden, 2010, pp. 1118–1127.

Razavian A S, Azizpour H, Sullivan J, et al. CNN features off-the-shelf: An astounding baseline for recognition. *Computer Vision and Pattern Recognition*, Washington, DC, USA, 2014, pp. 512–519.

Wang C, Mahadevan S. Heterogeneous domain adaptation using manifold alignment. *International Joint Conference on Artificial Intelligence*, Barcelona, Spain, July 16–22, 201, pp. 1541–1546.

Yosinski J, Clune J, Bengio Y, et al. How transferable are features in deep neural networks. *Neural Information Processing Systems*, Montreal, Canada, 2014, pp. 3320–3328.

Zeiler M D, Fergus R. Visualizing and understanding convolutional networks. *European Conference on Computer Vision*, 2014: 818–833.

Zhou J T, Pan S J, Tsang I W, et al. Hybrid heterogeneous transfer learning through deep learning. *National Conference on Artificial Intelligence*, Reykjavik, Iceland, 2014a, pp. 2213–2219.

Zhou J T, Tsang I W, Pan S J, et al. Heterogeneous domain adaptation for multiple classes. *International Conference on Artificial Intelligence and Statistics*, Reykjavik, Iceland, 2014b, pp. 1095–1103.

Zhu Y, Chen Y, Lu Z, et al. Heterogeneous transfer learning for image classification. *National Conference on Artificial Intelligence*, San Francisco, CA, 2011, pp. 1304–1309.

17 Grading System of Pear's Appearance Quality Based on Computer Vision

A system is constructed to realize the automatic grading of pear's appearance quality based on computer vision, which involves both hardware and software programs. It includes transport unit, control module, image acquisition module, and image processing and recognition module. It can perform automatic feature extraction. The features of identification include shape, color, and defect. According to the influence of spots on pear's surface on the defect detection, a spot removal method based on V component's dynamic threshold is put forward. According to the national standards, grading rules of fruit type and defect levels based on the above features are constructed. Furthermore, a grading model of pear is also constructed based on artificial neural network. The recognition rate of 630 images of pear reaches up to 90.3%. The system, equipment, and method in this chapter have a positive significance to online grading of pear's quality of appearance.

17.1 INTRODUCTION

Grading of pears is an important task before bringing the harvest to the market. It can greatly improve the fruit's added value, thus playing an important role in the trade and profit of the people involved in the pear industry. At present, the process of grading is carried out manually, that is, the fruits are graded according to visual perception of their color and size. This manual process of detection is susceptible to human errors. Therefore, there are efforts made to use computers in the grading process due to their high ability of data processing and accuracy of rapid and automatic analysis of the fruit's shape, size and quality based on computer vision. This will greatly improve the efficiency of grading process. Consequently, it has become the hot topic of relevant domains in recent years.

The research of computer's use in the automatic detection or grading of fruits has started in other countries in the beginning of the 1990s (Sarkar and Wolfe, 1990; Varghese et al., 1991; Laykin et al., 2002). However, the research started in China only in the mid- and late 1990s (Liu et al., 1996; Ying, 2001; Feng and Wang, 2002; Lin et al., 2005). In spite of the late beginning, we have attained some achievements. But the research is mostly made on the fruits having smooth face, such as apple, peach, tomato, and melon (Wang et al., 2011); less attention has been paid to the fruits having rough surface with pigments such as pear (Ying et al., 1999). At the same time, the research is made mostly on single feature such as shape or color. The number of extracted features is small, and so information of fruits' appearance is less considered. The images are mostly acquired in a static environment, and there is some difference between the real-time and online grading.

It is obvious that the detection of quality based on the appearance of spots is challenging in pears as the fruit already has natural pigments on its surface. This research is made in the following steps: (1) Construct the software and hardware environment to realize online grading of pear, (2) put forward a spots removal method based on V component's dynamic threshold for effective extraction of defects from the fruits having spots on their surface, (3) use comprehensively the features such as color and shape to extract more feature parameters, and (4) design a grading model based on artificial neural network to judge the shape, defect, and other features of pear's appearance according to the national standards.

17.2 SYSTEM DEVELOPMENT

17.2.1 FORMATION OF GRADING SYSTEM'S HARDWARE DEVICE

The grading system is composed of a hardware device and a software program. The hardware device is shown in Figure 17.1.

In the figure, 1 represents frequency controller of electric machine, 2 represents control line, 3 represents electric machine, 4 represents turbine deceleration controller, 5 represents chain, 6 represents roller, 7 represents hierarchical control device, 8 represents image collecting box, 9 represents soft light, 10 represents industrial camera, 11 represents image acquisition card, 12 represents host computer,

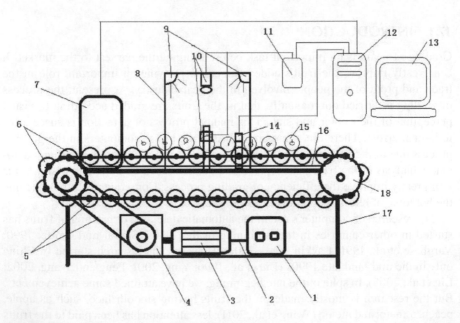

FIGURE 17.1 Hardware device schemes.

13 represents computer display, 14 represents infrared sensor, 15 represents pear, 16 represents rubber, 17 represents bracket, 18 represents gear, 19 represents trolley, 20 represents chain, 21 represents upper camera, 22 represents back camera, and 23 represents front camera.

The frequency controller of electric machine is linked to the electric machine, which is linked to the turbine deceleration controller, and the turbine deceleration controller drives the chaining trolley transport unit. Thus, the electric machine's rotation speed and the chaining trolley transport unit's propagation velocity can be controlled accurately. The infrared sensor and image collecting box are linked to the upper part of the chaining trolley transport unit. The soft light is fixed in the image collecting box to provide light. The infrared sensor linked to the single-chip processor is used to count pears and trigger the industrial camera to collect the images of pears. The industrial camera is linked to the image acquisition card, which is linked to the computer to get the images of pears. The image processing software in the computer will process the acquired images of pears to get the detection result of the pears' appearance quality. The hierarchical control device will grade according to the software's processed result.

17.2.2 CONCRETE IMPLEMENT

After all parts are installed according to Figure 17.1 and power is switched on, the device starts to work. The frequency controller controls the rotation of electric machine, and the electric machine's turbine deceleration controller drives the trolley to move horizontally, which drives the pears' horizontal movement.

When the pears move horizontally near the infrared sensor and come into the screen range, the switch of infrared sensor will send an electronic signal to the single-chip processor. When the single chip receives the signal, it will carry out corresponding process and send the result to the computer's software through the serial interface. After receiving the signal from the single-chip processor, the software will drive the cameras which are set above the pears facing downward to take photos.

The trolley not only moves horizontally under the driving of the chain, but it also rotates under the power of friction force, thus making also the pears to rotate. After the pears rotate each 120°, the camera on top will take a photo, so three photos are taken after a circuit. Five photos are taken including the two photos taken by the camera in the horizontal direction. The photos are transferred to the computer through the image acquisition card (image collection card). In application, the two cameras in the horizontal direction can be taken off to lessen the cost. Thus, only three photos can be acquired by the camera on top of the pears. The software will carry out corresponding process of pears' images, and then give the result of analysis and send out a controlling signal. The signal will be transported to the single-chip processor through the serial interface, which controls the hierarchical controlling device. The controlling machine will place the pears into the different boxes according to their grades.

17.2.3 TESTING OF REAL OBJECT

The real object of the hierarchical control device is shown in Figure 17.2. The real-time pears' images attained by the system are shown in Figure 17.3.

FIGURE 17.2   Physical figure of grading equipment.

FIGURE 17.3   Image of pears taken real time.

17.2.4   SOFTWARE SYSTEM

The software's operation interface is shown in Figure 17.4. The software provides the grading result of pears after comprehensive estimation of the images.

The interface is divided into four sections. They are section of controlling buttons, section of image monitoring, section of present pear's information, and section of

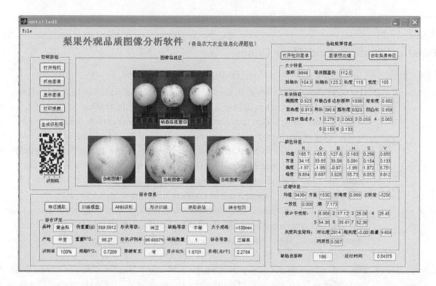

FIGURE 17.4 Graphical user interface (GUI) of the software.

comprehensive information. The functions of section of controlling buttons are mainly to take photos, display the photos, print report, and generate recognition code. The function of section of image monitoring is to display the real-time photos of the pears on the conveying belt and the real-time images of the pears' three sides attained in the process of rotation. The function of section of present pear's information is to display the features of present pear, which are shown in Table 17.1. The function of section of comprehensive information is to display the detection result of present pear.

17.3 IMPLEMENTATION OF ALGORITHM

17.3.1 DETECTION OF DEFECTS

There are always some spots present on the surface of pear. Removal of these spots is the difficulty and emphasis of this research. If the spots can't be removed completely, they will be mistaken as defects when extracted, thus imposing a great impact on the result of grading. After research, a spot removal method based on dynamic threshold of image's HSV color space's V component is put forward.

Images are divided into subimages of 15×15 pixel after the background is removed (as shown in Figure 17.5a). Then, the HSV color space's V component histogram is calculated, each subimage's threshold is dynamically set according to the V's doublet, and threshold segmentation is carried out; thus, the green section of pears' surface is removed and spots are left (shown in Figure 17.5b).

A square window with the side length of n is used to traverse the image of pears to remove the spots on the surface. After many experiments, a square window with the side length n of which is set at 7 pixel is used to traverse the whole image. If there is color in the inner layer and there is no color in the outer layer, the interior realm

TABLE 17.1
Statistical Features and Their Codes

Class	Features and Their Codes
Shape	There are eight features of size; they are area, the code of which is 1; length of long axis, the code of which is 2; length of short axis, the code of which is 3; length, the code of which is 4; width, the code of which is 5; circumference, the code of which is 6; diameter of equilateral circle, the code of which is 7; and convex area, the code of which is 8. There are five features of shape; they are ellipticity, the code of which is 9; rectangle ratio, the code of which is 10; circularity, the code of which is 11; compact ratio, the code of which is 12; and concave convex ratio, the code of which is 13; there are six Fourier descriptors, the codes of them are from 14 to 19.
Color	The code of the mean, variance, skewness, and kurtosis of the RGB color space's three components are, respectively, 20–22, 23–25, 26–28, and 29–31. The code of the mean, variance, skewness, and kurtosis of the HSV color space's three components are, respectively, 32–34, 35–37, 38–40, and 41–43.
Texture	The code of the mean of gray-level image is 44, the code of variance is 45, the code of smoothness is 46, the code of third-order moments is 47, the code of consistency is 48, and the code of entropy is 49. There are seven statistically invariant squares that reflect the number distributing characteristic of seeds image's gray value. The codes of them are 44–50. The gray-level co-occurrence matrix includes the features of contrast, correlation degree, energy, and homogeneity. The codes of them are, respectively, 57, 58, 59, and 60.

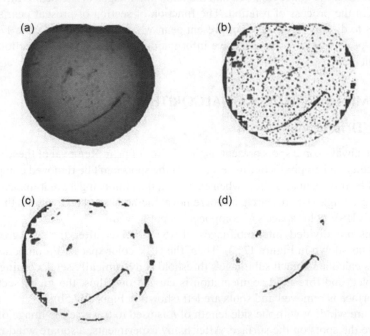

FIGURE 17.5 Process of removing spots on the surface of pear: (a) remove the background color, (b) remove the green color, (c) remove spots, and (d) final result.

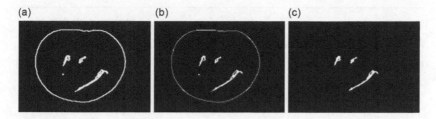

FIGURE 17.6 Defects detection combining morphological method: (a) dilation, (b) corrosion, and (c) removal of borderline.

will be set as achromaticity (as shown in Figure 17.5c). After the whole image is traversed, the V's mean of those removed spots is calculated, and then each pixel of the whole image will be traversed again; if the pixel is greater than the V's mean, it will be removed (as shown in Figure 17.5d). It is found in the experiment that adopting this method to remove the spots on the surface of pears can attain good results and preserve the defect on the pear's surface to a large extent.

The outline of defect part is acquired on the whole after the removal of spots on the pear's surface. Because the outline is not close, the method of morphological dilation is used to process the acquired image and more accurate part of defect can be acquired. Figure 17.6a, b, and c shows the images taken after dilation, corrosion, and removal of borderline, respectively.

17.3.2 FEATURE EXTRACTION

The features of pear and its defect part are extracted and listed in Table 17.1. These include 19 features of shape, 24 features of color, and 17 features of texture. The mean of 90 samples' corresponding features is calculated and listed in Table 17.1. The definitions of these features are available in the literature (Han and Yang, 2010; Han and Zhao, 2010).

The acquired samples need to be normalized (NY/T 440-2001). After that, they are used as the output of neural network.

17.3.3 GRADE JUDGMENT

As it allows for detection of pear's appearance quality according to the national standards (Liu et al., 2006), the back propagation (BP) artificial neural network algorithm is used as recognition model of variety and quality. BP neural network is a kind of multilayer feed forward neural networks trained by the BP algorithm. It is a widely used neural network model. The BP network can study and store a large quantity of mapping relationships in the mode of input and output, and doesn't need to apply any mathematics equations to describe these mapping relationships. Its study rule is using the steepest descent method to continuously adjust the network's weight and threshold through BP, thus minimizing the square sum of network's error. The adopted neural network model's topology includes three layers. They are input layer, hidden layer, and output layer (Table 17.2).

TABLE 17.2

Mean of Pear's Statistical Features

No.	Feature Value	No.	Feature Value	No.	Feature Value
1	6474.3	21	127.47	41	9.6277
2	96.736	22	103.52	42	4.5602
3	84.604	23	25.135	43	3.0822
4	90.711	24	25.686	44	26425
5	89.922	25	32.355	45	12106
6	90.085	26	−0.481	46	0.9994
7	0.4747	27	−0.481	47	−3E+07
8	6602.7	28	−0.221	48	0.0104
9	0.7828	29	3.0347	49	6.8877
10	0.9803	30	3.1187	50	6.6385
11	0.9299	31	2.8777	51	17.319
12	304.88	32	0.2377	52	25.299
13	4615	33	0.2556	53	26.13
14	0.2231	34	0.5191	54	53.709
15	0.1306	35	0.2131	55	35.545
16	0.1373	36	0.1589	56	52.476
17	0.1533	37	0.0998	57	1327.1
18	0.1574	38	2.5054	58	−0.002
19	0.16	39	1.0586	59	0.0002
20	129.61	40	−0.436	60	0.0798

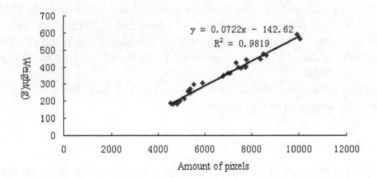

FIGURE 17.7 Relationship between pixel area and weight.

First, a recognition database of neural network is constructed to recognize the variety and producing area of 30 pears chosen from each variety (totaling three varieties: Huangjin, Fengshui, and Yali) are chosen. Using the different shapes of the 90 pears of the three varieties, a neural network is constructed to recognize the correctitude degree of pear's shape (correctitude, abnormality, and severe abnormality).

As pear's weight can be determined by its area, the regression relation between area and weight is constructed as shown in Figure 17.7. Thus, pear's weight can

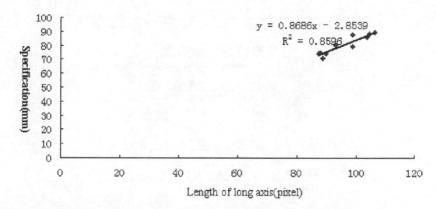

FIGURE 17.8 Relationship between images' long-axis length and size.

TABLE 17.3
Standards of Pear's Grade

Comprehensive Grade	Grade of Character	Grade of Defect
Top grade	Normal	None
Grade one	Normal	Middling
	Malformed	None
Grade two	Malformed	Middling
Grade three	Malformed	Severe
	Severely malformed	Middling
Off grade	Severely malformed	Severe

be calculated according to the regression relation. In Figure 17.8, the specification can be determined according to the relationship between long-axis length and size (Table 17.3).

At last, the pear's defect part is determined. According to the percentage of area of defect occupying the whole area, the grade of pear's defect is determined. If the percentage is less than 1%, there is no defect. If the percentage is between 1% and 5%, the defect is middling. If the percentage is more than 5%, the defect is severe.

Then, the comprehensive grade is determined according to the shape's correctitude and grade of defect.

Finally, the proposed selling price is given according to pear's weight, specification, and grade. The relevant information is given in the form of information table and information code. When the detection is finished, the report is printed when the controlling button is pressed. Thus, the detection result is presented in the report, which is shown in Figure 17.9.

17.4 RESULTS AND DISCUSSION

Partial and comprehensive experiment are, respectively, carried out aiming at 90 samples of each item totaling up to 630 samples, which are classified strictly

梨果检测结果报表

检测单位：青岛农业大学农业信息化项目组

品种	丰水梨	正确率	100%
产地	莱阳	正确率	100%
重量	571.6812	重量R^2	95.35
大小规格	>100mm	规格R^2	0.7096
形状等级	畸形	正确率	100%
缺陷等级	无	百分比	1.8701
缺陷数量	1	果梗有无	有
综合等级	一等果	建议价格	1.715

单位（盖章）：　　　负责人（签字）：

年　月　日

FIGURE 17.9 Detection results statements.

TABLE 17.4
Test Performance Table

Items	Detection by Human		Detection by Machine	
	Amount	Correct Rate (%)	Amount	Correct Rate (%)
Variety	90	93.5	90	100
Producing area	32	35.6	90	100
Specification	85	94.4	77	86.0
Weight	57	100	88	98.2
Shape	76	84.4	90	100
Defect	78	86.7	83	92.3
Comprehensive grade	79	87.8	82	90.6

according to the national standards. First five people detect the samples manually, and then the median of the detection result is taken as the final detection result attained by hand. Then, the samples are detected again using the system described in this chapter, and the result is shown in Table 17.4.

It can be seen that the correction rate of comprehensive grade detection by machine is up to 90.6%, which can meet the need of theory and practice. In the items, except that the correction rate of specification and weight are lower than that detected by man, the whole detection performance by machine is better than that by hand. The manual detection of specification and weight can only work by the use of specification set, such as balance and other tools. Moreover, there are mainly three factors leading to the detection error by machine. The first one is that there are fewer samples. The second one is that errors occur when the defect and outline are recognized. The third one is that defect is defined very strictly in the process of recognition, so even a little damage can influence the experiment result.

17.5 ENDING WORDS

In this chapter, a recognition system is constructed which includes a hardware device and a software program. It includes transport unit, control module, image acquisition module, and image processing and recognition module. It can extract features such as shape, color, and defect automatically. According to the influence of spots on the defect detection, a spot removal method based on V component's dynamic threshold is put forward. According to the national standards, grade rules of fruit type and defect level based on above features are constructed. Furthermore, grading model of pears based on artificial neural network is also constructed. Quantization research is carried out in the process of pear grading according to the agricultural standard of pear grading; the correction rate of the method referred in this chapter when used in the comprehensive quality grading is higher than 90%. This is better than the result of manual grading. Moreover, the grading method and system referred in this chapter has the advantages of high precision and automatization. Relevant conclusion and method in this research have positive significance to the industry of pear.

REFERENCES

Feng B, Wang M. Computer vision classification of fruit based on fractal color. *Transactions of the CSAE*, 2002, 18(2): 141–144.

Han Z, Yang J. Detection of embryo using independent components for kernel RGB images in maize. *Transactions of the CSAE*, 2010, 26(3): 222–226.

Han Z-Z, Zhao Y-G. Quality grade detection in peanut using computer vision. *Scientia Agricultura Sinica*, 2010, 43(18): 3882–3891.

Laykin S, Alchanatis V, Fallik E, Edan Y. Image-processing algorithm for tomato classification. *Transactions of the ASAE*, 2002, 45(3): 851–858.

Lin K, Wu J, Xu L. Separation approach for shape grading of fruits using computer vision. *Transactions of the Chinese Society for Agricultural Machinery*, 2005, 36(6): 71–74.

Liu H, Wang M. Study on neural network expert system for fruit shape judgment. *Transactions of the CSAE*, 1996, 12(1): 171–176.

Liu Y, Yin G Y, Fu X, et al. Automatic measurement system of fruit internal quality using near infrared spectroscopy. *Journal of Zhejiang University (Engineering Science)*, 2006, 40(1): 53–57.

NY/T 440-2001. Agricultural industry standard of the People's Republic of China: Standards for appearance grades of pears. *Standards Press of China*, Beijing, 2001, 2.

Sarkar N, Wolfe R R. Image processing for tomato grading. *Transactions of the ASAE*, 1990, 33(4): 564–572.

Varghese Z, Morrow CT, Heinemann PH, Sommer III HJ, Tao Y, Crassweller RM. Automated inspection of golden delicious apples using color computer vision. ASAE Paper 91-7002, 1991.

Wang S, Zhang J, Feng Q. Defect detection of muskmelon based on texture features and color features. *Transactions of the Chinese Society for Agricultural Machinery*, 2011, 42(3): 175–179.

Ying Y, Jing H, Ma J, et al. Shape identification of Huanghua pear using machine vision. *Transactions of the CSAE*, 1999, 15(1): 192–196.

Ying Y-B. Fourier descriptor of fruit shape. *Journal of Biomathematics*, 2001, 16 (2): 234–240.

18 Study on Defect Extraction of Pears with Rich Spots and Neural Network Grading Method

In order to study the feasibility of the computerized grading system of pears with rich spots, large number of pears are put in a light box, where images of different aspects of them are captured by cameras, which are connected to the computer. Additionally, two methods are proposed to remove the spots on the surface of pears in the image to reduce their effect on defect detection. The advantages and disadvantages of the two methods are discussed. One method works by applying an adaptive threshold and the other works by combining filtering with edge detection. In reference to the National Standard of China, grading based on shape and defect is studied. An ANN (artificial neural network) model, which colligates the information of shape, color, and defect, is established to grade the pears comprehensively. The method of the adaptive threshold has a better effect when it is used in processing pears with rich spots, such as Laiyang pear, whereas its executive efficiency is slightly lower. Yet, to the species on which the spots are not so conspicuous, such as the Huangjin pear and the Fengshui pear, the method that combines filtering with edge detection is supposed to be applied for better performance. The results of the grading by ANN model based on shape, defect, and comprehensive quality, respectively, reach the accuracy of 87.5%, 92.6%, and 90.3%. The methods proposed to remove spots and the model constructed for grading and recognition have positive significance to grading pears with rich spots by appearance.

18.1 INTRODUCTION

The grading of pears is an important part of commercial processing after picking and can greatly enhance a pear's added value. This is also directly related to the pear industry, trade, and the fruit growers' economic benefits. A pear's quality of appearance is important in grading and determines the pear's selling price to a great extent. At present, the work of grading is largely done by hand. A pear's appearance is estimated visually, and then the pears are divided into several grades according to color and size, and so on. This leads to the inherent human error of manual work. The grading of pears based on computer vision is to use the computer's high capability for data processing to realize rapid and automatic analysis and determine pear's appearance quality from its shape and size, and so on. This will greatly improve the efficiency of grading. Consequently, it has become the hot topic of relevant domain in recent years. Other countries have begun to research the computer's use in automatic detection or grading of fruits since the 1990s (Sarkar and Wolfe, 1990; Varghese et al., 1991; Laykin et al., 2002). Our country has begun much tracking research from the middle and later periods of 1990s (Liu et al., 1996; Ying, 2001; Feng and Wang, 2002; Lin et al., 2005). We have also attained some achievements despite the late beginning. But the research is mostly done on the fruit with smooth surfaces, such as apples, peaches, tomatoes, melons, and so on. Less research is done on fruit with rich spots on the surface, such as pears (Ying et al., 1999). But the spots' influence on pears' detection is great. Moreover, the above research has not fully considered the national standard's influence on pear's detection and still has some distance from actual grading. The objective detection of a pear's quality under the national norm is a challenging subject. This research intends to solve the following problems. (1) The method of image processing is to be improved. A full consideration of the influence of pear's spots on defect extraction and extraction method based on color space's dynamic threshold is to be put forward. The performance of this method will be compared to that of the traditional one. (2) In compliance with the national standard and comprehensive utilization of shape description, texture analysis, and other method, pears' appearance features will be quantified and feature parameters will be extracted. (3) Th recognition model based on an ANN will be designed, which can be used in shape judgment, recognition of defect variety, and judgment of pear's appearance's comprehensive grade. A grading experiment will be carried out.

18.2   MATERIAL AND METHOD

18.2.1   EXPERIMENTAL MATERIALS

Forty-three pears are chosen, which have been bought from a supermarket in 2010, including Laiyang pears, Huangjin pears, and Fengshui pears. The pears' appearance images will be collected. The image collecting system is shown in Figure 18.1. In order to avoid the instability of a natural light source, the process is implemented in a blocking light box. A ring light of 40 W is placed on the top of the box and shading cloth is lined on the box's wall in order to avoid reflection in the

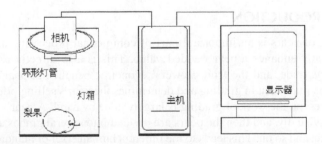

FIGURE 18.1   Image collecting system.

process of collecting images. The acquired images are stored in a computer through a transmission line for subsequent processing.

The camera used in the experiment is Olympus E300 SLR, and photos are taken under the macro mode of 1,728*2,304 pixels. The configuration of the computer is as follows: ThinkPad SL300, Intel(R) Core(TM)2 Duo CPU P8600@2.40 GHz, 2 GB EMS memory, 320 GB hard disk, Windows Vista(G) operating system. The image process is implemented by program developed by MATLAB® R2008a.

18.2.2 IMAGE PREPROCESSING

18.2.2.1 Background Removing and Outline Extraction

Before processing the collected images (showed in Figure 18.2), it is needed to remove the background. The H histogram is found to have a very strong doublet effect after the image's color space of RGB and HSV is studied. The image, the background of which is removed by the H component, is shown in Figure 18.3.

The object of outline extraction is to get the outer outline feature of target area. Images of the pear's appearance outline extracted by three edge detection operators are shown in Figure 18.4.

Figure 18.4a shows the result of edge detection with a Sobel operator. Figure 18.4b shows the result of edge detection with a Laplace operator. Figure 18.4c shows the result of edge detection with a Canny operator. After comparison, it can be seen that the Canny operator is the most effective one. But because of the influence of spots on the surface, there is some noise in pear's surface.

FIGURE 18.2   Pear appearance image.

FIGURE 18.3   Background removed.

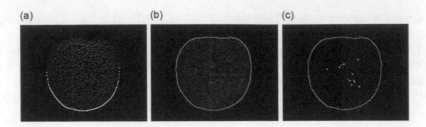

FIGURE 18.4　Result of three edge detection operators. (a) Edge detection with Sobel, (b) edge detection with Laplace, and (c) edge detection with Canny.

18.2.2.2　Removal of Spots on the Surface

On a pear's surface, there are always some spots of dark color and even size and shape. It is difficult to remove the spots. If they are not removed, they will be mistaken as defects during defect extraction and will greatly affect the grading result. After research, two methods have been found that can be used to remove the spots.

1. Dynamic threshold method based on the V component of image's HSV color space.

   After background removal, the image is divided into subimages of 15×15 pixel. The HSV color space's V component histogram of each subimage is calculated. Each subimage's threshold is dynamically set according to V's doublet. Threshold segmentation is carried out. The non-point region on the pear's surface is removed and the spots are left (shown in Figure 18.5b).

   A square window with a side length of n is used to traverse the pear's image to remove spots. After many experiments, a square window with a side length of 7 is used to traverse the whole image. If the inner layer of the window is colorful and the outer layer isn't colorful, the inner region is set white, shown in Figure 18.5c. After the whole image is traversed, the mean of V of the removed spots is calculated, and then each pixel of the image will also be traversed. If the pixel is larger than V's mean, it will be removed, shown in Figure 18.5d.

2. Method combining filtering with edge detection

   In order to enhance the efficiency of spot removal, an experiment is carried out in which, after background removal, the grayed image is

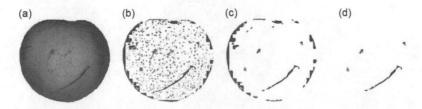

FIGURE 18.5　Process of removing spots on pear's surface. (a) Background removed, (b) non-point region removed, (c) spots removed, and (d) final result.

filtered and edge detection with a Canny operator is carried out (showed in Figure 18.6a) and edge detection with Canny operator is also carried out after morphology corrosion operation (showed in Figure 18.6b).

18.2.2.3   Extraction of Defective Parts

After spots are removed using the above methods, the outline of defective parts is attained. But some defects' outlines are not close and some outlines contain other outlines. More accurate defective parts can be acquired by the method of morphological dilation. The images after dilation, corrosion, and edge removal are shown in Figure 18.7a–c, respectively.

18.2.3   FEATURE EXTRACTION AND RECOGNITION

18.2.3.1   Scalarization of National Standard

A pear's appearance grade standard (NY/T 440-2001), which was promulgated by our country, is the authority guide of the judgment rules of a pear's appearance grade. In the national standard, a pear's appearance grade is divided into three grades. They are top grade, grade one, and grade two. Color, weight, standard of fruit type, bruising, abrasion, fruit rust, hail damage and insect damage, and several other aspects are described in the three grades. In the international standard, there is a grade of substandard pear. But there is no specific evaluation, and the three grades are inferior to China's grades.

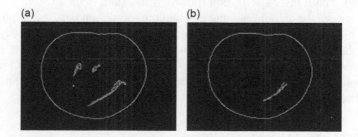

FIGURE 18.6   Method of filtering combined with edge detection. (a) Edge detection with Canny operator and (b) edge detection with canny operator after morphology corrosion operation.

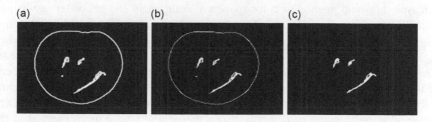

FIGURE 18.7   Defect detection using morphological dilation. (a) The images after dilation, (b) the images after corrosion, and (c) the images after edge removal.

18.2.3.2   Extraction of Fruit Type and Defective Part

Feature extraction of the fruit type is carried out on the pear's image, the background of which is removed. The features include three aspects. The first one is the basic geometry features of the surface (such as perimeter, area, and circularity). The second one is the features of color (the mean of color space of RGB or HSV). The third one is the shape features, such as moment invariants, Fourier descriptor, and so on. These features' definitions are referred in literatures (Han et al., 2009; Han and Yang, 2010; Han and Zhao, 2010; Zhao et al., 2009). They are used to judge the pear's correctitude. According to them, the fruit type's correctitude can be divided into four grades. They are standard, common, not so standard, and substandard.

Extraction of the defective part's features is carried out on the defective regions identified with the above methods. These features include the defect's geometrical features (such as circumference, area, and circularity of defect part) and color of defect and so on. According to them, the defects' size, quantity, and variety are estimated and the severity of defect is acquired. Defective varieties can be divided into scratches, scars, fruit rust, stems, and so on.

18.2.3.3   Judgment of ANN Grade

In this research, the severity of defect is divided into five grades, −1, −0.5, 0, 0.5, and 1. Standard of fruit type is divided into four grade, −1, −0.333, 0.333, and 1. At the same time, the HSV color information of pear's surface is also put into the input end of the back propagation (BP) neural network to implement the network's training and decision-making. In this process, the s-type function can be used as the network's each layer's activation transition function. Because there is little difference between the amount of input end and the former amount, the amount of the nodes of hidden layer is set at 12, the amount of the nodes of the input layer is set at 5, and the amount of the nodes of the output layer is set at 2. Thus, (0.001, 0.001), (0.001, 0.999), (0.999, 0.001), and (0.999, 0.999) are used, respectively, to represent the pear of top grade (T), pear of grade one (I), pear of grade two (II), and pear of substandard (F).

18.3   RESULTS AND ANALYSIS

18.3.1   Effect of Spot Removal and Defect Extraction

Research shows that in the defect detection of pears with obvious spots, using the dynamic threshold method of color space's V component can get better results and it can also perfectly keep the defective parts on the pear's surface. At the same time, it is observed that this method can't acquire high efficiency and can't be used when high efficiency is required. In the defect detection of pears on which there are no obvious spots, a method combining median filtering with a Canny operator can be used. Then morphologic operation is carried out to extract the defect parts, and defect parts can be extracted almost without being influenced by the spots on the surface, the information of defect parts on the pear's surface are preserved at the most, and the detected pear's outline is smooth. It is obvious that different methods should be used for different varieties. The method combining median filtering with a Canny operator can get better result and the information of defective parts can't be severely damaged when it is used in the varieties the spots of which are not obvious (such as Huangjin pears and Fengshui pears). The method of color space's V's dynamic threshold can be used to remove the spots in the variety with obvious spots (such as Laiyang pears).

18.3.2 Grade Judgment of Fruit Type and Defect

After feature extraction of large samples and acquisition of several features, the samples need to be normalized. Training of neural network and recognition is carried out with the extracted features. Compared to the result of manual judgment, the correct rate of judgment by computer is up to 87.5%.

The detection of defect variety is also based on neural network, and the features of samples for test are used as the input of neural network to carry out the training and recognition. Compared to manual judgment, the correct rate of judgment by computer is up to 92.6%.

18.3.3 Comprehensive Grade Judgments

In experiment of large samples, the correct rate of judgment result of BP neural network is up to 90.6%.

In the judgment result, there are three main reasons leading to error. The first one is that the number of samples is small. The second one is that error occurs in the defect or outline recognition. The third one is that defect is so strictly defined that little damage can affect the experiment result.

18.3.4 Influencing Factors of Grading of Comprehensive Quality

In the former experiments, much is considered about the pear's shape feature and little is considered about the appearance defect. Some researches of defect are only confined to the pear's severe damage on the surface, and little is described about pears' gall and fruit rust. The accuracy of the detection result needs to be enhanced. Only a few experimenters inspect the defect's severity, and the severity of defect is only defined by the experimenters themselves.

The above-mentioned conditions can lead to the experiments' results being incorrect, unsteady, or short of creditable basis. In order to solve these problems, it is necessary to use national trade standard to instruct the experiment process.

According to normal judgment method, in the complex process of judgment of organism features, on the basis of ensuring each aspect's accuracy, the judgment result can be calculated by converting to a linear mode. Moreover, it can be seen from the standard, the judgment of defect is the most important and concrete factor of judgment of samples for test. The correctness and stability of the result acquired by any grading method which ignores defect detection is to be checked.

Moreover, the trade standard has regulated the pear's fruit weight, size, and so on, so it is needed to convert the standard's unit to the manner that computer can recognize and express. Here the method of linear regression can be used to calculate the relationship between the pear's surface area and weight and the regression relationship between the pear's length and size. It is lucky that the relationship among these regression equations is linear (Figures 18.8 and 18.9). Thus, the pear's weight and specification can be correctly calculated from its area and length.

In addition, a pear's comprehensive quality includes not only appearance quality, but also internal quality (Liu et al., 2006). But a pear's appearance for sale is settled by appearance quality to a large extent. So, in this study, the internal quality is not considered.

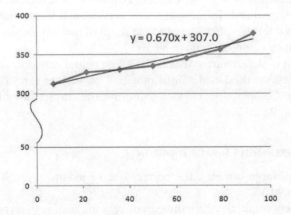

FIGURE 18.8   Regression curve of area and weight.

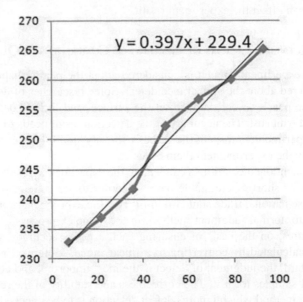

FIGURE 18.9   Regression curve of length and size.

18.4 DISCUSSIONS

The process of pears' appearance appraisement by computer is discussed in this chapter. Background removal, spots removal, defect extraction, shape description, and grading are researched and good results are acquired from practice. Through theory analysis and practice, conclusions have been attained as follows.

A stably lit environment is needed in the process of image collection. In this research, a self -made light box is used. According to the light box's bulk and illumination intensity, white color, which provides better contrast, is used as background color. Four images of each pear are collected. After analysis, binarization is carried out by changing the color space to get threshold value in the process of background removal. A perfectly obvious result is acquired after experiment. Two methods are compared in the process of removal spots. In the condition of giving priority to efficiency, the method combining morphologic filtering with edge detection realizes the effect. In the process of extraction defect on pear's surface, the method of seed filling and morphologic inflation is used, which can realize high sensitivity and exact extraction region. Grading quantization of agricultural standard of pear is researched and BP neural network classifier is used in recognition. The correct rate of grading of pear's defect and appearance quality is above 90%.

REFERENCES

Feng B, Wang M. Computer vision classification of fruit based on fractal color. *Transactions of the CSAE*, 2002, 18(2): 141–144.

Han Z, Yang J. Detection of embryo using independent components for kernel RGB images in maize. *Transactions of the CSAE*, 2010, 26(3): 222–226.

Han Z, Zhao Y. A cultivar identification and quality detection method of peanut based on appearance characteristics. *Journal of the Chinese Cereals and Oils Association*, 2009, 24(5): 123–126.

Han Z-Z, Zhao Y-G. Quality grade detection in peanut using computer vision. *Scientia Agricultura Sinica*, 2010, 43(18): 3882–3891.

Laykin, S, Alchanatis V, Fallik E, Edan Y. Image-processing algorithm for tomato classification. *Transactions of the ASAE*, 2002, 45(3): 851–858.

Liu H, Wang M. Study on neural network expert system for fruit Shape judgement. *Transactions of the CSAE*, 1996, 12(1): 171–176.

Lin K, Wu J, Xu L. Separation approach for shape grading of fruits using computer vision. *Transactions of the Chinese Society for Agricultural Machinery*, 2005, 36(6): 71–74.

Liu Y, Yin G Y, Fu X, et al. Automatic measurement system of fruit internal quality using near infrared spectroscopy. *Journal of Zhejiang University (Engineering Science)*, 2006, 40(1): 53–57.

NY/T 440-2001. Agricultural industry standard of the People's Republic of China: Standards for appearance grades of pears. *Standards Press of China*, Beijing, 2001, 2.

Sarkar N, Wolfe R R. Image processing for tomato grading. *Transactions of the ASAE*, 1990, 33(4): 564–572.

Varghese Z, Morrow CT, Heinemann PH, Sommer III HJ, Tao Y, Crassweller RM. Automated inspection of golden delicious apples using color computer vision. ASAE Paper 91-7002, 1991.

Ying Y, Jing H, Ma J, et al. Shape identification of Huanghua pear using machine vision. *Transactions of the CSAE*, 1999, 15(1): 192–196.

Ying Y-B. Fourier descriptor of fruit shape. *Journal of Biomathematics*, 2001, 16(2): 234–240.

Zhao C, Han Z, Yang J, et al. Study on application of image process in ear traits for DUS testing in maize. *Acta Agronomica Sinica*, 2009, 42(11): 4100–4105.

19 Food Detection Using Infrared Spectroscopy with k-ICA and k-SVM

Variety, Brand, Origin, and Adulteration

In the field of food testing, variety, brand, origin, and adulteration are four important testing projects. In this chapter, we propose a novel negentropy sorted kernel independent component analysis (k-ICA) method as a feature extraction method for Fourier-transform infrared (FTIR) spectroscopy. Since the independent components (ICs) obtained by ICA is random, it needs some criteria to sort these ICs. Here, we use negentropy as a criterion to measure the non-Gaussian of ICs in order to separate the IC from maximum negentropy first. Then we use a kernel support vector machine (k-SVM) as the classifier to distinguish different foods. We use four datasets to comprehensively investigate four kinds of problems (variety, brand, origin, and adulteration). The experimental result indicated that the k-ICA presents a superior performance than traditional method such as plusLrR, Fisher, and principal component analysis (PCA), also it is better than using original wavelength. The k-SVM model has the best performance, and it is better than back propagation-artificial neural network (BP-ANN) and partial least squares (PLS). The improved double kernel method (k-ICA and k-SVM) can detect food's variety, brand, origin, and adulteration simultaneously, and the recognition performance is steady, high, and efficient, and the recognition program works steadily efficiently. This conclusion has positive significance for food detection.

19.1 INTRODUCTION

In recent years, food safety problems have received more and more attention in food detection field. Illegal traders are shoddy, they use low-value food of other kinds, brands or regions as high-value food, and some products may be adulterated. Traditionally, many chemically analyzed methods are used to detect agriculture or food products, such as thin-layer chromatography (TCL) and high-performance liquid chromatography (HPLC). The result of these analytical methods is very accurate. But, they need to destroy the samples and are costly and time-consuming. In recent years, infrared spectroscopy which include near- and mid-infrared (MIR) spectroscopy technology has received more attention in the food detection field because of its ability to distinguish the small differences of different food, and they are nondestructive, fast, and accurate.

Associated with the detection of food industry, there are four key problems: variety, brand, origin, and adulteration. A feasibility study of MIR spectroscopy for authenticity problems in selected meats had been reported by Jowder et al. (1997). By obtaining FTIR spectroscopy with attenuated total reflectance (ATR) sampling, they proved that this technology can distinguish fresh chicken, pork, and turkey. Using 56 mid-infrared diffuse reflectance (MIR-DRIFT) spectra of lyophilized coffee produced from two species: Arabica (29 samples) and canephora var. Robusta (27 samples), Briandet et al. (1996) studied the problem about how to distinguish between Arabica and Robusta instant coffee by FTIR spectroscopy and chemometrics. Downey et al. (1997) studied the coffee varietal identification problem by near-infrared and MIR spectroscopy, respectively. Using 120 MIR spectra collected from 60 different authenticated extra virgin olive oils, supplied by the Institute of Food Research, UK, of the International Olive Oil Council, FTIR spectroscopy and multivariate analysis have been used to distinguish the geographic origin of extra-virgin olive oils (Tapp et al., 2003). In order to detect whether a strawberry has been adulterated, we use a 983 MIR spectra collected of two classes: "Strawberry" and "NON-Strawberry", Holland et al. (1998) used FTIR spectroscopy and PLS regression to detect the adulteration of strawberry purees.

However, one problem associated with this technology is how to deal with the huge amount of data acquired by the infrared spectroscopy device. Another problem is how to distinguish different food exactly and fast. In order to reduce the dimensionality of the data and distinguish these food, two key algorithms have been widely used in the papers mentioned above. They are PCA and PLS. However, the effect is not very good, and the universality performance of the algorithm is relatively poor. To solve the two problems aforementioned, this chapter proposed k-ICA and k-SVM to detect food. Like PCA, ICA can also be used to select optimal wavelengths to reduce data redundancy. PCA is based on second-order statistics, while ICA is based on fourth-order statistics.

Originally, ICA is proposed to solve Blind Signal Separation (BSS) (Hyviirinen et al., 2001). A fast ICA algorithm was proposed, which rapidly promotes ICA (Hyvärinen and Oja, 1997). Besides solving BSS problem, recently, ICA has also been proposed as a tool to optimal wavelength selection for differentiating walnut shells from meat using fluorescence hyperspectral imagery (Bin et al., 2007). Support vector machine (SVM) was first proposed by Cortes and Vapnik (1995). SVM has been widely used in many fields and can solve both linear and nonlinear multivariate calibration problems. In recent years, k-SVM is proposed, which uses the kernel function to improve the efficiency and speed of the problem identification, especially for small sample analysis. The kernel function used the idea of kernel function to transform multi-dimensional calculation to kernel space (Bach and Jordan, 2003), which can be used to improve the speed of computing. In this chapter, we use this method for food detection. In order to improve the performance of extract features by ICA, we propose negentropy as a novel criterion to measure the non-Gaussian of ICs in order to separate the IC with maximum negentropy firstly.

In the following section, we will describe materials and the spectral data collected. Then we will describe the improved ICA algorithm, and introduce the flow path of the method. In Section 19.3, we will report and discuss our experimental results. Finally, concluding remarks will be given in Section 19.4.

19.2 MATERIALS AND METHODS

19.2.1 MATERIALS

We selected four different agriculture products including meats of different varieties, coffee of different brands, oil from different original and adulteration of strawberry. Infrared spectroscopy, which was collected by the FTIR spectrometer, was provided by Analytical Sciences Unit, Institute of Food Research, UK.

The samples are summarized in Table 19.1, and the details of these data are illustrated as follows: 120 spectra which are duplicate acquisitions from 60 authenticated extra-virgin olive oils from 4 different countries of origin; raw data matrix size with a size of $[570 \times 120]$ obtained from FTIR spectroscopy with ATR sampling; 120 spectra of fresh minced meats – chicken, pork, and turkey. Duplicate acquisitions from 60 independent samples. Raw data matrix size is $[448 \times 120]$ obtained by FTIR spectroscopy with ATR sampling; 983 spectra of fresh fruit purees include: strawberry (authentic samples) and non-strawberry (adulterated strawberries and other fruits). Raw data matrix size is $[983 \times 235]$ obtained by FTIR spectroscopy with ATR sampling; 56 spectra of authenticated freeze-dried coffee samples (Arabica and Robusta species, respectively, 29 and 27 of each). Raw data matrix size is $[286 \times 56]$.

In order to illustrate the distribution of the infrared spectroscopy of different material and in order to show the different spectral response with respect to four kinds of sample categories, the spectral curves of each kind are given in Figure 19.1.

TABLE 19.1
Wavelength and Sample Number of Each Material

Object	Original		Variety		Adulteration		Brand	
Materials	Oil		Meat		Strawberry		Coffee	
Wavelength	799–1,897		1,005–1,868		899–1,802		810–1,910	
Total bands	570		448		235		246	
Kinds	Greece	10*2	FreshChicken	20	Strawberry	351	Arabica	29
	Italy	17*2	FreshPork	20	Non-strawberry	632	Robusta	27
	Portugal	8*2	FreshTurkey	20				
	Spain	25*2						

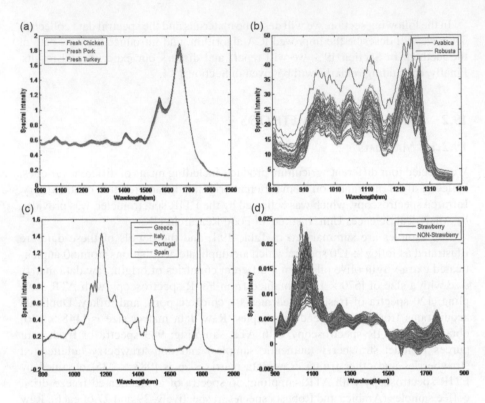

FIGURE 19.1 Infrared spectroscopy of the materials. (a) Meat, (b) coffee, (c) oil, and (d) strawberry.

19.2.2 Algorithm

ICA can be expressed as the problem that a latent random vector X can be recovered from observations of unknown linear functions of that vector. The components of X are assumed to be independent of each other. And, an observation Y is modeled as:

$$Y = AX \text{ here}, X = (x_1, x_2, \ldots, x_m), Y = (y_1, y_2, \ldots, y_m) \qquad (19.1)$$

where x is a latent random vector with ICs, and A is a parameter matrix of $m*m$. Given N independently, identically distributed observations of y, we hope to estimate A and, thereby, to recover the latent vector x corresponding to any specific y by solving a linear problem.

We can obtain a parametric model which can be estimated via maximum likelihood by specifying distribution for the components xi. With $W = A-1$ as the parameterization, one can easily obtain a gradient or fixed-point algorithm that yields an estimate $\hat{W}$ and provide estimates of the latent components via $\hat{X} = \hat{W}Y$. Hyvärinen et al. have proposed an algorithm named fast fixed-point algorithm for ICA.

Unfortunately, it is difficult to approximate and optimize the mutual information based on a finite sample. In this chapter, we provide a new solution to the ICA problem based on an entire function space of candidate nonlinearities instead of a single nonlinear function. Especially, these functions have been dealt with in a reproducing kernel Hilbert space (RKHS), which we can use "kernel trick" to search efficiently. It is the use of the function space that makes it possible to adapt to all kind of sources and make algorithms more robust on various source distributions as depicted as follows.

Bach et al. defined a contrast function that can do a rather direct measurement of the dependence of a set of random variables. For simplicity, we assume x_1 and x_2 are two univariate random variables and F is a vector space and F-correlation ρ_F is the maximal correlation between the random variables $f_1(x_1)$ and $f_2(x_2)$, where f_1 and f_2 range over F:

$$\rho_F = \max_{f_1,f_2 \in F} \mathrm{corr}(f_1(x_1), f_2(x_2)) = \max_{f_1,f_2 \in F} \frac{\mathrm{cov}(f_1(x_1), f_2(x_2))}{\left(\mathrm{var}\, f_1(x_1)\right)^{1/2} \left(\mathrm{var}\, f_2(x_2)\right)^{1/2}} \quad (19.2)$$

It is clear that the F-correlation would equal to zero if the variables are independent. Furthermore, the converse is also true if F is big enough (Wang and Chang, 2006a,b).

We use the idea of RKHS to get a computationally manipulable implementation of the F-correlation. Let F be an RKHS on R, $K(x,y)$ be the associated kernel, and $\Phi(x) = K(\cdot, x)$ be the feature map, where $K(\cdot, x)$ is a function in F for each x. Then we have the famous reproducing property.

$$f(x) = \langle \Phi(x), f \rangle, \quad \forall f \in F, \quad \forall x \in \mathbb{R} \quad (19.3)$$

This implies:

$$\mathrm{corr}(f_1(x_1), f_2(x_2)) = \mathrm{corr}\left(\langle \Phi(x_1), f_1 \rangle, \langle \Phi(x_2), f_2 \rangle\right) \quad (19.4)$$

Consequently, between one-dimensional linear projections of $\Phi(x_1)$ and $\Phi(x_2)$, the F-correlation is the maximal possible correlation that is exactly the definition of the first canonical correlation between $\Phi(x_1)$ and $\Phi(x_2)$, which suggests that the computation of a canonical correlation can be based on an ICA contrast function in a function space.

The separated ICs were disordered using traditional ICA. The first separated ICs may not be important. Therefore, it needs some criteria to sort these ICs. In this chapter, we use negentropy as a criterion to measure the non-Gaussian of ICs. Then, the IC with maximum negentropy will be separated first. Negentropy is given by:

$$N_g(Y) = H(Y_{\mathrm{Gauss}}) - H(Y) \quad (19.5)$$

Y_{Gauss} is a random Gauss variable and has the same variance as Y, and $H(\cdot)$ is the differential entropy of the random variable.

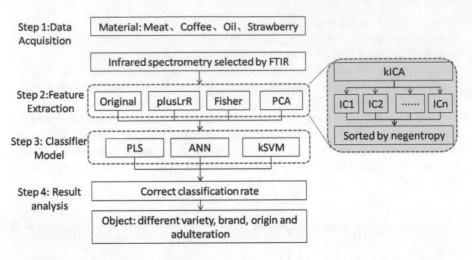

FIGURE 19.2   Flowchart and research framework.

19.2.3 FLOWCHART

In order to investigate the contribution of all wave bands of infrared spectrometry on the performance of classifiers, we utilized k-ICA and k-SVM as feature selection method and classifier for each material. Meanwhile, the main method of PCA and PLS in the previous study and some other methods, such as plusLrR, Fisher, and ANN, are also used in this study. Figure 19.2 depicts a general flowchart of this study. There are four steps of the whole procedures which can be described as follows.

In order to classify the varieties of the meat, the brands of the coffee, the origination of the oil and the adulteration of the strawberry, the raw infrared spectrometry of the four kinds of material was obtained by infrared spectrometer (FTIR) in the first step.

In the second step, the food samples were divided into calibration groups and prediction groups according to a ratio of 4:1. First, we did feature extraction using k-ICA. By comparison, PCA, plusLrR, and Fisher, which were widely used, had been realized in this chapter. Meanwhile, all the wavelengths of spectrometry have been used as the raw feature also.

In the third step, we use three classifier models to test the classifier performance on different feather set. One fist model is k-SVM, and the other two models are PLS and ANN. Optimal identification model was selected by comparison in terms of the identification power (correct classification rate, CCR).

Finally, based on the recognition results, we analyze the results of the models for identification of different kinds, brands, origins, and adulteration in detail.

19.3   RESULTS AND DISCUSSION

In this chapter, we use K-fold cross validation technique to evaluate the generalization performance. In the machine learning community, Wong (2017) suggests that K is usually selected as 5 or 10. Therefore, in this chapter, we set K as 5.

Our dataset is randomly divided into five disjoint folds. Four of them are used for training and validation purposes (calibration set), and the remaining was used as the test set for our predictive model (validation set). To compute the average accuracy rate, the process for each fold was repeated to get test results from the five folds. The optimal identification model was selected by comparison in terms of the identification power (CCR).

19.3.1 DIFFERENT FEATURES SELECTION METHOD

The main goal of features selection and optimization in this work was to reduce the spectral dimensionality of the spectrum, which typically contains highly correlated information in neighboring bands. Minimizing data dimensionality was a necessary first step for the discriminant technique that followed (Vijay Kartik et al., 2017).

The aim of effective features selection is to seek a subset of features as small as possible to cover the full wavelengths. The subsets of features, as the substitution of the full spectral features, are equally or more efficient because they reduce the dimensionality of raw data which makes the identification less time-consuming, sometimes it can also improve the recognition rate of the object. Plus L reduce R (plusLrR) and Fisher algorithms are used in this chapter (Han et al., 2016) as feature selection methods to identify the most significant wavelengths. With the reduced spectral bands, it will be possible to construct a simple machine vision system for food detection.

Feature optimization methods seek a linear or a nonlinear mapping from all bands' wavelength to a feature space, whose size is smaller than the size of the original bands' wavelength. We use principle component analysis (PCA) and ICA for feature detection. ICA has been used to identify the single component spectra in glucose (Braig et al., 2016). As a kind of improvement, k-ICA is also a kind of feature detection method essentially (Wang et al., 2016). They can be carried out to identify the most significant feature mapping from feature set. The improved k-ICA proposed in this chapter uses negentropy as a criterion to measure the non-Gaussian of ICs. The maximum negentropy IC will be first separated. This will be very useful for classification.

In addition, to evaluate the classification performance of these methods above, we compared reduced features using these methods with original features. For a fair comparison, we unify the size of the feature subsets to 10 and unify k-SVM as recognition model. The recognition results are listed in Table 19.2. In any table cell, the bottom figure is the correct recognition rate (CRR) of training set and the top figure is the CRR of testing set. A black body identifies the best result of the rank. The recognition results are the mean value of fivefold verification.

Table 19.2 shows overall accuracy rates using different feature selection methods and k-SVM method as classifier on four feature sets. Bold values indicate best feature selection methods. Table 19.2 indicates that feature optimization can improve the classification performance generally. In most case, k-ICA method outperforms the other methods. Even though PCA, 100% accuracy rate in one case (Oil) can be achieved. k-ICA still shows the best performance. The result of plusLrR and Fisher methods has been undesirably proved. Meanwhile, using original features,

TABLE 19.2
Result of the Extracted Features with Different Feature Selection Methods by k-SVM Classifier

Methods	Meats	Coffee	Oil	Strawberry
Original	**97.50**	78.57	89.17	91.96
	100.0	**100.0**	92.50	**99.29**
plusLrR(10)	67.50	57.15	42.50	64.29
	86.67	75.00	45.00	89.32
Fisher(10)	77.50	71.43	86.67	81.89
	95.00	83.93	75.83	92.68
PCA(10)	95.00	96.43	81.67	96.64
	98.96	98.21	**100.0**	97.15
k-ICA(10)	95.83	**98.21**	**100.0**	**97.76**
	100.0	**100.0**	**100.0**	97.86

Note: Bold values indicate maximum recognition rate.

we can also get a good performance, but the size of the feature set is large, and the computational process is time-consuming. To summarize, k-ICA, especially after we improved it, is the best approach of all the feature selected and optimized methods.

19.3.2 DIFFERENT RECOGNITION MODELS

Besides feature selection and feature optimization methods, another important factor that affects the recognition and detection results is the method of classification models.

PLS regression is a kind of multiple statistic data analytical method which is produced from the chemistry field. The outstanding characteristic of PLS is that it can make the multiple linear regression analysis, and it is often used in conjunction with PCA.

In recent years, SVM has been widely used in many fields and can solve both linear and nonlinear multivariate calibration problems. SVM classification is a non-parametric supervised classification technique. Often SVM-based classifiers are shown to perform well in small sample research. The salient feature of SVM is the ability to handle data with very high dimensions with very little training samples. SVM can map the low dimension space where the data is inseparable to high dimension space, so that the data can be divided. A kernel function is often used to transform the data to higher dimension to make classes separable by hyperplane. The radial basis function (RBF) was used as the kernels in consideration of its excellent performance. There is a free A Library for Support Vector Machines (LIBSVM) toolbox (Chang and Lin, 2011).

Another popular classification method is ANN. In the ANN, it is still an open problem on how to determine the optimal number of neurons in the hidden layer. Specifically, the Rapid Miner tool (2015) suggested the number of neurons in the hidden layer could be half the number which is the number of entering features plus number of output categories.

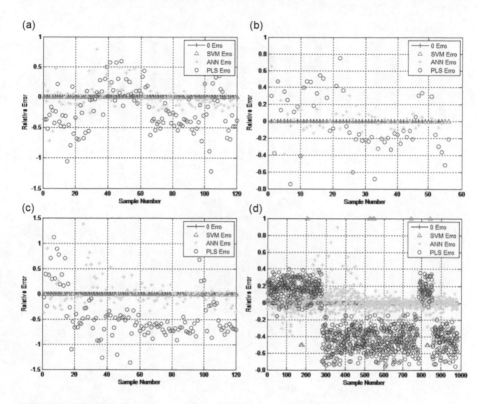

FIGURE 19.3 Identification results of different models by k-ICA. (a) Meat, (b) coffee, (c) oil, and (d) strawberry.

Figure 19.3 showed the performance of the three methods, and it could be found that k-SVM showed it performs much better than ANN and PLS. k-SVM achieves the excellent testing performance in four foods. As shown in Figure 19.3, the results of neural network recognition are not stable, which may be due to the random weights generated by the system. The identification results of PLS are the worst, and it cannot be used in food detection. Here we use k-ICA as the feature selection method.

19.3.3 DIFFERENT SAMPLES OR FEATURES

For further evaluation of the performance of the proposed approach, additionally, we found the number of samples or features which may affect the identification results, and we will further discuss these factors below. Here, we just use the best model that is k-ICA and k-SVM.

For total of 1,279 samples of different meat, coffee, oil and strawberry, we employ k-ICA as the selected optimal features method and test the top 2, 4, 6, 8, and 10 optimal features. The relationship between the number of selected features and the recognition rate for each kind of food is plotted in Figure 19.4.

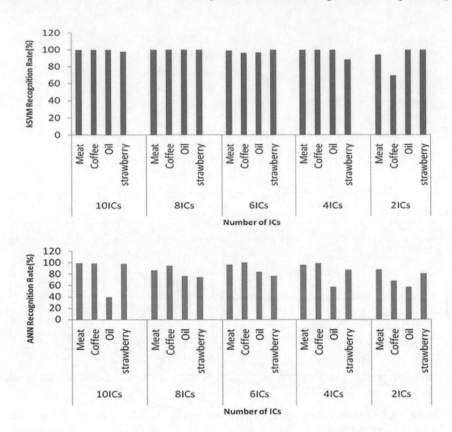

FIGURE 19.4   Result of different ICs for SVM and ANN.

It was shown that as the number of features (ICs) increases, the recognition rate of k-SVM increases too. When the number of features is larger (above 6), the recognition rate is higher, and when the number of features is lower (under 4), the recognition rate is lower too. But the decline is not obvious. This shows that the previous few ICs contain most of the information of the testing data. The degree of dependence for different datasets is different. For the coffee dataset, using two ICs and k-ICA, the recognition rate is only about 60%. The reason may be that there are too few coffee samples. It has been reported in some literature, such as Xiao et al., (2016) point out that even in small sample analysis, the sample size should be above 100 when using k-SVM. Here, the coffee dataset is only 56, so the recognition result isn't very good. Using four ICs, the recognition rate of the strawberry dataset is the worst of the four. That may be the number of strawberry sample is the largest (983). When the sample size is large, it is difficult for the ICA to find valuable information from the data. So, the size of the sample should be considered carefully when you use the k-ICA. But in general, the recognition rate of this method (k-ICA and k-SVM) is high, with an average of more than 90%, and the recognition result is relatively stable, which shows that this method is an effective method in food testing area. The overall recognition rate of ANN is low, and the result is not always accurate.

19.4 SUMMARY

This reviewed chapter deals with an interesting topic – developing new algorithms based on ICA and SVM for food authentication including variety, brand, origin, and adulteration. IR spectroscopy was selected as an instrumental technique. ICA was used as a feature extraction method as well as the classification approach. Before this study, ICA has been used as feature extraction method for nuclear magnetic resonance (NMR) in wine (Monakhova et al., 2015), fruit juice (Cuny, 2007; Cuny et al., 2008), etc. In this chapter, we improve the ICA model and use it in NIR spectroscopy analysis area. In addition, our research is more comprehensive, including not only the viscous liquid (Woodcock et al., 2007) but also the paste solid and the powder solid; not only the variety and the origin but also the brand and adulteration.

This chapter aims to use a double kernel way (k-ICA and k-SVM) to identify different kinds of food such as meat, coffee, oil, and strawberry of different variety, brand, origin, and adulteration, respectively. Firstly, we based on the infrared spectroscopy data with the traditional methods (PCA and PLS), and then, we use improved k-ICA as feature selection method combined with k-SVM as identification model. In order to further verify the effectiveness of this method, finally, we compared some other methods, such as plusLrR, Fisher, and ANN. By comparing the performance of different feature extraction methods, the performance of different recognition models, the related factor of different number of datasets, and different number of feature sets, we find that the combined model of k-ICA and k-SVM is the best model. This method can effectively identify the brand, origin, variety, and adulteration of agricultural product or food.

REFERENCES

Bach F R, Jordan M I. Kernel independent component analysis. *The Journal of Machine Learning Research*, 2003, 3: 1–48.

Bin Z, Lu J, Fenghua J, Lei Q, Abby V, Yang T. Walnut shell and meat differentiation using fluorescence hyperspectral imagery with ICA-kNN optimal wavelength selection. *Sensing and Instrumentation for Food Quality and Safety*, 2007, 1: 123–131.

Braig J R, Keenan R, Rule P, et al. Fluid component analysis systems and methods for glucose monitoring and control. U.S. Patent 9,414,782, 2016.

Briandet R, Katherine Kemsley E, Wilson R H. Discrimination of Arabica and Robusta in instant coffee by Fourier transform infrared spectroscopy and chemometrics. *Journal of Agricultural and Food Chemistry*, 1996, 44(1): 170–174.

Chang C C, Lin C J. LIBSVM: A library for support vector machines, 2011. www.csie.ntu.edu.tw/-cjlin/libsvm.

Cortes C, Vapnik V. Support-vector networks. *Machine Learning*, 1995, 20: 273–297.

Cuny M, Evolving window zone selection method followed by independent component analysis as useful chemometric tools to discriminate between grapefruit juice, orange juice and blends. *Analytica Chimica Acta*, 2007, 597(2): 203–213.

Cuny M, Vigneau E, Le Gall G, et al. Fruit juice authentication by 1H NM(2015)R spectroscopy in combination with different chemometrics tools. *Analytical and Bioanalytical Chemistry*, 2008, 390(1): 419–427.

Downey G, Briandet R, Wilson R H, et al. Near- and mid-infrared spectroscopies in food authentication: Coffee varietal identification. *Journal of Agricultural and Food Chemistry*, 1997, 45(11): 4357–4361.

Han Z, Wan J, Deng L, et al. Oil adulteration identification by hyperspectral imaging using QHM and ICA. *PLoS One*, 2016, 11(1): 1–11.

Holland J K, Kemsley E K, Wilson R H. Use of Fourier transform infrared spectroscopy and partial least squares regression for the detection of adulteration of strawberry purees. *Journal of the Science of Food and Agriculture*, 1998, 76: 263–269.

Hyviirinen A, Karhunen, J, Oja E. *Independent Component Analysis*. Hoboken, NJ: Wiley and Sons, 2001.

Hyvärinen A, Oja E. A fast fixed-point algorithm for independent component analysis. *Neural Computation*, 1997, 9(7): 1483–1492.

Jowder A O, Kemsley E K, Wilson R H. Mid-infrared spectroscopy and authenticity problems in selected meats: A feasibility study. *Food Chemistry*, 1997, 59: 195–201.

Monakhova YB, Godelmann R, Kuballa T, et al. Independent components analysis to increase efficiency of discriminant analysis methods (FDA and LDA): Application to NMR fingerprinting of wine. *Talanta*, 2015, 141: 60–65.

Rapid Miner Tool. Neural net learner (rapid miner class documentation), 2015. http://rapid-i.com/api/rapidminer-4.4/com/rapidminer/operator/learner/functions/neuralnet/NeuralNetLearner.html.

Tapp HS, Defernez M, Katherine Kemsley E. FTIR spectroscopy and multivariate analysis can distinguish the geographic origin of extra virgin olive oils. *Journal of Agricultural and Food Chemistry*, 2003, 51(21): 6110–6115.

Vijay Kartik S, Carrillo R E, Thiran J P, et al. A Fourier dimensionality reduction model for big data interferometric imaging. *Monthly Notices of the Royal Astronomical Society*, 2017, 468(2): 2382–2400.

Wang H, Xu W, Guan N, et al. Fast kernel independent component analysis with Nyström method. *2016 IEEE 13th International Conference on Signal Processing (ICSP)*, 2016, pp. 1016–1020.

Wang J, Chang C I. Applications of independent component analysis in endmember extraction and abundance quantification for hyperspectral imagery. *IEEE Transactions on Geoscience and Remote Sensing*, 2006a, 44(9): 2601–2616.

Wang J, Chang C I. Independent component analysis-based dimensionality reduction with applications in hyperspectral image analysis. *IEEE Transactions on Geoscience and Remote Sensing*, 2006b, 44(6): 1586–1600.

Wong T T. Parametric methods for comparing the performance of two classification algorithms evaluated by k-fold cross validation on multiple data sets. *Pattern Recognition*, 2017, 65: 97–107.

Woodcock T, Downey G, Kelly J D, et al. Geographical classification of honey samples by near-infrared spectroscopy: A feasibility study. *Journal of Agricultural and Food Chemistry*, 2007, 55(22): 9128–9134.

Xiao M, Zhong Z, Shan-hai X. Spectrum quantitative analysis based on bootstrap-SVM model with small sample set. *Spectroscopy and Spectral Analysis*, 2016, 36(5): 1571–1575.

20 Study on Vegetable Seed Electrophoresis Image Classification Method

To verify the performance of crop classification and variety clustering based on electrophoretogram and investigate vegetable seed genetic relationships, three kinds of breeder seeds' samples are collected. They are bell pepper, Chinese cabbage, and cucumber. Thirty varieties of each kind of crop have been collected. Standard electrophoretograms of them are prepared by the method of protein ultrathin isoelectric focusing electrophoresis. Then, the digital images of them are obtained by scanners. Using these images, a pattern recognition model of crop recognition based on principal component analysis (PCA) and support vector machine (SVM) is constructed. Test results obtained by the leaving-one method show that, more than 97% crops can be recognized correctly. In addition, through k-means clustering analysis, clustering trees of these three kinds of crops with different varieties are established and the paternities of partial varieties based on clustering tree are discussed. This study has positive significance to automatic seed inspection by computer based on electrophoresis and selection of breeding direction.

20.1 INTRODUCTION

In recent years, frequent incidents of fake and shoddy seeds have caused great economic losses to truck growers and seed administrators. The quality of vegetable seeds and authenticity of variety are important factors in vegetables' increased production. In order to decrease or avoid the economic loss of seeds' incident, it is needed to test seeds' variety and seeds' purity quickly and efficiently.

Ultrathin-layer isoelectric focusing is one of the methods with low cost and rapid test speed and has been written into the International Seed Testing Association (ISTA) seed test regulations (TeKrony, 2006) and the national standard of crop seed test regulations (GB/T 3543, 1995).

While the final test result is dependent on the later image analysis to a large extent, and nothing but accurate recognition, comparison and quantification of protein spots can effectively analyze the protein data of each gel, and thus, significant biological information can be obtained. Protein atlas analysis is an important application of pattern recognition of digital image processing in biomedicine domain. At present, the atlas is obtained by manual recognition, and the result is not accurate at some extent because of the testers' knowledge and some other deviations.

Pattern recognition test based on digital image process by computer is a new method with high test speed, strong resolving ability, and high repeatability and can test large quantities tirelessly. This method has been widely used in the variety recognition and quality test of corn (Zhao et al., 2009; Han and Zhao, 2010), peanuts (Han et al., 2010; Han and Zhao, 2009), wheat (Sakai et al., 1996), rice (Dubey et al., 2006), and other varieties. Present cluster analysis of electrophoretogram is from the aspect of biology (Wang et al., 2004) or simply studies protein spots' distribution rule (Shen et al., 2005) by analysis software. Little is researched on the automatic seed test by computer.

Standard electrophoretograms of main cultivation varieties of bell pepper, Chinese cabbage, and cucumber are constructed by technology of protein ultrathin isoelectric focusing electrophoresis. On the basis of these electrophoretograms, image process method is tried in crop classification and genetic relationships to provide a new, efficient, rapid, and correct test method of seed variety's authenticity and purity.

20.2 MATERIAL AND METHOD

20.2.1 Experimental Materials

Thirty seeds (Table 20.1) of each variety of the three, bell pepper, Chinese cabbage and cucumber, are collected. These almost contain the main varieties of these crops' main producing area. Standard electrophoretograms of the crop (bell pepper, Chinese cabbage, and cucumber) seeds are prepared by the method of protein ultrathin iso-electric focusing electrophoresis. The equipment of collecting electrophoresis film are isoelectric focusing electrophoresis slot, low-temperature circulating water bath slot, high bistable electrophoresis instrument power, horizontal shaker, vortex mixer, removing liquid instrument, glass plate, and Gel-Fix polyester film. The electrophoresis film is obtained according to the protein ultrathin isoelectric focusing electrophoresis technology's operation regulation. In the process of acquisition, the request of protein ultrathin isoelectric focusing electrophoresis technical manual is fully considered (GB/T 3543, 1995) and electrophoretograms are obtained after the aired films are scanned through scanner (Figure 20.1).

The type of the scanner used in the experiment is a Canon CanoScan 8800F Flatbed CCD scanner with optical resolution 4,800 dpi*9,600 dpi and maximum resolution 19,200 dpi, scan range 216*297 mm. The type of the computer is a Lenovo ideaCentre Kx 8160 with CPU of Intel Core 2 Quad Q8300 2.5 GHz, memory DDRIII4G; Flash 1G, HDD 500 GB; Windows XP operating system. The image recognition process is implemented by program developed by MATLAB® R2008a.

20.2.2 Method

The feature optimization method referred this chapter is primary component analysis. The recognition pattern is SVM algorithm. Preprocessing of image includes some routine methods, such as images' enhancement, de-noising, median filtering, and gray processing.

Recognition of image is based on features. In engineering, an image is equated to a matrix, and the value of each pixel composes the original features called pixel features. But with the increase of image's resolution, the composing data

TABLE 20.1
Experimental Material

No.	Bell Pepper	Cucumber	Chinese Cabbage
1	Jinxing bell pepper	Yulong	Jiaoyan 9
2	Juxing bell pepper	Changlv 1	Jiaoyan 65
3	Julong bell pepper	Zhongnong 20	Jiaoyan 58
4	Zili bell pepper	Zhongnong 106	Jiaoyan 55
5	Meitian 5	Cuilong	Jiaobai 4
6	Zhudi	Huangfen 2	Jiaobai 6
7	Meimeng	Fuhua 3	Improved Zaoshu 2
8	Huangtaifei	Fuhua 1	Improved Chengza 5
9	Hongsusan	Jingxuanguifei	Jiaoyanhuangxinbai F1
10	Guoqinghong	Jinong 3	Jiaoyanchunxianfeng
11	Hongluodan	Jinong 8	Jiaoyanchunfeng
12	Bulanni	Jinongbaiye 3	Jiaochun 2
13	Suofeiya	Lvzuanshi	Jiaochun 1
14	Aoteman	Jinong 12	Chunxiu
15	Leiao	Fuyang 1	Jiaoyanzaoxiawang
16	Weigeli	Xiaobailong	Jiaoyanxiaxing
17	Libeite	Bafute	Jiaoyanxiajin 2
18	Zimeigui	Fuyuan 3	Jiaoyanxiaqiuwang
19	Jinhuang	Oubao 1617	Jiaoyanxiaqiu 65
20	Jincheng	Huangjinfeng	Jiaoyan 10
21	Jinhong	Jinlongjiecheng	Jiaoyan 3
22	Baiyuxue	Xiaoqiubanbai	Jiaoyanqiuzao
23	Qiaokeli	Lushu 21	Jiaoyanqiusheng
24	Hongweier	Jinyou 12	Jiaoyanjinqiu
25	Huangfeite	Jinyou 35	Jiaoyangaokang 78
26	Juxiya	Jinyou 36	Jiaobai 8
27	Jinzi	Jinyou 48	Jiaobaijinqiu
28	Jinkaidi 179	Jinlv 4	Jinqiubai 1
29	Honglina 116	Jinchun 4	Jiaobaiqiushou
30	Zhongjiao 8	Jinchun 2	Jiaochun 2

becomes huge, and dimensional catastrophe occurs. In order to improve computer's performance, dimension reduction and feature optimization become necessary. Different optimization results can be obtained on the basis of different methods of data dimension reduction and optimization. The optimization results are called optimization features.

PCA is an effective method of analyzing data in statistics. The purpose of it is to find out a set of vectors to interpret the data's variance as much as possible from the data space, to map data from the former space of R dimension to the space of M dimension $(R > M)$ through dimension reduction and keep the main information of the data, thus, it is easier to deal with the data. The method of PCA is to

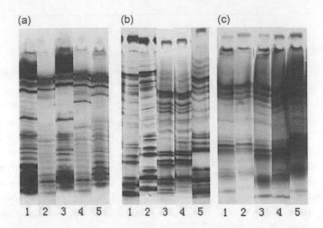

FIGURE 20.1 Some electrophoretograms of three vegetable crops. (a) Bell pepper, (b) cucumber, and (c) Chinese cabbage.

look for some mutually orthogonal axles along the direction of the biggest variance of dataset.

The method of SVM is on the basis of Vapnik-Chervonenkis dimension (VC dimension) and structural risk minimum principle of statistical learning theory. It is to find out the best compromise between the model's complexity (that is the precision of learning aiming at specific sample) and ability of learning (that is, the ability of recognition of any sample without error) according to the limited information and to obtain the best ability of generalization. SVM is to map vector to a higher dimension space and make a big gest interval hyperplane in the space. On the two sides of the hyperplane separating the data, two hyperplanes are built in parallel. An appropriately built separating hyperplane can make the space of the two hyperplanes, which is equivalent to it being the biggest. It is summed that the classifier's total error margin is less when the space or gap among the paralleling hyperplanes is bigger. In the process of classification, there are many kernel functions for selection, such as linear, polynomial, and radial basis function (RBF) kernel functions.

20.3 RESULTS AND ANALYSIS

20.3.1 RECOGNITION OF CORP

Classification recognition of the electrophoretograms of the three crops is carried out on the basis of SVM based on the grayed pixel features, PCA data dimension reduction, and RBF kernel function. The test result of leaving-one method aiming at the three crops is up to 97.62%. At the same time, it is found that the sample's amount can influence the recognition rate to some extent, and if the amount is larger, the recognition is higher. The function relationship between sample's amount x and recognition rate y is $y = -0.0006x^2 + 0.0306x + 0.5948$.

When the sample's amount is above 10, the whole recognition rate is above 90% and the recognition effect is good. In the experiment, 20 primary components are collected in PCA. PCA can be used not only in data dimension reduction but also in looking for optimized feature combinations appropriate to all varieties.

Different kernel functions' recognition effect and computation time are listed in Table 20.2, from which it can be seen that RBF kernel function has the best recognition performance (Figure 20.2).

Classification recognition effect of the first two primary components (Figure 20.3) and the first three primary components (Figure 20.4) of the three crops are particularly researched. It can be seen that the three crops have clustered in two principal components' space and aliasing has occurred in some part. This phenomenon has alleviated in the three principal components' space and appropriate use of hyperplane can isolate the three crops to a great extent.

TABLE 20.2

Recognition Effect by SVM with Different Kernel Functions

Feature Type	Kernel Function	Recognition Rate	Computation Time
Pixel feature	Linearity	78.22	2.7″
	Polynome	90.54	4.6″
	RBF	92.23	4.7″
PCA feature	Linearity	82.13	2.7″
	Polynome	93.75	4.6″
	RBF	97.62	4.7″

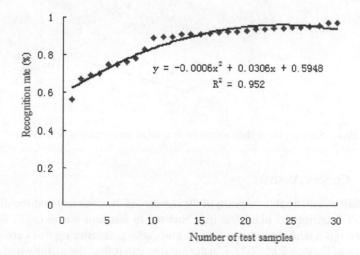

FIGURE 20.2 The influence of test sample amount on recognition rate.

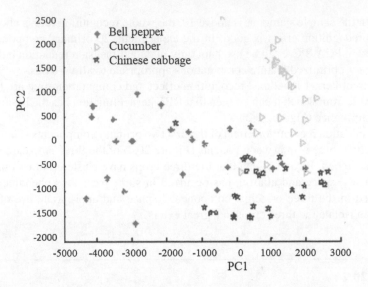

FIGURE 20.3 Scatter plot of three crops in first two principal components.

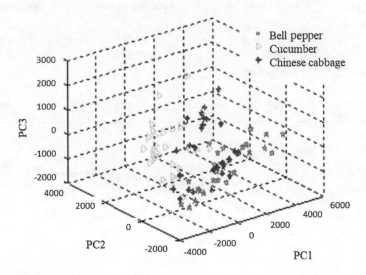

FIGURE 20.4 Scatter plot of three crops in first three principal components.

20.3.2 CLUSTER ANALYSIS

On the basis of electrophoretogram of 30 samples of each variety and the difference of their PCA-optimized pixel features, method of k-means clustering is used, and the three crops (cucumber, sweet pepper, and cabbage) clustering trees are obtained and shown in Figures 20.5–20.7. Clustering tree can reflect the affinity and heredity relationship among crop varieties to a great extent and sometimes is called pedigree diagram.

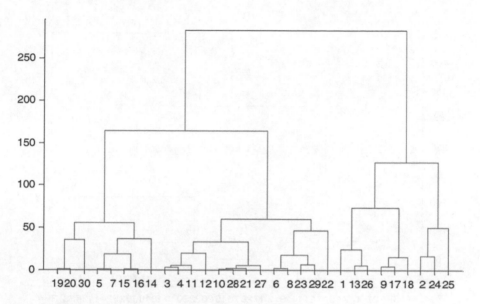

FIGURE 20.5 K-means clustering tree of cucumbers.

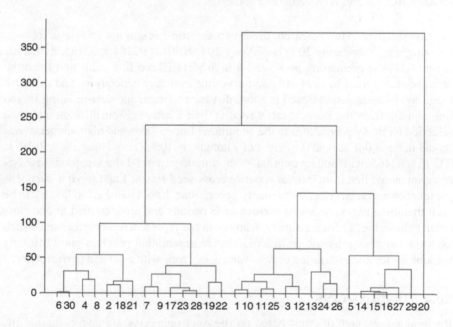

FIGURE 20.6 K-means clustering tree of sweet peppers.

Cluster analysis can help to analyze the genetic relationship among varieties. For example, when the two varieties with the code of 3 and 4 on the clustering tree (Zhongnong 20 and Zhongnong 106) are researched, they are first classified as one class, and it can be concluded that there is very near genetic relationship between

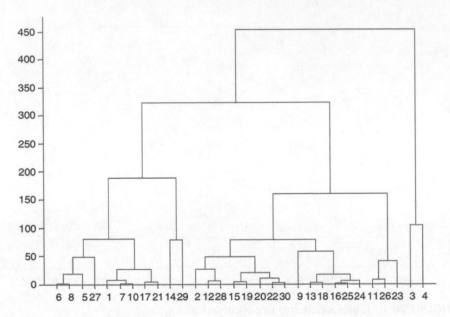

FIGURE 20.7   K-means clustering tree of cabbages.

the two varieties. After research, these two varieties' origin are shown as follows. The origin of Zhongnong 20 (ZhongNong 20#, 2010) is 02484 × 02465; its female parent 02484 is premature, good, and a high-yield inbred line cultivated from the cross seed imported in 1994 after self-crossing from any generations and directive breeding; its male parent 02465 is good, disease-resistant, and heat-resisting inbred line coming from the internal cross seed of Jiyu 8 after self-purification and seed selection for many generations in the condition of open fields and high temperatures. Its old name is 04C8. And the origin of Zhongnong 106 (ZhongNong 106#, 2010) is 04796 × 02465; its female parent 04796 and male parent 02465 are good, disease-resistant inbred line from interior superior cross-seed Jinyou 1 and Jinyu 8 after self-purification and seed selection for many generations. Its old name is 04C38. It can be seen that these two varieties have close male parents and are classified to one class during clustering. There are similar features in the representation performance such as good and strong disease resistance. This representation provides good thinking for looking for and cultivation of new variety of crops with a certain performance.

20.4   DISCUSSIONS

Recognition research of crops based on the electrophoretogram adopts mainly the method of pixel points. Through PCA, the first 20 primary components are attained, and using these primary components as features, method of SVM is used to recognize and classify crops. It has been found that this method is effective and the whole recognition rate is up to above 90%. There is great similarity among vegetable seeds. Correct recognition means when other crop seeds are used to imitate the true seeds, their images will be identified automatically. It should be noted that features acquired by the method of pixel points can distinguish different kinds of crops, but it will meet some difficulty when it is used to distinguish the different varieties of the same crop. This is because that the different varieties of one crop belong to the same genus and similarity exists in the normal electrophoretogram. More features, such as texture of electrophoretogram, should be attained in order to recognize the different varieties of one crop.

In this research, normal electrophoretogram is attained and the standard of it is manually judged by the experts' knowledge and method of image analysis is not considered. If the method of image analysis is used to standardize the electrophoretogram of the same batch of varieties or looking for the clustering center of the electrophoretogram, then normal electrophoretogram can be attained automatically. This problem needs further research. Moreover, it should be noted that electrophoretogram's swimming lane information is very important and extraction of relevant swimming lane can provide good reference value for further variety identification, pedigree analysis, and breeding new variety. This will be the emphasis of further research.

In clustering experiment, each variety can be clustered, and the result indicates at some extent that there is genetic relationship among varieties. Seeds with same male parent or female parent are first clustered, and then seeds with same grandparent are clustered. But because of restrictions on experimental conditions, the accurate standardizing of each electrophoretogram can't be secured, so there is some error in clustering. Moreover, the collected pixel features can't truly represent the difference of variety pedigree chart, and some information is lost during clustering. So, it is difficult to construct one crop's genealogy system by method of cluster analysis which needs much germ plasm resource of the crop and long-time experiment. Of course, mutation of some extent will occur in the growth of crop which can be embodied in the electrophoretogram. In this research, only genetic relationship of two varieties (Zhongnong 20 and 106) are analyzed. The breeding process needs to be made clear, and an information base of mass germ plasm resource is also needed when the relationships of other varieties are to be tested. But at present, our country does not have a relevant wholesome information base, so there is much work to do. The work of this research is a fundamental attempt at crops' variety classification and variety kinship clustering. In order to construct an automatic seed test method and system based on electrophoretogram by computer, there is much work to do.

Seed's electrophoretogram has an important function in distinguishing variety, crop, variety's purity, and authenticity. Fully excavate the biological information of the electrophoretogram can provide reference to the breeding of new variety.

20.5 CONCLUSIONS

To verify the performance of crop classification and variety clustering based on electrophoresis and investigate vegetable seed genetic relationships, three kinds of breeder's seeds samples are collected. They are bell pepper, Chinese cabbage, and cucumber. Each crop has 30 varieties. And standard electrophoretograms of them are prepared by method of protein ultrathin isoelectric focusing electrophoresis. Then, the digital images of them are obtained by scanners. Using these images, a pattern recognition model of crop recognition based on PCA and SVM is constructed. Test results by leaving-one method show that, more than 97% crops can be recognized correctly. The amount of training set and testing set affects the recognition rate. On the whole, the recognition rate can be improved with the increase of samples. When the number of samples is above 10, the recognition is up to above 90%. This phenomenon has some stability. In addition, through k-means clustering analysis, the clustering process of different varieties of the same crop is researched. The genetic relationship of varieties and the far and near of the relationship can be found according to the clustering chart. This study has positive significance to automatic seed inspection by computer based on electrophoretogram and selection of germ plasm resource in the breeding of new variety.

REFERENCES

Dubey B P, Bhagwat S G, Shouche S P, et al. Potential of artificial neural networks in varietal identification using morphometry of wheat grains. *Biosystems Engineering*, 2006, 95(1): 61–67.

GB/T 3543. Crop seed test procedures, National bureau of technical supervision of P.R.C, 1995, 8.

Han Z-Z, Zhao Y-G. A cultivar identification and quality detection method of peanut based on appearance characteristics. *Journal of the Chinese Cereals and Oils Association*, 2009, 24(5): 123–126.

Han Z-Z, Zhao Y-G. Quality grade detection in peanut using computer vision. *Scientia Agricultura Sinica*, 2010, 43(18): 3882–3891.

Han Z-Z, Zhao Y-G, Yang J-Z. Detection of embryo using independent components for kernel RGB images in maize, *Transactions of the CSAE*, 2010, 26(3): 222–226.

Sakai N, Yonekawa S, Matsuzaki A. Two-dimensional image analysis of the shape of rice and its application to separating varieties. *Journal of Food Engineering*, 1996, 27(4): 397–407.

Shen P, Fan X-H, Zeng Z, et al. Proteome two-dimensional electrophoresis analysis Based on image feature and mathematical morphology. *Science in China Series B: Chemistry*, 2005, 35 (1): 44–50.

TeKrony D M. Seeds. *Crop Science*, 2006, 46(5): 2263–2269, International Seed Testing Association, ISTA.

Wang Y-W, Wei L-J, Ainiwaer, et al. Analysis of principal component and cluster in spring wheat high-molecule-weight glutenin subunits and gliadin bands and main quality traits. *Journal of Shihezi University (Natural Science)*, 2004, 22(4): 277–282.

Zhao C-M, Han Z-Z, Yang J-Z, et al. Study on application of image process in ear traits for DUS testing in maize, *Acta Agronomica Sinica*, 2009, 42(11): 4100–4105.

ZhongNong 20#. http://baike.baidu.com/view/4682387.htm, 2010, 11.

ZhongNong 106#, http://baike.baidu.com/view/46823.htm, 2010, 11.

21 Identifying the Change Process of a Fresh Pepper by Transfer Learning

In this chapter, we studied the feasibility of using transfer learning to identify the change process of a fresh pepper. It is very meaningful to identify the process of pepper change, especially in the degree of freshness of fruits and vegetables. As far as we know, no one judges the relationship between the quality of fruits and vegetables and time by deep learning. We tried to identify the change process of the pepper with the help of pretrained model (Googlenet). The fresh pepper collected on the same day was placed in a shaded environment, and the change of the pepper for 20 days was identified. The accuracy of 1-day, 2-day, 3-day, and 4-day, respectively, are 55.3%, 69%, 82.4%, and 96.2%. To test the generalization of the model, we also applied the model to the classification of dried pepper. In order to expand the application of transfer learning to image recognition, more studies on the changing process of fruit and vegetable varieties are needed in the future.

21.1 INTRODUCTION

There are many methods to identify the quality of fruits and vegetables. For example, the image processing technology has been widely used in agricultural products inspection and gradings, such as cereal grain color, grain shape, and type identification; fruit shape and defects; and other exterior quality inspection (Liao et al., 1994; Yarnia et al., 2012). Others use spectroscopy to identify the quality of fruits and vegetables, such as apple quality, lychee fruit, citrus fruit, and so on (Tang et al., 2011; Lorente et al., 2013; Guo et al., 2017; Juntao et al., 2018).

However, these methods are only to identify the quality of fruits and vegetables. And what we're going to do is correlate the quality of fruits and vegetables with how long they have been stored. A pepper was used as the experimental material, and the quality of the pepper was related to the storage time. We used the method of transfer learning to identify the change process of pepper storage.

In many practical applications, reorganizing training data and rebuilding models are expensive. It is necessary to reduce the need and workload of reorganizing training data. In this case, knowledge transfer or transfer learning is required between task domains. As big data repositories become more and more popular and use relevant but not identical existing datasets, interest in the target domain makes transfer learning of solutions an attractive approach. Transfer learning has been successfully applied to image classification (Duan et al., 2012; Kulis et al., 2011; Zhu et al., 2011), text emotion classification (Wang and Mahadevan, 2011), human

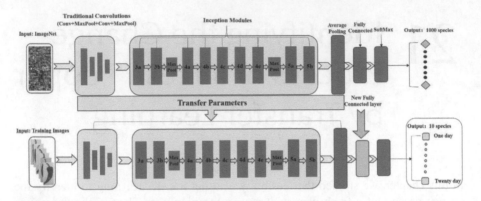

FIGURE 21.1 Network model structure diagram.

disease detection (Tan et al., 2018), multilingual text classification (Carneiro et al., 2015; Prettenhofer and Stein, 2010; Zhou et al., 2014), software defect classification (Nam et al., 2017), human activity classification (Harel and Mannor, 2011), and plant disease identification (Ramcharan et al., 2017). Transfer learning has greatly reduced the workload of image acquisition, enabling the small datasets to achieve higher recognition accuracy (Razavian et al., 2014; Donahue et al., 2014; Yosinski et al., 2014). The model is pretrained by using image-net dataset, and the data is imported into the model. Replace the last layer of the model with the full connection layer for local fine-tuning to improve the recognition accuracy (Figure 21.1).

21.2 MATERIALS AND METHOD

21.2.1 IMAGE CHARACTERISTICS

The variety of pepper used in the experiment was a Hang pepper, and 100 fresh peppers were picked on the same day (Figure 21.2). The camera used in the experiment is a Canon DS126491. These fresh peppers were observed in a shady environment for 20 days. An A4 sheet was used as the background of the pepper. The image collection time is 7 o'clock every day. Then the pepper was put back in its place.

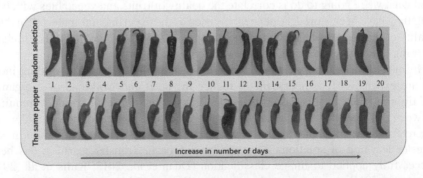

FIGURE 21.2 Pepper images.

21.2.2 Workflow Diagram

First, the processed images were divided into a training set and test set in the ratio of 7:3. Secondly, the training set is imported into the pretrained GoogLeNet model (change the parameters of the model's final full connection layer for training), and the training is stopped after a specific number of steps. By testing the model, the recognition accuracy and loss of the model are obtained.

21.2.3 Convolutional Neural Network

Deep learning constitutes a recent, modern technique for image processing and data analysis, with promising results and large potential. As deep learning has been successfully applied in various domains, it has recently also entered the domain of agriculture (Kamilaris and Prenafetaboldu, 2018). Meanwhile, the convolutional neural network (CNN) has achieved good results in cultivated land information extraction (Lu et al., 2017), plant species classification (Lee et al., 2015), and plant diseases by leaf image classification (Sladojevic et al., 2016). Therefore, we carried out transfer learning on the basis of CNN network and achieved excellent results.

In this study, we used the famous CNN model GoogLeNet architecture (Szegedy et al., 2015). GoogLeNet, proposed by Szegedy, won the 2014 ImageNet Large Scale Visual Recognition Challenge. GoogLeNet uses an average pool rather than a fully connected layer, and this serves to reduce the number of parameters. The overall number of layers (independent building blocks) of GoogLeNet, which is 22 layers deep (or 27 layers deep if we also count pooling) deep when counting only layers with parameters, is approximately 100 (Yurtsever and Yurtsever, 2018). The network structure of GoogLeNet based on CNN is composed of more than 6.7 million parameters (Szegedy et al., 2015). In addition, GoogLeNet uses nine inception modules. The architecture of the standard GoogLeNet algorithm is shown in Figure 21.3, and the flow chart of the experiment is shown in Figure 21.4.

Previously, the standard activation functions used the tanh or sigmoid functions, but with gradient descent training, these saturated (saturating) nonlinear activation

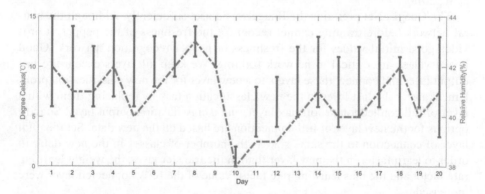

FIGURE 21.3 Changes of temperature and relative humidity during the test.

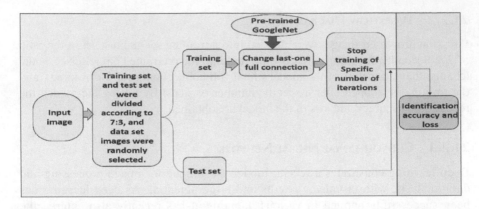

FIGURE 21.4   Experimental flowchart.

functions would cause gradient disappearance and reduce the training speed. This problem can be avoided by using unsaturated nonlinear activation function ReLu. Overlapping pooling improves accuracy and reduces the probability of overfitting. Dropout is adopted to prevent over-fitting in full connection layer. The network adopts small-size convolutional layer with three to five convolutional layers, because the small-size convolutional layer is more nonlinear than a large-size convolutional layer, making the function more accurate and reducing the number of parameters (Liao et al., 2019). Because the depth of the network and the small size of the convolution kernel help to achieve implicit regularization, the network starts to converge after a few iterations. Therefore, we chose this network to complete the task of determining the freshness of the pepper.

21.2.4   TRANSFER LEARNING

Transfer learning is not training a completely blank network; the model can recognize the distinguishing features of a specific category of images much faster and with significantly fewer training examples and less computational power (Kermany et al., 2018).

The GoogLeNet pretrains the ImageNet dataset sufficiently. Although the neural network before training cannot recognize the freshness of the pepper, it provides good initial values for the freshness of pepper recognition network. Good initial values are critical to network training. We keep all layers except the last output layer and connect these layers to a new layer for the new classification problem. Transfer the last layer to the new classification task by replacing them with a layer of full connection, a softmax layer, and a classification output layer. Specify options for the new layer of full connection are based on the new data. Set the total layer of connection to the same size as the number of classes in the new data. In order to learn faster in the new layer than in the transfer layer, the weight learning rate factor and the bias learning rate factor value of the fully connected layer were increased.

21.3 THE PERFORMANCE OF THE MODEL

The curve of recognition accuracy and cross loss (Figure 21.5) is drawn to illustrate the performance of our model. The curve of the training process was processed by smoothing with a smoothing factor of 0.6, and the variation trend was easily observed. We also add a part of the image with high-noise interference in the training set to avoid the occurrence of the overfitting phenomenon, which is conducive to the generalization of the classification model. On the precision curve (Figure 21.5a), the training dataset is a solid line and the verification dataset is a dashed line. When the cross-entropy loss is plotted, the training dataset is a solid line and the verification dataset is a dashed line.

The model training process proceeded for 7,000 iterations for the problem of recognizing the days of fresh pepper, and the recognition accuracy was 55.3%.

To make it easier to observe the performance of the predicted results, we draw the real point and the predicted point. If the two points coincide, the prediction is correct and vice versa. As shown in Figure 21.6a, most of the error points are very close to the real point. At the same time, we expanded the classification time range of the model, including 1 day before and 2 days after, 3 days before and 3 days after, and 4 days before and 4 days after. According to our statistics, the accuracy of the expanded classification range was 55.3%, 69%, and 82%. We calculated that the accuracy rate after expanding the time range was 55.3%, 69%, 82%, and 96.5%. We present the confusion matrix in the model at the same time, through the

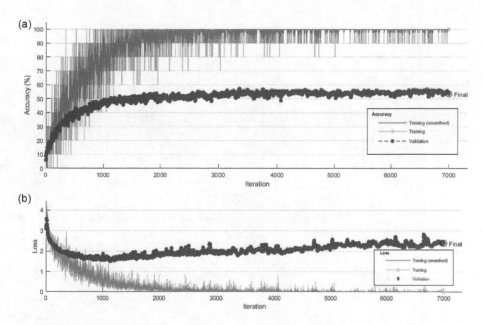

FIGURE 21.5 Accuracy and loss curve during training. (a) The precision curve in the process of fresh pepper identification training for 20 days and (b) the loss curve in the process of fresh pepper identification training for 20 days.

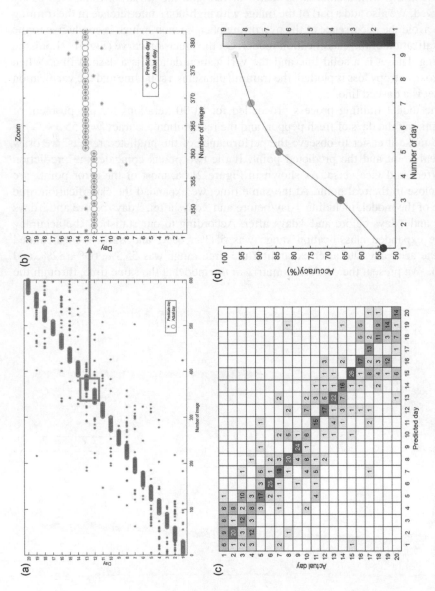

FIGURE 21.6 The model performance of fresh pepper. (a) True hit chart, (b) zoom true hit map, (c) the confusion matrix of the model, and (d) the accuracy of predicting a wider range of days.

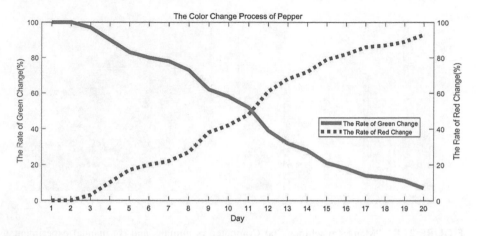

FIGURE 21.7   The color change process of pepper.

TABLE 21.1

The Accuracy of Three-Category Recognition

Items	Number of Errors Identified	Number of Correctly Identified	Total Number of Category Recognition	Accuracy
1	9	21	30	70%
10	5	25	30	83.3%
20	1	29	30	96.6%
Total image	15	75	90	83.3%

observation data distribution in the matrix, and find that the recognition rate is low when pepper changes in the early and late stage, and the recognition rate is more accurate in the middle change stage. So, we can estimate changes of chili pepper: in both early and late, the change is slow, and in the middle stage the change is faster and it quickly changed from green to red.

As time goes on, the peppers began to turn red. In order to observe the change of pepper, we used the image processing method to count the change rule of a solid line and dotted line. The experimental results are shown in Figure 21.7.

In order to further test the model performance, we identified three types, namely day 1, day 10, and day 20. The overall recognition accuracy was 83.3%. The experimental results are shown in Table 21.1.

21.4   COMPUTER vs HUMANS

To test the performance of our model, we held a "man-machine war". In the test, four people were asked to judge 600 images. They can refer to the image of model training. The people who took part in the experiment went crazy after identifying 600 photos, especially in the later stages of recognition, because it was difficult for them to observe

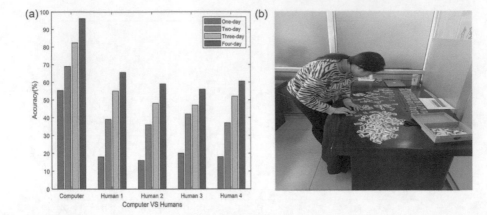

FIGURE 21.8 "Men vs machines". (a) Computer vs humans and (b) manual experiment scene diagram.

TABLE 21.2
Comparison of Model and Artificial Recognition Accuracy

	Accuracy (%)			
Class	1 Day	2 Days	3 Days	4 Days
Model	55.3	69	82.4	96.2
Human 1	18	39	55	65.5
Human 2	16	36	48	59
Human 3	20	42	47	56
Human 4	18	37	52	60.5

the features of the photos and then make judgments (Figure 21.8b). Finally, the results of "man-machine war" are as follows (Figure 21.8a). Analysis of the experimental results clearly shows that the performance of the model far exceeds that of humans (Table 21.2).

21.5 EXPANDING APPLICATION

In order to study the generalization of our model, we also applied the model to the classification of dried chili (Figure 21.9). Three men divided 1,000 peppers into 10 groups with 100 in each group; the proportion of training set and test set is the same as the above experiment. The recognition accuracy of the model after 3,100 iterations is 49.5% (Figure 21.10), and the experimental accuracy is not high. However, this does not mean that the performance of the model is not good enough. There is a great controversy on the manual judgment of the quality grade of dried chili. We speculate that if the model is applied to the identification of pepper varieties, the accuracy of the model will be greatly improved.

FIGURE 21.9 Image of a dried chili sample.

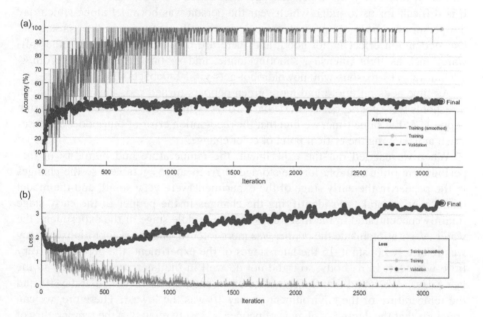

FIGURE 21.10 Accuracy and loss curve during training. (a) The precision curve in the process of dried chili identification training for ten kinds of maize and (b) the loss curve in the process of dried chili identification training for ten kinds of maize.

21.6 DISCUSSION

At present, some people use the electrochemical method to analyze the freshness of fruits and vegetables. It is found in the experiment that in the aging process, the open circuit's potential value of the measured fruits and vegetables presents an upward trend (http://www.wseas.us/e-library/conferences/2005argentina/papers/503-246. pdf). This method involves destructive detection. In the experiment, the probe needs to be inserted into the fruit and vegetable for detection, which is still very diffi- cult in practical engineering application. Some people use chlorophyll fluorescence measurement to measure the freshness of vegetables containing leaves (Qiu et al., 2017). The above method also involves destructive detection, which is difficult to be used in practice.

Our model was not very accurate in identifying the specific day of the pepper quality. However, when we collated the experimental data, we found that there were significant differences between individual peppers. For example, some peppers turned red on the third day after picking, and some pepper turned red on the 17th day. This poses a huge challenge to our model. We will do further research on this issue in the future. If we broaden the identification time period, the performance of the model will be greatly improved. This suggests that most pepper freshness trends are similar. There are too many factors influencing individual and individual change, and there is no very precise limit for an individual change. Similarly, when we judge a person's age, we can tell what age group the person is. It is difficult for us to judge which year the person was born, let alone which day the person was born. At present, there is much selfie software in the market, but the recognition effect is not good, and there are many factors affecting the software, such as light intensity, shooting angle, and so on (https://medium.com/@ yushunwang/10-reasons-why-howoldrobot-gets-viral-1a0eb5cb0d96).

As time goes on, more and more green peppers turned red. The red and green thresholds were found between 11 and 12 days, which could be useful for storing peppers. At the same time, we find that the recognition error of our model increases when approaching the critical point of color change.

When we carried out the experiment, the temperature and humidity of that period were quite suitable for the storage of fruits and vegetables, so the changes in the pepper in the early stage of the experiment were very small, and the model made many mistakes in identifying the changes in the pepper in the early stage (mainly concentrated in the first 5 days). In the middle stage of the experiment, the water evaporation inside the pepper was more intense, and the recognition accuracy was relatively constant. In the later stages of the experiment, the pepper had very little water left in its body, so it did not do well in the later stages (mainly in the last 5 days). We found that the error rate of the model was very high at day 10, and the temperature of the environment on day 10 was the lowest. Therefore, we can conclude that the degree of change of pepper is closely related to the temperature of the environment (Rao et al., 2011). In the future study, we will conduct the research in the environment with large temperature variation differences, so as to find out the most suitable environmental temperature for the model to identify the freshness of the pepper.

The difference in the selection of experimental fruits and vegetables may affect the experimental results (Ludikhuyze et al., 2003). A pepper was used in this experiment, and the change in the pepper was relatively slow from the experimental process. So, we conjecture if we were to replace peppers with tropical fruits like bananas, will the results of our experiment improve? This will be the research content of our next experiment.

Although transfer learning allows the training of a highly accurate model with a relatively small training dataset, its performance would be inferior to that of a model trained from a random initialization on an extremely large dataset of images, since even the internal weights can be directly optimized for maize feature detection (Kermany et al., 2018). In this experiment, we adopted the method of local fine-tuning model weight and bias to rapidly train the model. We suspect that the global fine-tuning of the model weight and bias method can further improve the performance of the model, and the time spent on training will increase accordingly. However, this method will take less time than traditional CNN model. In the next step, we will study the case of fine-tuning the global to prove our conjecture.

21.7   CONCLUSIONS AND FUTURE WORK

Our model did not do very well in determining which changes the pepper experienced on which days. However, the performance of the model is stable in a certain period of time. We found that there was a close relationship between the temperature and the change of the pepper. Only image recognition is far from enough, with other advanced tools, and we believe that the future of fruit and vegetable changes can be real-time monitoring.

1. We extended the model to other types of fruit and vegetable freshness recognition. Study the general law of the change of most fruits and vegetables.
2. We will apply the model to different environments to further improve the robustness of our model.

REFERENCES

Carneiro G, Nascimento J C, Bradley A P, et al. Unregistered multiview mammogram analysis with pre-trained deep learning models. *International Conference on Medical Image Computing and Computer-Assisted Intervention*, Springer, Cham, 2015, pp. 652–660.

Donahue J, Jia Y, Vinyals O, et al. DeCAF: A deep convolutional activation feature for generic visual recognition. *Proceedings of the 31st International Conference on Machine Learning, PMLR*, Beijing, China, June 21–26, 2014, pp. 647–655.

Duan L, Xu D, Tsang I W, et al. Learning with augmented features for heterogeneous domain adaptation. *Proceedings of the 29th International Conference on Machine Learning, ICML*, Edinburgh, Scotland, 2012, pp. 711–718.

Guo Z, Chen Q, Zhang B, et al. Design and experiment of handheld near-infrared spectrometer for determination of fruit and vegetable quality. *Transactions of the Chinese Society of Agricultural Engineering*, 2017, 33(8): 245–250.

Harel M, Mannor S. Learning from multiple outlooks. *Proceedings of the 28th International Conference on International Conference on Machine Learning* Bellevue, Washington, USA. Omnipress, 2011, pp. 401–408.

Juntao X, Rui L, Rongbin B, et al. A micro-damage detection method of litchi fruit using hyperspectral imaging technology. *Sensors*, 2018, 18(3): 700.

Kamilaris A, Prenafetaboldu F X. Deep learning in agriculture: A survey. *Computers and Electronics in Agriculture*, 2018, 147: 70–90.

Kermany D, Goldbaum M H, Cai W, et al. Identifying medical diagnoses and treatable diseases by image-based deep learning. *Cell*, 2018, 172(5): 1122–1131.

Kulis B, Saenko K, Darrell T, et al. What you saw is not what you get: Domain adaptation using asymmetric kernel transforms. *Proceedings of the IEEE Conference on Computer Vision and Pattern Recognition (CVPR 2011)*, Colorado Springs, CO, 2011, pp. 1785–1792.

Lee S H, Chan C S, Wilkin P, et al. Deep-plant: Plant identification with convolutional neural networks. *2015 IEEE International Conference on Image Processing (ICIP)*, Quebec City, Canada. IEEE, 2015, pp. 452–456.

Liao K, Paulsen M R, Reid J F. Real-time detection of colour and surface defects of maize kernels using machine vision. *Journal of Agricultural Engineering Research*, 1994, 59(4): 263–271.

Liao W X, Wang X Y, An D, et al. Hyperspectral imaging technology and transfer learning utilized in haploid maize seeds identification. *2019 International Conference on High Performance Big Data and Intelligent Systems (HPBD&IS)*, Shenzhen, China. IEEE, 2019, pp. 157–162.

Lorente D, Blasco J, Serrano A J, et al. Comparison of ROC feature selection method for the detection of decay in citrus fruit using hyperspectral images. *Food and Bioprocess Technology*, 2013, 6(12): 3613–3619.

Lu H, Fu X, Liu C, et al. Cultivated land information extraction in UAV imagery based on deep convolutional neural network and transfer learning. *Journal of Mountain Science*, 2017, 14(4): 731–741.

Ludikhuyze L, Van Loey A, Indrawati, et al. Effects of combined pressure and temperature on enzymes related to quality of fruits and vegetables: From kinetic information to process engineering aspects. *Critical Reviews in Food Science and Nutrition*, 2003, 43(5): 527–586.

Nam J, Fu W, Kim S, et al. Heterogeneous defect prediction. *IEEE Transactions on Software Engineering*, 2017, 44(9): 874–896.

Prettenhofer P, Stein B. Cross-language text classification using structural correspondence learning. *Proceedings of the 48th Annual Meeting of the Association for Computational Linguistics*, Uppsala, Sweden, 2010, pp. 1118–1127.

Qiu Y, Zhao Y, Liu J, et al. A statistical analysis of the freshness of postharvest leafy vegetables with application of water based on chlorophyll fluorescence measurement. *Information Processing in Agriculture*, 2017, 4(4): 269–274.

Ramcharan A, Baranowski K, Mccloskey P, et al. Using transfer learning for image-based cassava disease detection. *Frontiers in Plant Science*, 2017, 8, doi:10.3389/fpls.2017.01852.

Rao T V, Gol N B, Shah K K, et al. Effect of postharvest treatments and storage temperatures on the quality and shelf life of sweet pepper (Capsicum annum L.). *Scientia Horticulture*, 2011, 132: 18–26.

Razavian A S, Azizpour H, Sullivan J, et al. CNN features off-the-shelf: An astounding baseline for recognition. *Proceedings of the 2014 IEEE Conference on Computer Vision and Pattern Recognition Workshops*, Washington, DC, USA, 2014, pp. 512–519.

Sladojevic S, Arsenovic M, Anderla A, et al. Deep neural networks based recognition of plant diseases by leaf image classification. *Computational Intelligence and Neuroscience*, 2016, Article ID 3289801, 1–11.

Szegedy C, Liu W, Jia Y, et al. Going deeper with convolutions. *Proceedings of the IEEE Conference on Computer Vision and Pattern Recognition*, Las Vegas, USA 2015, pp. 1–9.

Tan T, Li Z, Liu H, et al. Optimize transfer learning for lung diseases in bronchoscopy using a new concept: Sequential fine-tuning. *IEEE Journal of Translational Engineering in Health and Medicine*, 2018, 6, pp. 1–8.

Tang C X, Li E B, Zhao C Z, et al. Quality detection and specie identification of apples based on multi-spectral imaging. *Advanced Materials Research*, 2011, 301–303: 158–164.

Wang C, Mahadevan S. Heterogeneous domain adaptation using manifold alignment. *Twenty-Second International Joint Conference on Artificial Intelligence*, Barcelona, Spain, July 16–22, 2011, pp. 1541–1546.

Yarnia M, Farajzadeh E, Tabrizi M. Effect of Seed priming with different concentration of GA 3, IAA and Kinetin on Azarshahr onion germination and seedling growth. *Journal of Basic and Applied Scientific Research*, 2012, 2(3): 2657–2661.

Yosinski J, Clune J, Bengio Y, et al. How transferable are features in deep neural networks. *Proceedings of the 27th International Conference on Neural Information Processing Systems—Volume 2*, Montreal, Canada, 2014, pp. 3320–3328.

Yurtsever M, Yurtsever U. Use of a convolutional neural network for the classification of microbeads in urban wastewater. *Chemosphere*, 2018, 216: 271–280.

Zhou J T, Tsang I W, Pan S J, et al. Heterogeneous domain adaptation for multiple classes. *International Conference on Artificial Intelligence and Statistics*, Reykjavik, Iceland, 2014, pp. 1095–1103.

Zhu Y, Chen Y, Lu Z, et al. Heterogeneous transfer learning for image classification. *Twenty-Fifth AAAI Conference on Artificial Intelligence*, San Francisco, CA, 2011, pp. 1304–1309.

22 Identifying the Change Process of Fresh Banana by Transfer Learning

Bananas are a very delicious and nutritious fruit. Every year there are a large number of bananas sold at home and abroad and loved by the people. It is necessary and meaningful to study the freshness of bananas. Understanding the process of change will provide important guidance. Manual identification of change processes is time-consuming, expensive, and requires experienced experts, which is usually limited. This chapter analyzed the feasibility of using transfer learning to identify the changing process of fresh banana and established the relationship between fruit quality and time. We use a pretrained GoogLeNet model to identify the process of banana change. The experiment was carried out at room temperature for 11 days to determine the change process. The result shows that the correct recognition rate of this method is 92%, which is higher than human level. In order to make the experimental model universal, we also used the model to detect the change process of strawberries. In order to expand the application of transferred learning image recognition, we will conduct more research on the change process of fruit and vegetable varieties in the future, so as to establish a complete fruit and vegetable freshness detection system.

22.1 INTRODUCTION

Nowadays, the fruit must be fresh and healthy, meeting the standard to get a good price and be accepted by the people. The production and application of various automatic separators can ensure that these standards are met. In order to improve the economic value of large tonnage of fruit varieties, the corresponding automatic sorting machine equipment is needed. There are many ways to identify the color and size of fruits and vegetables. For example, the use of digital image processing technology to complete the defect detection and quality grading of some agricultural products (Donahue et al., 2014), apple defect detection, category identification and quality grading (Chen and Ting, 2004), carrot appearance without root, and shape detection. In addition, some devices use hyperspectral technology to analyze and identify the quality and substance content of fruits and vegetables, such as the quality of apples and the sugar content of mangoes (Guo et al., 2017; Juntao et al., 2018).

However, these devices and methods are used to check the quality of fruits and vegetables. What we wanted to do in our experiment was to correlate the quality of fruits and vegetables with their freshness, that is, to correlate the quality of fruits and vegetables with their storage time. The quality of banana was evaluated by storage time. In this experiment, transfer learning was used to identify the changing process of banana storage.

With the development of transfer learning, the combination of transfer learning and image recognition has become the focus of research and application in recent years. Mohanty et al. (2016) used transfer learning approach which is the process of using pretrained AlexNet for classification of new categories of image where in this case it is used for disease classification. It was able to classify 26 different diseases in 14 crop species using 54,306 images with a classification accuracy of 99.35%. Simonyan and Zisserman (2014) use the deep neural network to carry on the large-scale image recognition, and its model effect is far higher than the traditional method. Additionally, it has been widely applied in many sectors of the world such as business, agriculture, and automotive industry in object detection and image classification (Sladojevic et al., 2016; Dyrmann et al., 2016; Reyes et al., 2015). However, in these applications, it is very time-consuming and expensive to reorganize the training pretraining data and reconstruct the evaluation model. In order to reduce the workload, transfer or transfer learning is usually carried out between work domains, so that experiments can be completed quickly and accurately. With the advent of the era of big data, it has become a popular method to adopt migration learning as a solution for the perceptual part of the target field with the existing datasets that are practical and relevant but different. Transfer learning has been successfully applied to product classification (Zhu et al., 2011), product defect detection, text language classification (Wang and Mahadevan, 2011), human disease prediction (Tan et al., 2018), and plant disease recognition. Migration learning greatly reduces the workload of data collection and enables small data and high-recognition accuracy. The ImageNet dataset was used for preprocessing, and the data was imported into the used model. A series of fine tuning was performed on the model to improve the execution time and accuracy of the model. The sections which explain our study in detail are as follows; the whole process of the experiment and the methods adopted were mentioned in Section 22.2. In Section 22.3, the results of the studies were presented. In Section 22.4, we discuss their positive and negative aspects. Finally, we have concluded our study in general in Section 22.5.

22.2   MATERIALS AND METHOD

22.2.1   Image Dataset

In this experiment, we selected 103 bananas of two kinds that were transported from the production area on that day (Figure 22.1). Of these, 71 are from Good Farmer and 32 are from Chiquita. The bananas were stored at room temperature for 11 days. We placed the bananas on A4 sheet and took pictures with a hand-held Canon DS126491 camera, keeping the camera parameters unchanged. When collecting pictures of bananas, we kept the bananas in the same position and take pictures in the same order every other day. The placement of the bananas is random, which makes is more likely how they would be stored in the real world. The sequence number of the bananas is shown in Figure 22.2a. The shooting time is from 9:30 to 10:00 every morning. The shooting process at the temperature and humidity of is shown in Figure 22.2b.

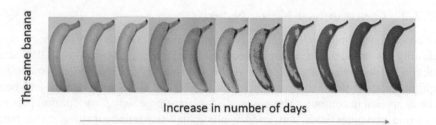

FIGURE 22.1 Example diagram of a banana sample.

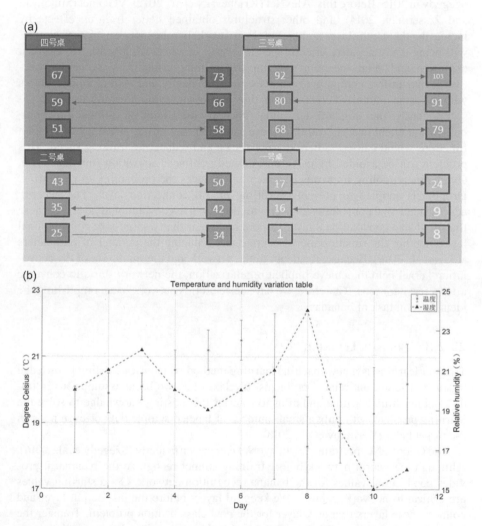

FIGURE 22.2 The sequence of the banana shots (a) and changes in temperature and relative humidity during testing (b).

22.2.2 Convolutional Neural Network

As an emerging and modern image processing and data analysis technology, deep learning has broad application prospect and development potential. With the introduction of deep learning theory and the improvement of numerical computing equipment, convolutional neural network (CNN) has developed rapidly and been widely applied in computer vision, natural language processing, agricultural product detection, and other fields, with good results achieved. At the same time, on the basis of CNN analysis of learning, better experimental results can be obtained.

In this study, we used GoogLeNet (Szegedy et al., 2015) architecture, a prominent CNN model. GoogLeNet is a brand deep learning structure proposed by Christian Szegedy in 2015. Before this, AlexNet (Krizhevsky et al., 2012), VGG net (Simonyan and Zisserman, 2014), and other structures obtained better training effects by increasing the depth (layer number) of the network, but the increase of layer number will bring many negative effects, such as overfit, gradient disappearance, gradient explosion, and so on. Inception may improve training results from another perspective: it may utilize computing resources more efficiently and extract more features with the same amount of computation, thus improving training results.

Previously, the standard activation functions used tanh or sigmoid functions, but with gradient descent training, these saturated (saturating) nonlinear activation functions would cause gradient disappearance and reduce the training speed. This problem can be avoided by using unsaturated nonlinear activation function ReLu. Overlapping pooling improves accuracy and reduces the probability of overfitting. Dropout is adopted to prevent overfitting in full connection layer. The network adopts small-size convolutional layer with three to five convolutional layers, because the small-size convolutional layer is more nonlinear than a large-size convolutional layer, making the function more accurate and reducing the number of parameters (Liao et al., 2019). Because the depth of the network and the small size of the convolution kernel help to achieve implicit regularization, the network starts to converge after a few iterations. Therefore, we have selected this network in accomplishing the identification task of bananas.

22.2.3 Transfer Learning

Transfer learning is a new machine learning method which uses existing knowledge to solve different but related problems. It relaxes the two basic assumptions in traditional machine learning and aims to transfer the existing knowledge to solve the learning problem with only a small number of labeled sample data or even none in the target field. (Kermany et al., 2018).

The GoogLeNet pretrains the ImageNet dataset sufficiently (Szegedy et al., 2016). Although the neural network before training cannot recognize the banana, it provides good initial values for the banana recognition network. Good starting values are critical to network training. We keep all layers before the last output layer and connect these layers to a new layer for the new classification problem. Transfer the last layer to the new classification task by replacing them with a layer of full connection, a softmax layer, and a classification output layer. Specify options for the new layer of full connection are based on the new data. Set the total layer of connection to the same size as the number of classes in the new data. In order to learn faster in the new layer than in the transfer layer, the weight learning rate factor and the bias learning rate factor value of the fully connected layer were increased.

22.2.4 EXPERIMENTAL SETUP

In this experiment, using the neural network toolbox provided in MATLAB® 2018b, transfer learning (Bengio, 2012) was applied to CNN (GoogLeNet) trained. GoogLeNet architecture was selected because of its superior performance in the identification of fruits and vegetables (Mohanty et al., 2016; Ferentinos, 2018). Its network structure is shown in Figure 22.3. After fine-tuning, the parameters used for training network are determined as follows: basic learning rate 0.001; power, 0.9; small batch, 10; and number of times, 10. All experiments were performed using the NVIDIA GEFORCE GTX950 graphics processing unit (GPU). All images were resized prior to training to meet the GoogLeNet's input dimension requirement ($224 \times 224 \times 3$ pixels).

First, we divide the preprocessed dataset into training set and test set in the ratio of 7:3. Secondly, the training set is imported into the GoogLeNet model before training (fine-tuning some parameters of the model), and the training stops after appropriate steps. By testing the model, the recognition precision and the loss degree are obtained.

The performance of the classifier is compared by using the confusion matrix. Each row of the obfuscation matrix represents an instance of the number of predicted banana days, while each column represents an instance of the number of actual

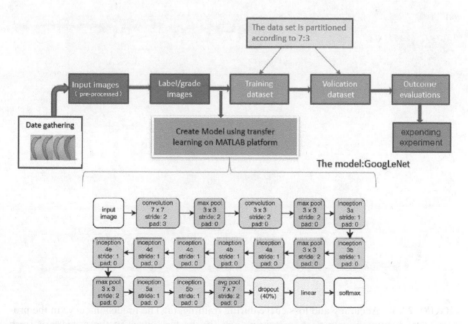

FIGURE 22.3 Experimental flow chart and GoogLeNet model structure chart.

banana days. Based on the confusion matrix, three metrics, precision, sensitivity, and Fl score are used to evaluate the method's performance.

22.3 THE PERFORMANCE OF MODEL

22.3.1 EXPERIMENTAL RESULT

The curve of recognition accuracy and cross loss (Figure 22.4) is drawn to illustrate the performance of our model. The curve of the training process was processed by smoothing with a smoothing factor of 0.6, and the variation trend was easily observed. We also add some interfered images in the training set to avoid the over-fitting phenomenon and facilitate the generalization of classification model. When drawing the precision curve, the training dataset is a solid line and the verification dataset is a dashed line. When the cross-entropy loss is plotted, the training data set is a solid line and the verification dataset is a dashed line.

The model training process proceeded for 2,150 iterations on the problem of recognizing the days of fresh pepper, and the recognition accuracy was 94.62%. To make it easier to observe the performance of the predicted results, we draw the real point and the predicted point. If the two points coincide, the prediction is correct and vice versa. Most of the error points are very close to the correct point. At the same time, we expanded the classification time of the model, from the previous time interval of 1 day to 2 days, and we got the prediction accuracy of the model of 99.19%. As can be clearly seen from confusion matrix (Table 22.1), most of the error recognition is focused on the correct one before or after, without too much deviation. We measured

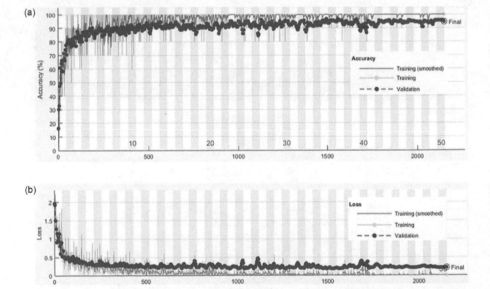

FIGURE 22.4 Accuracy and loss curve during training. (a) The precision curve in the process of fresh banana identification training and (b) the loss curve in the process of fresh banana identification training.

TABLE 22.1

Confusion Matrix of Bananas

Day	1	3	5	7	9	11	Precision(%)	Sensitivity(%)	F1 Score
1	31	1	0	0	0	0	96.88	100	98.42
3	0	30	3	0	0	0	90.90	96.77	93.74
5	0	0	27	0	0	0	100	87.10	93.11
7	0	0	1	31	0	0	96.88	100	98.42
9	0	0	0	0	30	4	88.23	96.77	92.30
11	0	0	0	0	1	27	96.42	87.10	91.52
Accuracy (%)	94.62								

TABLE 22.2

The Accuracy at Different Time Intervals

	1 Day	2 Day	3 Day	4 Day
Accuracy	68.33%	94.86%	99.14%	

the accuracy of time intervals of 1 day, 2 days, 4 days, and 5 days. The results were shown in Table 22.2.

22.3.2 COMPUTER VS HUMANS

In order to test the superiority of the performance of our model, we design a comparative experiment of computer and human recognition. In the test, four people were asked to identify 120 images, and the tester trained them before the recognition to make their game with the computer fairer. At the beginning of the experiment, each participant was able to recognize pictures accurately. However, as the recognition time increased, they would become agitated and their judgment accuracy would decline to different degrees. The final result of "man-machine war" is shown in Figure 22.5. It is obvious from the experimental results that the performance of this model is much better than that of human beings.

22.3.3 EXPANDING APPLICATION

In order to study the general adaptability of the model, we also applied the model to the classification of fresh strawberries. A total of 103 samples were collected. After the same operation steps and 1,050 iterations, the recognition accuracy of the model is 92.47%. This experiment is slightly less accurate than the banana experiment. The change in the strawberry is harder to spot than the banana because the strawberry itself is red. When it goes bad, the color changes to a deep red, which is less obvious. However, our model can still achieve high accuracy. The results show that the model has good adaptability. It can be used for a variety of fruit tests (Figure 22.6; Table 22.3).

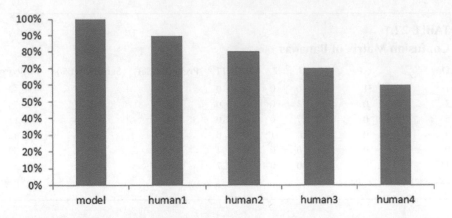

FIGURE 22.5 Comparison of GoogLeNet with human experts.

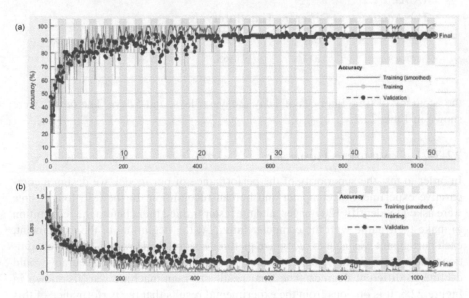

FIGURE 22.6 Accuracy and loss curve during training. (a) The precision curve in the process of fresh pepper identification training and (b) the loss curve in the process of fresh pepper identification training.

TABLE 22.3

Confusion Matrix of Strawberries

Day	1	3	5	Precision (%)	Sensitivity (%)	F1 Score
One	29	3	0	90.62	93.54	92.06
Three	2	27	1	90.00	87.09	88.52
Five	0	1	30	96.77	96.77	96.77
Accuracy(%)	92.47					

22.4   DISCUSSION

We put forward the idea of combining the freshness of fruit with storage time. This idea uses transfer learning to learn banana representation from the banana dataset and then transfers the knowledge to the recognition of fruit freshness. GoogLeNet can effectively identify the freshness of different types of fruits, because it has the inception module and multidimensional convolution reaggregation module, which are new deep learning structures. They outperform human experts. In reality, people's evaluation of the freshness of a certain fruit mainly depends on individual subjective judgment. This model can judge the freshness of fruit more accurately and provide decision support for people. The judgment camera of human and computer makes the judgment more accurate and effective.

Currently, some people use electrochemical methods to analyze the freshness of fruits and vegetables. This approach involves lossy detection. In the experiment, it is necessary to insert the probe into the fruit and vegetable for detection, which is still very difficult in the practical engineering application. Some people use chlorophyll fluorescence method to measure the freshness of leafy vegetables (Qiu et al., 2017). The method mentioned above also involves lossy detection, which is difficult in practical application.

Our model was not very accurate in determining the dates of the banana qualities. At present, it is difficult for all transfer learning models to accurately determine a specific day by traditional photos. When we collated the experimental data, we found that there was a big difference between different bananas. For example, some bananas go bad on the fifth day, and some bananas go bad on the ninth day after they are picked. At the same time, we also found that the process of change of the two kinds of bananas was not consistent (the time of deterioration of the Goodfarmer was shorter, while the Chiquita was longer), which posed a great challenge to our model. We will do further research on this issue in the future. If our model liberalizes the identified time period, the performance of the model will be greatly improved. This indicates that the freshness trend of most bananas is the same.

Our data prove the effectiveness of migration learning method in detecting the freshness of fruits, which is better than people's subjective judgment. In order to better apply this technology to the reality, this fast and scalable method can be deployed in mobile devices and has a high practical guiding significance. With the widespread use of smart phones, we can develop corresponding mobile phone app, so that we can get the state of fresh fruit at any time. This will be a very convenient process with low cost and simple operation.

One defect of our experiment is that the number of datasets is still too small. Only 104 bananas were used in this experiment, which is still far from the large-scale experiment. Therefore, we will expand the number and types of datasets to verify the accuracy of our experiment. Although the specific change process (the first day and the next day's change) is still lacking, we have done the experiment and found that in the data collection days to 1 day later, the range of accuracy of the model has fallen slightly, with an accurate rate around 70%*l*. We will discuss the number of days in the future on the result of recognition and the corresponding solutions. In addition, this experiment adopts the method of local fine-tuning model's weight and deviation to train the model quickly. We believe that global fine-tuning of the model weight and deviation method can further improve the performance of the model, and the training time will increase accordingly. However, this method still takes less time than the traditional CNN model. The next step is to look at fine-tuning global variables to prove our guess.

22.5 CONCLUSIONS

In this chapter, we use two types of banana to train and test transfer learning model. Recognition accuracy is 68.33%, higher than most human experts and traditional neural network model. In order to verify the universality of the model, we used the fresh strawberry model for training and testing, and the accuracy rate was 63.98%. Our model did not do very well in determining which day the banana process was specific. However, the performance of the model is stable in a certain period of time. In the future, we will extend the model to other types of fruit and vegetable freshness recognition and study the general law of the change of most fruits and vegetables. We will try to further improve the robustness of the model by using a technique designed to handle images taken by different devices and people under different environmental conditions.

REFERENCES

Bengio Y. Deep learning of representations for unsupervised and transfer learning. *Proceedings of the 2011 International Conference on Unsupervised and Transfer Learning Workshop*, 2012, 27, pp. 17–37.

Chen H H, Ting C H. The development of a machine vision system for shiitake grading. *Journal of Food Quality*, 2004, 27(5): 352–365.

Donahue J, Jia Y, Vinyals O, et al. DeCAF: A deep convolutional activation feature for generic visual recognition. *International Conference on Machine Learning*, Beijing, China, June 21–26, 2014, pp. 647–655.

Dyrmann M, Karstoft H, Midtiby H S. Plant species classification using deep convolutional neural network. *Biosystems Engineering*, 2016, 151(2005): 72–80.

Ferentinos KP. Deep learning models for plant disease detection and diagnosis. *Computers and Electronics in Agriculture*, 2018, 145: 311–318.

Guo Z, Chen Q, Zhang B, et al. Design and experiment of handheld near-infrared spectrometer for determination of fruit and vegetable quality. *Transactions of the Chinese Society of Agricultural Engineering*, 2017, 33(8): 245–250.

Juntao X, Rui L, Rongbin B, et al. A micro-damage detection method of litchi fruit using hyperspectral imaging technology. *Sensors*, 2018, 18(3): 700.

Kermany D, Goldbaum M H, Cai W, et al. Identifying medical diagnoses and treatable diseases by image-based deep learning. *Cell*, 2018, 172(5): 1122–1131.

Krizhevsky A, Sutskever I, Geoffrey EH. ImageNet classification with deep convolutional neural networks. *Advances in Neural Information Processing Systems*, 2012, 25(NIPS2012): 1–9. doi:10.1109/5.726791.

Liao W X, Wang X Y, An D, et al. Hyperspectral imaging technology and transfer learning utilized in haploid maize seeds identification. *2019 International Conference on High Performance Big Data and Intelligent Systems (HPBD&IS)*, Shenzhen, China. IEEE, 2019, pp. 157–162.

Mohanty S P, Hughes D P, Salathe M. Using deep learning for image-based plant disease detection. *Frontiers in Plant Science*, 2016, 7, Article ID 1419.

Qiu Y, Zhao Y, Liu J, et al. A statistical analysis of the freshness of postharvest leafy vegetables with application of water based on chlorophyll fluorescence measurement. *Information Processing in Agriculture*, 2017, 4(4): 269–274.

Reyes A K, Caicedo J C, Camargo J E. Fine-tuning deep convolutional networks for plant recognition. *CEUR Workshop Proceedings 1391*, Aachen, Germany, 2015.

Simonyan K, Zisserman A. Very deep convolutional networks for large-scale image recognition. arXiv preprint arXiv:1409.1556, 2014.

Sladojevic S, Arsenovic M, Anderla A, Culibrk D, Stefanovic D. Deep neural networks based on recognition of plant diseases by leaf image classification. *Computational Intelligence and Neuroscience*, 2016(June), Article ID 3289801, 1–11.

Szegedy C, Liu W, Jia Y, et al. Going deeper with convolutions. *Proceedings of the IEEE Conference on Computer Vision and Pattern Recognition*, Las Vegas, USA, 2015, pp. 1–9.

Szegedy C, Vanhoucke V, Shlens J. Rethinking the inception architecture for computer vision. *Proceedings of the IEEE Conference on Computer Vision and Pattern Recognition*, Las Vegas, NV, 2016, pp. 2818–2826.

Tan T, Li Z, Liu H, et al. Optimize transfer learning for lung diseases in bronchoscopy using a new concept: sequential fine-tuning. *IEEE Journal of Translational Engineering in Health and Medicine*, 2018, 6, 1–8.

Wang C, Mahadevan S. Heterogeneous domain adaptation using manifold alignment. *International Joint Conference on Artificial Intelligence*, Barcelona, Spain, July 16–22, 2011, pp. 1541–1546.

Zhu Y, Chen Y, Lu Z, et al. Heterogeneous transfer learning for image classification. *National Conference on Artificial Intelligence*, San Francisco, CA, 2011, pp. 1304–1309.

23 Pest Recognition Using Transfer Learning

Pest recognition is very important to crops growing healthily, and this in turn affects crop yields and quality. At present, it is a great challenge to obtain accurate and reliable pest identification. In this study, we put forward a diagnostic system based on transfer learning for pest detection and recognition. On ten types of pests, the transfer learning method have achieved an accuracy of 93.8%. We compared the transfer learning method with human experts and a traditional neural network model. Experimental results show that the performance of the proposed method is comparable to human experts and a traditional neural network. To verify the general adaptability of this model, we used our model to recognize two types of weeds: Sisymbrium Sophia and Procumbent Speedwell, and achieved an accuracy of 99.46%. The proposed method can provide evidence for the control of pests and weeds and the precise spraying of pesticides.

23.1 INTRODUCTION

Pest recognition is very important to crops growing healthily, and this in turn affects crop yields and quality. The current recognition technology for vegetable pests mainly relies on artificial statistics, which exists many shortages such as needing a large amount of labor, low efficiency, feedback delay, artificial faults, and no decision support for pesticide spraying (Xiao et al., 2018). The pest problem is very complicated because of differences in soil type, weather conditions, cultivar, and so on (Alptekin, 2011). Today, people's requirements on the quality of agricultural products have become higher and higher. However, the presence of pests and disease on crops has hampered the quality of agricultural produce (Faithpraise et al., 2013). Pests are usually inspected manually by agricultural experts. This requires continuous observation and recording of crop pests. For farmers, this approach needs a huge cost (Al-Hiary et al., 2011). Pesticides can be used to control pests, but they are harmful to plants and humans if used uncontrollably and in excessive amounts (Dey et al., 2016). If pests are detected in the early stages and prevented accordingly, then big losses can be avoided.

With the development of machine learning, combining machine learning with image recognition has evolved into a hotspot for research and application in recent years. Fuentes et al. (2017) used deep convolutional neural networks to effectively recognize nine types of diseases and pests for tomato. Li et al. (2009) proposed a detection method based on different color features between pest and plant leaves to identify the pest. Faithpraise et al. (2013) utilized a k-means clustering algorithm and correspondence filters to detect and recognize pests. It can achieve rotational invariance of pests up to angles of 360°. Dey et al. (2016) used the statistical feature to extract features and the k-means clustering method to detect the white fly pest. Roldán-Serrato et al. (2015) used a random subspace classifier to build a special neural network for the Colorado potato beetle, and its accuracy rate is 85%. Wen et al. (2015) utilized improved pyramidal stacked de-noising auto-encoder (IpSDAE) for pest estimation-dependent identification and achieved a recognition accuracy of 96.9%. Cheng et al. (2017) used deep residual learning in complex background for pest identification. For ten classes of agricultural pests, recognition accuracy reached 98.67%. Xiao et al. (2018) presented a model based on the bag-of-words model and support vector machine (BOF-SVM) to classify and recognize four vegetable pests in south of China. Experiments showed that the average recognition accuracy reached 91.56%.

Transfer learning is currently very popular in the field of machine learning. From the relevant but different source domain, transfer learning can learn to apply knowledge and it can improve learning level on the target domain. In recent years transfer learning has aroused people's great concern. And the application of knowledge transfer learning from related and different source fields has improved the study of the target field, which has aroused considerable concern (Bengio, 2012). Among them, the two most representative models of transfer learning are AlexNet and GoogLeNet. Many people focus on choosing the relevant source domain instances or features to facilitate learning. However, in each source instance, only some of the attributes may be relevant or contribute to the transfer. This technique allows the last outer layer (classification layer) and uses the remaining structure to retrain and get new weight corresponding classes of interest-damaged kernels. Compared with a completely blank network, transfer learning can retrain the weight of the upper levels faster with fewer training examples and less computational power (Kermany et al., 2018).

In this study, we tried to develop an effective transfer learning algorithm to deal with pest images. The primary task of this technique involved digital image processing images, and the model was also used in two types of weeds: Sisymbrium Sophia and Procumbent Speedwell to validate the generalizability of this model.

23.2 MATERIALS AND METHODS

23.2.1 Materials

We investigated ten types of pests that mainly affect tea plants, which are: Locustamigratoria, Parasalepida, Gypsy moth larva, Empoascaflavescens, Spodopteraexigua, Chrysochuschinensis, Laspeyresiapomonella larva, Spodopteraexigua larva, Atractomorphasinensis, and Laspeyresiapomonella, as show in Figure 23.1, and the number of each type is shown in Table 23.1. These images are from Deng et al. (2018) among these samples, some were gathered from online resources, such as Insert Image, IPM images, and so on. The others were taken outdoors using a digital Single Lens Reflex (SLR) camera, which have been uploaded to Mendeley Data (Deng et al., 2018). We only use part of the dataset, because some URLs have expired. So, there are 484 images of ten types of pests. The sample images show great variation in scale, position, viewpoint, lighting conditions, and backgrounds. In this chapter, we train transfer learning models with 70% and test with 30% of these images.

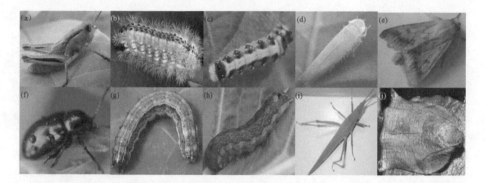

FIGURE 23.1 Ten types of pest image. (a) Locustamigratoria, (b) Parasalepida, (c) Gypsy moth larva, (d) Empoascaflavescens, (e) Spodopteraexigua, (f) Chrysochuschinensis, (g) Laspeyresiapomonella larva, (h) Spodopteraexigua larva, (i) Atractomorphasinensis, and (j) Laspeyresiapomonella.

TABLE 23.1
Number of Ten Types of Pests

Name	Number	Name	Number
Locustamigratoria	69	Chrysochuschinensis	46
Parasalepida	54	Laspeyresiapomonella larva	41
Gypsy moth larva	34	Spodopteraexigua larva	45
Empoascaflavescens	36	Atractomorphasinensis	57
Spodopteraexigua	56	Laspeyresiapomonella	46

23.2.2 INTRODUCE OF THE MODEL

In this chapter, the transfer learning model we used is AlexNet. AlexNet is a Convolutional Neural Network (CNN) which won the ImageNet Large-Scale Visual Recognition Challenge (ILSVRC), an annual challenge that is intended to evaluate algorithms for object detection and image classification. The model is built on more than 1 million images and can be categorized into 1,000 object categories. AlexNet achieved a top-five error around 16%, which was an extremely good result back in 2012. To put it into context, until that year, no other classifier had been able achieve results under 20%. AlexNet was also more than 10% more accurate than the runner-up. AlexNet is composed of eight trainable layers, five convolution layers, and three fully connected layers. All the trainable layers are accompanied by a ReLu activation function, except for the last fully connected layer, where the Softmax function is used. Besides the trainable layers, the network also has three pooling layers, two normalization layers, one dropout layer, which is used for training to reduce the overfitting. In this chapter, workflow diagram and the architecture of AlexNetare are shown in Figure 23.2. All experiments in this chapter were conducted in the MATLAB® 2018a and images processing using Intel HD Graphics 6000 1536MB Graphics Processing Unit (GPU).

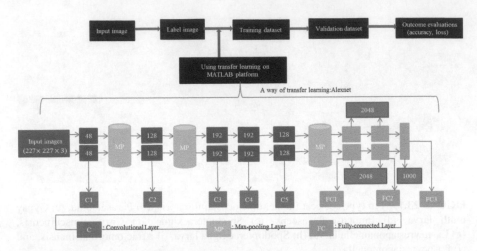

FIGURE 23.2   Workflow diagram and the architecture of AlexNet.

23.3   RESULT AND ANALYSIS

23.3.1   Pests Recognition Result by Transfer Learning Model

In order to recognize the object rapidly and accurately, we input these images of insect pests into this model. Through the eight data process steps of the AlexNet model, we can get the accuracy of the output. We achieved an accuracy of 93.84%. Training processes are shown in Figure 23.3.

We made the thermodynamic diagram of obfuscation matrix using the output result, as shown in Figure 23.4. Through the obfuscation matrix, we can find the recognition accuracy rate is 100% for Locustamigratoria, Parasalepida, Gypsy moth larva, Empoascaflavescens, Chrysochuschinensis, and Spodopteraexigua larva. The recognition accuracy rate of Laspeyresiapomonella larva, Atractomorphasinensis, and Laspeyresiapomonellaisover 80%. But the recognition accuracy rate of Spodopteraexigua is the lowest, only 76.47%.

23.3.2   Comparison of the Model with Traditional Methods

In the traditional methods of image classification, the features which were extracted manually were used to train a traditional classifier (Goldbaurn et al., 1996).

To assess its effectiveness in recognition performance, we compared our model with the traditional neural network model, such as SIFT-HMAX and Convolutional Neural Network (CNN). The recognition accuracy is illustrated in Table 23.2. Among these three methods, the recognition accuracy of transfer learning is the highest. SIFT-HMAX recognition rate is the lowest.

23.3.3   Comparison of the Model with Human Expert

Those pest images were also recognized by human experts. The recognition result is shown in Figure 23.5. The experimental results show that the proposed method

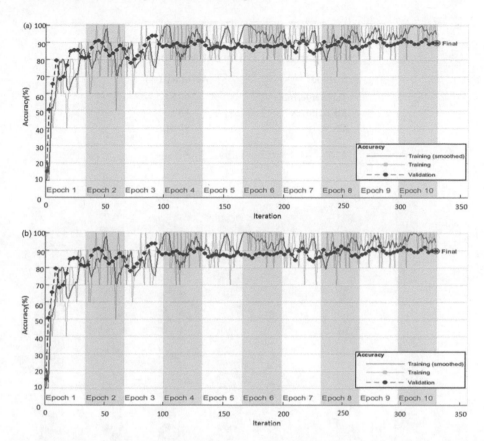

FIGURE 23.3 Training process. (a) The accuracy rate and (b) the loss rate.

Predicted species

	Locusta migratoria	Parasa lepida	Gypsy moth larva	Empoasca flavescens	Spodoptera exigua	Chrysochus chinensis	Laspeyresia pomonella larva	Spodoptera exigua larva	Atractomorpha sinensis	Laspeyresia pomonella
Locusta migratoria	100	0	0	0	0	0	0	0	0	0
Parasa lepida	0	100	0	0	0	0	0	0	0	0
Gypsy moth larva	0	0	100	0	0	0	0	0	0	0
Empoasca flavescens	0	0	0	100	0	0	0	0	0	0
Spodoptera exigua	0	0	0	0	76.47	5.88	0	0	0	19.65
Chrysochus chinensis	0	0	0	0	0	100	0	0	0	0
Laspeyresia pomonella larva	0	0	0	0	0	0	83.33	16.67	0	0
Spodoptera exigua larva	0	0	0	0	0	0	0	100	0	0
Atractomorpha sinensis	5.88	0	0	0	0	0	0	0	94.12	0
Laspeyresia pomonella	0	0	0	0	14.29	0	0	0	0	85.71

True species

FIGURE 23.4 Thermodynamic diagram of obfuscation matrix.

TABLE 23.2

Comparison of Different Methods

Methods	Accuracy (%)
Transfer learning	93.84
SIFT-HMAX	85.50
CNN	90.41

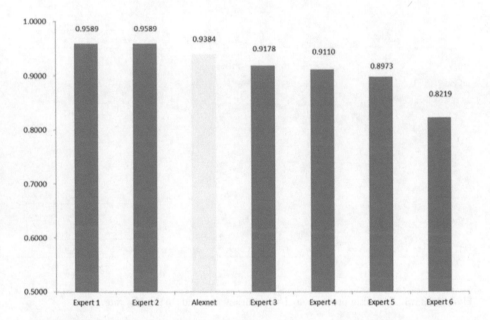

FIGURE 23.5 Comparison of AlexNet with human experts.

is significantly better than four human experts. We also made the thermodynamic diagram of obfuscation matrix by the human expert results, as shown in Figure 23.6. By analyzing the results of human experts, we found that Locustamigratoria, Parasalepida, Gypsy moth larva, Empoascaflavescens, Chrysochuschinensis, and Atracto morphasinensis recognition rate is very high. But for Laspeyresiapomonella larva, Spodopteraexigua, Spodopteraexigua, and Laspeyresiapomonella, human experts have made more mistakes in the process of recognition.

23.3.4 UNIVERSAL OF THE TRANSFER LEARNING MODEL

In order to validate the universal adaptability of our model, we performed experiments on two types of weeds: Sisymbrium Sophia and Procumbent Speedwell, as shown in Figure 23.7. There are totally 622 images and the number of weeds is shown in Table 23.3. We also used 70% of these images to train this model and 30% to test the model. The training process is shown in Figure 23.8. Recognition accuracy is 99.46%.

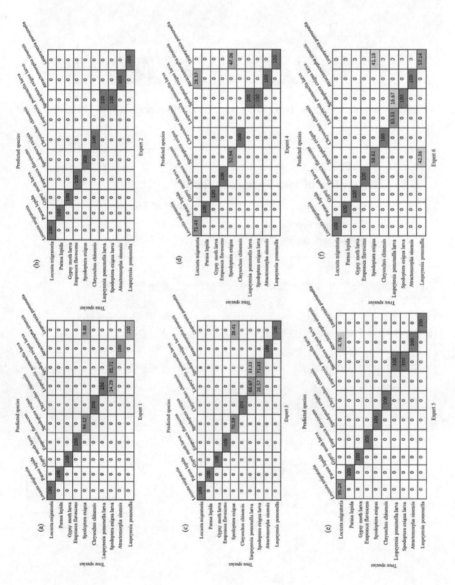

FIGURE 23.6 (a–f) The obfuscation matrices of experts 1–6.

FIGURE 23.7 Sample image of two plants. (a) Sisymbrium Sophia and (b) procumbent speedwell.

TABLE 23.3
Number of Two Types of Weeds

Species	Number of Sample
Sisymbrium Sophia	125
procumbent speedwell	497

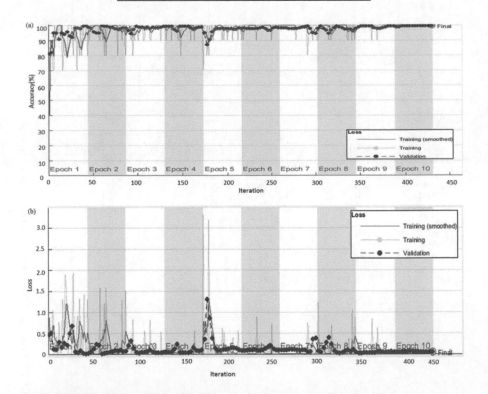

FIGURE 23.8 Training process. (a) The accuracy rate and (b) the loss rate.

23.4 DISCUSSION

23.4.1 IMAGE NUMBERS AND IMAGE CAPTURE ENVIRONMENT

In this study, our model only uses fewer than 500 images, and the recognition accuracy is 93.84%. This dataset is very small compared to the traditional neural network model. Ferentinos (2018) utilized 87,848 images of plant disease to train the CNN model, with the best performance reaching 99.53%. More than 82,000 images were used, but the accuracy only improved by about 5.7%. So, transfer learning gets a better model than the CNN model. Due to the transfer of pretrained models trained on millions of images, a more accurate model can be produced when retraining layers for other pest classifications, thus making it more accurate. We don't need to collect too many samples and can have a better result.

In addition, image capture environment is also very important for machine learning. Most researchers perform experiments in a controlled lab environment; however, in field conditions, complex environments bring a huge challenge for image capture (Deng et al., 2018). The success rate fell from 99% to 68% when a net trained with field-condition images was used to determine laboratory-condition images (Ferentinos, 2018). So, those images captured under actual cultivation are essential for the development of useful disease recognition tools.

23.4.2 IMAGE BACKGROUND AND SEGMENTATION

Target detection is the key issue to pest identification, which is traditionally achieved by image segmentation. Traditional machine learning methods can be adversely affected by the image background, especially when it contains other leaves or soil (Barbedo, 2017). In reality, pest images often have complex background, which makes it difficult to isolate them from the background (Barbedo, 2018). The segmentation algorithm under complex background is very hard, and the practicability and operability are poor. Therefore, the existing method is not particularly suitable for the detection of pests under natural conditions. Barbedo (2018) used deep learning for plant disease recognition and the accuracy of the original image is about 76%, and the accuracy of the removal background image is 79%. This may have a complex background and an element that mimics some of the disease's characteristics, so the network will eventually learn them, leading to learning errors.

23.4.3 SIMILAR OUTLINE DISTURB

When we analyzed the obfuscation matrix of the experts, we find that Laspeyresiapomonella larva and Spodopteraexigua larva are difficult to recognize. The images are presented in Figure 23.8a and b. None of the six experts correctly identified the Laspeyresiapomonella larva and Spodopteraexigua larva. Expert 2, expert 4, and expert 5 classified Laspeyresiapomonella larva and Spodopteraexigua larva into one category. The best result is expert 6. He misclassified 16.67% Laspeyresiapomonella larva images to Spodopteraexigua larva, like the result of transfer learning. For expert 1, he misclassified 35.71% Spodopteraexigua larva images to Laspeyresiapomonella larva. For expert 3, he misclassified 33.33% Laspeyresiapomonella larva images to Spodopteraexigua larva and misclassified 28.57% Spodopteraexigua larva images to Laspeyresiapomonella larva. So, the transfer learning is preferable to manual classification. The reason may be that the key features extracted during the transfer learning training process are more accurate than those extracted manually. And as with any activity carried out by human beings, the manual approach is also affected by psychological and cognitive phenomena that can lead to prejudice, visual illusion and, ultimately, errors. But for transfer learning, there is no interference by any external factors. For Laspeyresiapomonella larva and Spodopteraexigua larva, the model can extract very accurate feature to distinguish the two species. This suggests that transfer learning is better than humans in some ways.

In addition, Spodopteraexigua and Laspeyresiapomonella also have similar appearance features, as presented in Figure 23.8c and d. There are have similar outline, for example, wings, tentacle, color, and so on. These similar features are difficult to recognize for experts. For expert 1, he misclassified 5.88% Spodopteraexigua images to Laspeyresiapomonella. For expert 3, he misclassified 29.41% Spodopteraexigua images to Laspeyresiapomonella. For expert 4, he misclassified 47.06% Spodopteraexigua images to Laspeyresiapomonella. For expert 6, he misclassified 41.18% Spodopteraexigua images to Laspeyresiapomonella, and 42.86% Laspeyresiapomonella images to Spodopteraexigua. But experts 2 and 5 correctly identified as Spodopteraexigua and Laspeyresiapomonella. For transfer learning model, it misclassified 5.88% Spodopteraexigua images to Chrysochuschinensis, 19.65% images to Laspeyresiapomonella, and 14.29% Laspeyresiapomonella images to Spodopteraexigua. Transfer learning is compared with expert, only two expert recognition rates are higher than transfer learning. The possible reason is that in the process of feature extraction, transfer learning eliminates key feature

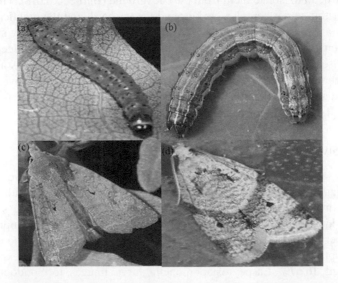

FIGURE 23.9 Sample images of similar species. (a) Laspeyresia pomonella larva, (b) Spodoptera exigua larva, (c) Spodoptera exigua, and (d) Laspeyresia pomonella.

points. However, experts have conquered the transfer learning model with relevant knowledge accumulation and experience. So, in the next work, we need to adjust the parameters of transfer learning model and increase layers of the neural network to improve the accuracy of recognition (Figure 23.9).

23.5 CONCLUSION

In this chapter, we use ten types of pests to train and test transfer learning model. Recognition accuracy is 93.84%, higher than most human experts and a traditional neural network model. In order to validate the wide adaptability of this model, we utilize two types of weeds to train and test the model, and the accuracy rate is 99.46%. It is a good illustration of the wide adaptability of this model. In future, we will try to use a technique that is designed to be used to actually handle images captured by different devices and people, at different angles and light conditions, and under different environmental conditions. These images acquired under actual cultivation conditions are essential for the development of effective pest identification tools. With the development of technology, some limitations can be overcome, but much work remains to be researched.

REFERENCES

Al-Hiary H, Bani-Ahmad S, Reyalat M, et al. Fast and accurate detection and classification of plant diseases. *International Journal of Computer Applications*, 2011, 17(1), 31–38.

Alptekin Y. Integrated pest management of potatoes. *Agricultural Sciences*, 2011, 2(3): 297.

Barbedo J G A. Automatic image-based detection and recognition of plant diseases—A critical view. *EmbrapaInformáticaAgropecuária-Artigoemanais de congresso (ALICE). CONGRESSO BRASILEIRO DE AGROINFORMÁTICA*, 11., 2017, Campinas. Ciência de ados na era da agricultura digital: anais. Embrapa Informática Agropecuária, Campinas, Editora da Unicamp, 2017.

Barbedo J G A. Factors influencing the use of deep learning for plant disease recognition. *Biosystems Engineering*, 2018, 172: 84–91.

Bengio Y. Deep learning of representations for unsupervised and transfer learning. *Proceedings of ICML Workshop on Unsupervised and Transfer Learning*, 2012, pp. 17–36.

Cheng X, Zhang Y, Chen Y, et al. Pest identification via deep residual learning in complex background. *Computers and Electronics in Agriculture*, 2017, 141: 351–356.

Deng L, Wang Y, Han Z, et al. Research on insect pest image detection and recognition based on bio-inspired methods. *Biosystems Engineering*, 2018, 169: 139–148.

Dey A, Bhoumik D, Dey K N. Automatic detection of whitefly pest using statistical feature extraction and image classification methods. *International Research Journal of Engineering and Technology (IRJET)*, 2016, 3(9): 950–959.

Faithpraise F, Birch P, Young R, et al. Automatic plant pest detection and recognition using k-means clustering algorithm and correspondence filters. *International Journal of Advanced Biotechnology and Research*, 2013, 4(2): 189–199.

Ferentinos K P. Deep learning models for plant disease detection and diagnosis. *Computers and Electronics in Agriculture*, 2018, 145: 311–318.

Fuentes A, Yoon S, Kim S C, et al. A robust deep-learning-based detector for real-time tomato plant diseases and pests recognition. *Sensors*, 2017, 17(9): 2022.

Goldbaum M, Moezzi S, Taylor A, et al. Automated diagnosis and image understanding with object extraction, object classification, and inferencing in retinal images. *Proceedings of the International Conference on Image Processing, 1996*. IEEE, 1996, 3, pp. 695–698.

Kermany D S, Goldbaum M, Cai W, et al. Identifying medical diagnoses and treatable diseases by image-based deep learning. *Cell*, 2018, 172(5): 1122–1131.e9.

Li Y, Xia C, Lee J. Vision-based pest detection and automatic spray of greenhouse plant. *IEEE International Symposium on Industrial Electronics,* ISIE 2009. IEEE, 2009, pp. 920–925.

Roldán-Serrato L, Baydyk T, Kussul E, et al. Recognition of pests on crops with a random subspace classifier. *4th International Work Conference on Bioinspired Intelligence (IWOBI)*, 2015. IEEE, 2015, pp. 21–26.

Wen C, Wu D, Hu H, et al. Pose estimation-dependent identification method for field moth images using deep learning architecture. *Biosystems Engineering*, 2015, 136: 117–128.

Xiao D, Feng J, Lin T, et al. Classification and recognition scheme for vegetable pests based on the BOF-SVM model. *International Journal of Agricultural and Biological Engineering*, 2018, 11(3): 190–196.

24 Using Deep Learning for Image-Based Plant Disease Detection

Plant disease recognition is very important to crops' healthy growth. Deep learning is quickly becoming one of the most important tools for image classification. This technology is now beginning to be applied to the tasks of plant disease classification and recognition. In this chapter, transfer learning model (Alexnet, GoogLeNet, VGG16) was developed to perform tomato disease detection and diagnosis using simple leaves' healthy and diseased images. Training of the models was performed with the use of an open database of 18,160 images, containing healthy and nine different diseases of tomato. Three model architectures were trained, with the best performance reaching a 97.98% success rate in identifying the tomato diseases using GoogLeNet network. To verify the general adaptability of the model, we use other eight plant images to test this model. The test accuracy of the network is above 95%. The approach of training transfer learning method on increasingly large and publicly available image datasets presents a clear path toward smartphone-assisted crop disease diagnosis on a massive global scale.

24.1 INTRODUCTION

Plant diseases can cause significant damage to agriculture crops which decrease the production significantly. Plant disease recognition is very important to crops' healthy growth, and this in turn affects crop yields and quality. Prophylactic actions are not always enough to prevent outbreaks, thus constant monitoring is essential for early detection and consequent application of control measures (Barbedo, 2018). Modern technologies have given human society the ability to produce enough food to meet the demand of more than 7 billion people. However, food security remains threatened by a number of factors including climate change (Tai et al., 2014). Plant diseases are not only a threat to food security at the global scale but can also have disastrous consequences for smallholder farmers whose livelihoods depend on healthy crops. Plant disease diagnosis through optical observation of the symptoms on plant leaves incorporates a significantly high degree of complexity. Due to this complexity and the large number of cultivated plants and their existing phytopathological problems, even experienced agronomists are consequently led to mistaken conclusions and treatments (Ferentinos, 2018).

Deep learning is currently a remarkably active research area in machine learning and artificial intelligence and has been extensively and successfully applied in numerous fields (Too et al., 2018). Sladojevic et al. (2016) use CaffeNet model to recognize plant disease images, and the accuracy rate achieved precision between 91% and 98%, for separate class tests, on average 96.3%. Amara et al. (2017) use the LeNet architecture for banana leaf diseases classification, the results demonstrate the effectiveness of the proposed approach even under challenging conditions such as illumination, complex background, different resolution, size, pose, and orientation of real scene image. Brahimi et al. (2017) use a convolutional neural network (CNN) to recognize tomato leaves infected with nine diseases, the results are reaching 99.18% of accuracy. DeChant et al. (2017) use CNN model to recognize northern leaf blight (NLB) in maize, the model achieved 96.7% accuracy on test set images not used in training. Fuentes et al. (2017) used deep CNNs to effectively recognize nine types of diseases and pests for tomato. Liu et al. (2017) design a novel architecture of a deep CNN based on AlexNet to detect apple leaf diseases, and the model achieves an overall accuracy of 97.62%. Lu et al. (2017) use CNN model to identify ten common rice diseases, under the tenfold cross-validation strategy, achieves an accuracy of 95.48%. Karmokar et al. (2015) use an ANN model to recognize diseases of the tea leaf, and the testing process 91% of accuracy was found.

Transfer learning is currently very popular in the field of machine learning. From the relevant but different source domain to apply knowledge transfer learning and improve the target domain learning level, in recent years, transfer learning has aroused people's great concern. And the application of knowledge transfer learning from related and different source fields has improved the study of the target field, which has aroused considerable concern (Bengio, 2012). The conjunction of deep learning and transfer learning, together with the development of Graphics Processing Units (GPUs), has provided a powerful tool for classification and recognition of diseases in plants (Ferentinos, 2018).

In this chapter, we use four transfer learning models—AlexNet, GoogLeNet, VGG16, and ResNet to recognize nine tomato diseases that detail in Section 24.2. In Section 24.3, we use eight plant disease images to test the best model in Section 24.2. In Section 24.4, we will discuss in detail the factors that affect the efficiency of identification.

24.2 MATERIALS AND METHODS

24.2.1 Dataset

In order to develop accurate image classifiers for the purposes of plant disease diagnosis, we needed a large, verified dataset of images of diseased and healthy plants. Until very recently, such a dataset did not exist, and even smaller datasets were not freely available. To address this problem, the Plant Village project has begun collecting tens of thousands of images of healthy and diseased crop plants (Hughes and Salathé, 2015) and has made them openly and freely available.

Plant Village has released more than 50,000 expertly curated images of healthy and infected leaves of 14 different crops (apple, blueberry, corn, grape, etc.) and 12 of them also have healthy leaves and a total number of 26 different diseases. We use part dataset to train and test our models. The number images of crops are shown in Table 24.1. And ten kinds of tomato leaves are shown in Figure 24.1.

TABLE 24.1
Number Images of Crops

Crops	Disease	Number	Crops	Disease	Number
Tomato	Healthy	1,591	Grape	Healthy	423
	Bacterial spot	2,127		Lack rot	1,180
	Early blight	1,000		Esca	1,383
	Late blight	1,909		Leaf blight	1,076
	Leaf mold	952	Cherry	Healthy	854
	Septoria leaf spot	1,771		Powdery mildew	1,052
	Spider mites	1,676	Peach	Healthy	360
	Target spot	1,404		Bacterial spot	2,297
	Mosaic virus	373		Bell healthy	1,478
	Yellow leaf curl virus	5,357		Bell Bacterial spot	997
Apple	Healthy	1,645	Potato	Healthy	152
	Scab	630		Early blight	1,000
	Black rot	621		Late blight	1,000
	Cedar apple rust	275	Strawberry	Healthy	456
Corn	Healthy	1,162		Leaf scorch	1,109
	Cercospora leaf spot	513			
	Common rust	1,192			
	Northern leaf blight	985			

FIGURE 24.1 Examples of different disease in tomatoes. (a) Bacterial spot, (b) early blight, (c) healthy, (d) late blight, (e) leaf mold, (f) Septoria leaf spot, (g) spider mites, (h) target spot, (i) mosaic virus, and (j) yellow leaf curl virus.

24.2.2 Introduced Model

In recent years, CNNs have shown great results in many image classification tasks, which give researchers the opportunity to improve classification accuracy in many fields, including agriculture and plant disease classification. Transfer learning is an important part of CNN. Transfer learning is a model developed for a task and then reused as the starting point for a model on a second task. It is a popular approach in deep learning, where pretrained models are used as the starting point on computer vision and natural language processing tasks given the vast compute and time resources required to develop neural network models on these problems and from the huge jumps in skill that they provide on related problems. At present, transfer learning mainly includes AlexNet, GoogLeNet, and VGG network.

AlexNet is composed of eight trainable layers, five convolution layers, and three fully connected layers and has 60 million parameters and 650,000 neurons (Krizhevsky et al., 2012). GoogLeNet network was designed with computational efficiency and practicality in mind. The network is 22 layers deep when counting only layers with parameters (or 27 layers if we also count pooling). The overall number of layers (independent building blocks) used for the construction of the network is about 100 (Szegedy et al., 2015). The architecture is shown in Figure 24.2. VGG is CNN model devised by Simonyan and Zisserman (2015) for the ILSVRC-2014 challenge. This model is with only 3 × 3 convolutional layers stacked on top of each other in increasing depth. Max pooling handles reducing the size of the volume (down sampling). Additionally, two fully connected layers each with 4,096 nodes and a softmax classifier are shown in their work (Simonyan and Zisserman, 2015).

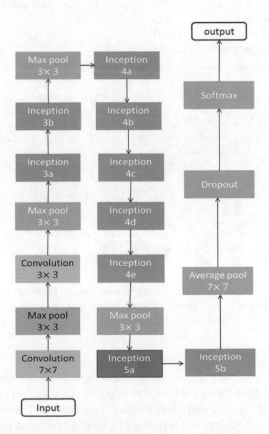

FIGURE 24.2 The architecture of GoogLeNet.

TABLE 24.2

Experiment Environment of Hardware and Software

No	Hardware and Software	Characteristics
1	Memory	8 GB
2	Processor (CPU)	Intel(R) Core(TM) i7-7700HQ CPU @ 2.8 GHz
3	Graphics (GPU)	NVIDIA GeForce GTX 1060 4 GB
4	Operating system	Windows 10 Family Chinese version

24.2.3 EXPERIMENT ENVIRONMENT

Transfer learning was applied using the Neural Network Toolbox provided by MATLAB® 2018a. At first, three different networks were trained. The parameters used to train the network were the following: Mini Batch Size, 10; Max Epochs, 10; Initial Learn Rate, 0.0001; Validation Frequency, 3. All experiments were run using a NVIDIA GeForce GTX 1060 GPU. And other experiment environment of hardware and software is shown in Table 24.2.

24.2.4 EXPERIMENT RESULT

We used three CNN architectures AlexNet, VGG16, and GoogLeNet based on two learning strategies: train and test. In this study, the accuracy results using 70%–30% train-test distribution. From the results, we observe that the recognition accuracy of the VGG16 network is the highest, but the training time is also the longest and GoogLeNet network training time is shortest, and the recognition accuracy was only 1% lower than VGG16. We use the scoring system to measure the recognition capability of each model. In terms of training time, GoogLeNet model takes the least time, score 1, the second is AlexNet model, score 2, and the last is VGG16 model, score 3. In terms of accuracy, VGG16 model has the highest recognition accuracy, score 1, the second is GoogLeNet model, score 2, and the last is AlexNet model, score 3. GoogLeNet model total score is 3, VGG16 model total is 4, and AlexNet model total is 5. As shown in Table 24.3. So, considering the factors of training time and recognition accuracy, we decided to use the GoogLeNet model for the following research. The training process of GoogLeNet is shown in Figure 24.3.

24.3 MODEL UNIVERSAL ADAPTABILITY

24.3.1 BIG DATASETS VALIDATION

There are four types of apple datasets, four types of corn datasets, and four types of grape datasets in Plant Village. Detailed data are shown in Table 24.1. Examples of apples, corn, and grapes are shown in Figure 24.4. We use these datasets to train GoogLeNet model and to prove the general adaptability of the model, under the same conditions in Section 24.2. The test results are shown in Table 24.3. Apple had

TABLE 24.3
Network Training Time, Accuracy, and Score

Network	Training Time (h)	Score	Accuracy	Score	Total Score
AlexNet	46.13	2	97.25%	3	5
VGG16	87.85	3	98.99%	1	4
GoogLeNet	39.52	1	97.98%	2	3

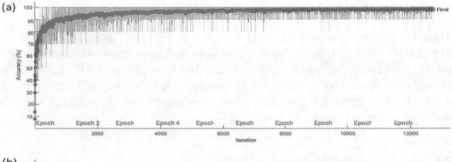

FIGURE 24.3   Training process. (a) The accuracy rate and (b) the loss rate.

the most recognition accuracy, with 99.89%; the second is grape, with recognition accuracy of 99.34%; and the last is corn, recognition accuracy is 97.32%.

24.3.2 SMALL DATASETS VALIDATION

In order to verify GoogLeNet model general adaptability under small datasets, we used the remaining datasets from Plant Village to train and test the model (include cherry, peach, potato, and strawberry). Detailed data are shown in Table 24.1. But there is a data imbalance in these datasets. Such as a healthy potato has 152 images, an early blight potato and a late blight potato has 1,000 images. In order to solve this problem, we decided to randomly pick 100 images per sample. For example, for a healthy cherry, we randomly pick 100 images, and for a powdery mildew we also randomly pick 100 images. We used GoogLeNet model, under the same conditions in Section 24.2. The recognition accuracy is shown in Table 24.4. Peaches have the highest recognition accuracy of 1, and peppers have the lowest recognition accuracy of 95.00%. But the average recognition rate is 97.44%.

FIGURE 24.4 Examples of apples, corn, and grapes. (a) Apple with scab, (b) apple with black rot, (c) apple with rust, (d) healthy apple, (e) corn with gray leaf spot, (f) corn with rust, (g) healthy corn, (h) corn with NLB, (i) grape with black rot, (j) grape with black measles, (k) healthy grape, and (l) grape with leaf blight.

TABLE 24.4
The Recognition Accuracy of GoogLeNet

Plant Type	Accuracy (%)	Time(s)
Cherry	98.33	171
Peach	100.00	169
Pepper	95.00	168
Potato	95.56	279
Strawberry	98.33	178

24.3.3 ARTIFICIAL RECOGNITION

In order to test GoogLeNet model, we chose six experts to detect apple, grape, pomato, and tomato. We randomly chose 100 images of every disease for people recognition. And also used GoogLeNet for detection. Recognition results are shown in Table 24.5.

From Table 24.5, the time of GoogLeNet model is much less than that of artificial recognition, and the recognition rate is much higher than that of artificial recognition. Artificial recognition of grapes is more accurate than that of apples and takes less time. Artificial recognition of a potato is more accurate than that of a tomato and takes less time. So, with the increase of time, the accuracy of artificial recognition is reduced.

TABLE 24.5
Artificial and GoogLeNet Recognition Result

Plant	Expert	Time (s)	Average Time (s)	Accuracy (%)	Average Accuracy
Apple	1	566	1028	77	73.50
	2	1203		45	
	3	872		91.67	
	4	1022		75.67	
	5	830		56.33	
	6	1675		95.33	
	GoogLeNet	295		94.44	
Grape	1	544	846.33	65.33	80.06
	2	852		59.67	
	3	1114		81.67	
	4	633		98	
	5	790		97.67	
	6	1145		78	
	GoogLeNet	272		96.67	
Potato	1	350	407.5	63	78.92
	2	324		63.5	
	3	401		99	
	4	312		97	
	5	473		93	
	6	585		58	
	GoogLeNet	175		98.33	
Tomato	1	300	428.67	70	69.67
	2	312		60.5	
	3	735		86.5	
	4	317		64.5	
	5	365		80	
	6	543		56.5	
	GoogLeNet	179		93.33	

24.4   DISCUSSION

24.4.1   Effect of Data Size

CNNs usually require a very large number of samples to be trained. However, in many real-world applications, it is expensive or unfeasible to collect the training data needed by the models (Pan and Yang, 2010). In Section 24.2 of this chapter, we use 18,160 images of tomatoes to train and test the GoogLeNet network. The recognition accuracy was 97.98%. But in Section 24.3, we use 3,171 images of apples, 3,852 images of corn, and 4,062 images of grapes to train and test GoogLeNet network, the recognition accuracy was 99.89%, 97.32%, and 99.34%. Therefore, it is not the case that more the data, the higher the recognition rate of network train and test. This may be because the Plant Village dataset is unbalanced, where some classes have more images than others, which could be very misleading and could lead to overfitting if not trained carefully. The overfitting problem of deep learning models appears when the model describes random noise or errors rather than the underlying relationship (Liu et al., 2017). There are a few ways to avoid overfitting. If transfer learning is being applied, freezing the first layers and retraining only the last few ones can prevent those layers from overfitting to the new dataset. Also, dataset augmentation operations such as image rotation, mirror symmetry, brightness adjustment, and principal component analysis (PCA) jittering increase the diversity of the training set and cause the model to generalize better (Liu et al., 2017). Mohanty et al. (2016) addressed the issue of overfitting by varying the train and test set ratio. They argued that if the degradation of performance is not significant when the size of the training dataset is decreased, this indicates that the model is not overfitting the data.

24.4.2 IMAGE BACKGROUND

Traditional machine learning methods may be adversely affected by the image background, especially if it contains other leaves or soil (Barbedo, 2016). These objects detect algorithms trained using simple images and adapt them to localize and classify diseases in complex images. This type of algorithm is used in literature in many contexts. Because of that, placing some kind of panel behind the leaf is a common requirement during image capture. Deep neural nets are known for learning the objects of interest even in busy images (Krizhevsky et al., 2012), so in theory, the requirement for leaf segmentation could be relaxed. Indeed, experiments conducted by Mohanty et al. (2016) indicated that better results could be achieved by keeping the background intact. Barbedo (2018) used deep learning for plant disease recognition and the accuracy of the original image is about 76%, and the accuracy of the removal background image is 79%. This is likely due to the fact that the backgrounds have elements that mimic the characteristics of certain diseases, so the network ends up learning them as well, leading to error. This seems to indicate that, if the background is busy and has elements that share characteristics with leaves and symptoms, removal may be useful.

In addition, there are many factors that affect deep learning-based tools when they are used under real field conditions, but in most cases, those are only briefly discussed (or not considered at all). A large fraction of the Plant Village dataset is images of leaves in a controlled environment with a simple background. Adding images with different qualities and complex backgrounds in the training and validation dataset could improve accuracy and produce a classifier more useful for practical usage.

24.4.3 SYMPTOM VARIATIONS

Symptoms and signs in plant leaves may come from a wide variety of sources, which include diseases, nutritional deficiencies, pests, phytotoxicity, mechanical damage, and cold and heat damage. As a result, a truly comprehensive diagnosis system should be able to deal with a classification problem having many classes. Although deep learning techniques have a remarkable ability to classify a large number of classes, currently there is not enough data available to make such a comprehensive diagnosis system feasible. In practice, only a few more common and relevant diseases are usually considered. As a consequence, when tools for disease recognition are used under real-world conditions, they have to find the class that best explains the symptoms among the limited subset of disorders for which they were trained, often leading to incorrect diagnosis. In this case, the learning system, used in data analysis, must deal with complex image containing several leaves and maybe many diseases.

The stage of the disease (symptom severity) is arguably the most important source of variability, as symptoms may range from very mild and barely visible in the beginning, to causing widespread necrosis in the most advanced stages of infection. As a result, diseases may be easier or harder to identify, depending on the stage of infection. If all variations expected for a given symptom are included in the training dataset, deep learning-based tools can properly deal with the challenges caused by such diversity. In practice, it is hard to predict the full extent of symptom variability associated to a given disease and even more difficult to obtain enough images to represent each one of those situations.

24.4.4 MACHINE RECOGNITION AND HUMAN RECOGNITION

As any activity carried out by humans, this approach is subject to psychological and cognitive phenomena that may lead to bias, to optical illusions, and ultimately to error (Bock et al., 2010). Inexperienced farmers can gain intuition about disease and symptoms used by the classifier. Similarly, agriculture experts and experienced farmers can evaluate the classifier decision by showing its classification mechanism. Also, their experts can exploit the transparency of the classifier to discover new symptoms or to localize known symptoms that are difficult to see with the human eye. Most studies in deep learning for plant diseases classification have only focused on analyzing images containing one leaf taken in a controlled environment. Although these approaches can classify a disease accurately in one image taken by a human agent, they are unable to find disease regions automatically in large fields. In general, better known and, as a consequence, more easily recognized by farmers and farmworkers. Rarer disorders have a higher probability of being misidentified and mishandled. Fortunately, in most cases, rare diseases tend to have low impact on crop production. On the other hand, misclassification may result in the application of pesticides that will be innocuous against that disorder, unnecessarily increasing costs.

In addition, a human agent is unable to monitor a large field and detect the earlier symptoms of diseases in order to take a picture. For this reason, a practical disease detection system should automate the monitoring of fields to interact with plant diseases in due course. This automatic monitoring that leads to early detection of diseases can considerably reduce the damages on crops.

REFERENCES

Amara J, Bouaziz B, Algergawy A. A deep learning-based approach for banana leaf diseases classification. *BTW (Workshops)*, 2017, pp. 79–88.

Barbedo J G A. A review on the main challenges in automatic plant disease identification based on visible range images. *Biosystems Engineering*, 2016, 144: 52–60.

Barbedo J G A. Factors influencing the use of deep learning for plant disease recognition. *Biosystems Engineering*, 2018, 172: 84–91.

Bengio Y. Deep learning of representations for unsupervised and transfer learning. *Proceedings of ICML Workshop on Unsupervised and Transfer Learning*, 2012, pp. 17–36.

Bock C H, Poole G H, Parker P E, et al. Plant disease severity estimated visually, by digital photography and image analysis, and by hyperspectral imaging. *Critical Reviews in Plant Sciences*, 2010, 29(2): 59–107.

Brahimi M, Boukhalfa K, Moussaoui A. Deep learning for tomato diseases: classification and symptoms visualization. *Applied Artificial Intelligence*, 2017, 31(4): 299–315.

DeChant C, Wiesner-Hanks T, Chen S, et al. Automated identification of northern leaf blight-infected maize plants from field imagery using deep learning. *Phytopathology*, 2017, 107(11): 1426–1432.

Ferentinos K P. Deep learning models for plant disease detection and diagnosis. *Computers and Electronics in Agriculture*, 2018, 145: 311–318.

Fuentes A, Yoon S, Kim S C, et al. A robust deep-learning-based detector for real-time tomato plant diseases and pests recognition. *Sensors*, 2017, 17(9): 2022.

Hughes D P, Salathé M. An open access repository of images on plant health to enable the development of mobile disease diagnostics. *Computers and Society*, 2015. arXiv:1511.08060.

Karmokar B C, Ullah M S, Siddiquee M K, et al. Tea leaf diseases recognition using neural network ensemble. *International Journal of Computer Applications*, 2015, 114(17). 27–30.

Krizhevsky A, Sutskever I, Hinton G E. ImageNet classification with deep convolutional neural networks. *Proceedings of the Annual Conference on Neural Information Processing Systems*, 2012, pp. 1106–1114.

Liu B, Zhang Y, He D J, et al. Identification of apple leaf diseases based on deep convolutional neural networks. *Symmetry*, 2017, 10(1): 11.

Lu Y, Yi S, Zeng N, et al. Identification of rice diseases using deep convolutional neural networks. *Neurocomputing*, 2017, 267: 378–384.

Mohanty S P, Hughes D P, Salathé M. Using deep learning for image-based plant disease detection. *Frontiers in Plant Science*, 2016, 7, Article ID 1419, 1–11.

Pan S J, Yang Q. A survey on transfer learning. *IEEE Transactions on Knowledge and Data Engineering*, 2010, 22(10): 1345–1359.

Simonyan K, Zisserman A. Very deep convolutional networks for large-scale image recognition. *Sixth International Conference on Learning Representations*, 2015, pp. 1–14.

Sladojevic S, Arsenovic M, Anderla A, et al. Deep neural networks based recognition of plant diseases by leaf image classification. *Computational Intelligence and Neuroscience*, 2016: 11 Article ID 3289801.

Szegedy C, Liu W, Jia Y, et al. Going deeper with convolutions. *Proceedings of the IEEE Conference on Computer Vision and Pattern Recognition*, Las Vegas, USA, 2015: 1–9.

Tai A P, Martin M V, Heald C L. Threat to future global food security from climate change and ozone air pollution. *Nature Climate Change*, 2014, 4: 817–821. doi: 10.1038/nclimate2317

Too E C, Yujian L, Njuki S, et al. A comparative study of fine-tuning deep learning models for plant disease identification. *Computers and Electronics in Agriculture*, 2019, 161: 272–279.

25 Research on the Behavior Trajectory of Ornamental Fish Based on Computer Vision

The trajectory of fish is an important index reflecting the behavioral characteristics of fish, which provide important information for studying the living habits of fish. In this chapter, two color RGB cameras were used to build the shooting platform to film the fish's long-term motion. Through preprocessing, background interframe difference, three-dimensional reconstruction, and other methods, the three-dimensional trajectory of the fish was obtained, and the behavior characteristics of the fish were analyzed and studied. We found that fish have phototaxis, prefer to swim in mild light and the surrounding environment close to the impact of objects; and have tendency to edge, prefer to swim around and stick the tank. Moreover, the fish have a tendency to deepen and are accustomed to staying in the underwater environment below 0.25 m. Through long-term monitoring, we also found that fish have indirect "resting" behaviors similar to those of human beings around 8:00, 14:00, and 21:00. The purpose of this study is to expand the research on ornamental fish culture and fish behavior.

25.1 INTRODUCTION

It is important to evaluate the behavior of fish. From an economic point of view, farmers have a strong interest in the behavioral characteristics of fish. Compared with manual observation, quantitative measurement by computer provides valuable information in an efficient and noninvasive way. Detection of fish behavior, such as individual feeding or swimming speed, can provide useful information for improving production management (Oppedal et al., 2011) and help farmers to observe fish behavior as an indicator for taking relative measures (Zion, 2012). Aquatic organisms are sensitive to changes in the surrounding environment, and they will respond to these variables with unique movements or behaviors (Mancera et al., 2008). For example, the habits of fish will change under the influence of human factors, and the vertical distribution will change when artificial light is used (Oppedal et al., 2011). In addition, fish can also be used as cheap objects in experiments for laboratory studies, and the corresponding water quality environment can be monitored by analyzing the behavior of fish. For example, the chemical reagent in water can be detected by the abnormal tail-beat frequency of crucian carp (Xiao et al., 2015). Or it can reflect the abnormality of water quality through multiple characteristic parameters such as abnormal movement of fish (Leihua, 2017). It can also be used in the research and development of innovative drugs such as central nervous system (CNS), for example, to test neuroactive drugs and test their pharmacological mechanism by observing the reaction of fish (Michael et al., 2015).

There are a variety of conventional detection methods for fish monitoring, such as direct visual monitoring, including manual labeling (Shengmao et al., 2017). However, this traditional method is not fast and real-time monitoring and is inefficient in processing large lots of data. Video tracking was carried out by computer, and an easy-to-detect tag was used on fish, or a harmless and easy-to-monitor elastomer was implanted under the skin of fish (Delcourt et al., 2013), or a fluorescent tag was carried out on fish fins, and positioning was conducted by monitoring the fluorescent tag (Marti-Puig et al., 2018). But in some cases, these methods are invasive and can lead to stressful behavior in fish. A video-tracking software can be directly used for location such as EthoVision XT7 (Cachat et al., 2015). In addition, sound images are acquired through mechanical scanning sonar imaging, and characteristics of fish behavior are studied through image processing (Shih-Liang et al., 2019). In recent years, with the wide application of deep learning, these methods are gradually applied to the research on fish. For example, features of fish head are used to identify fish under the framework (Zhiping and Xi, 2017).

All the above methods are based on the tracking of a two-dimensional environment. However, most organisms live in a three-dimensional environment, which leads to the loss of information such as vertical motion during the tracking trajectory. One method is to obtain 3D information by using reflected light intensity of near-infrared camera when taking images (Aliaksandr et al., 2015). Another method is to use a Kinect depth camera and other methods to directly obtain 3D information of fish (Saberioon and Cisar, 2016), and the postures of pigs were studied by deep learning (Chan et al., 2018). However, in order to track objects in 3D, multiple cameras are usually used to reconstruct 3D scenes. For example, the 3D environment was reconstructed by using the images taken by two cameras, and the position of individual zebrafish was artificially recorded to understand the neural phenotype of adult zebrafish in the 3D environment (Cachat et al., 2011). A stereoscopic vision system was used to track the four to six groups of giant Danio aequipinnatus (Viscido et al., 2004). Another proposed a method of tracking multiple bees with two cameras based on motion information (Veeraraghavan et al., 2006). In addition, two cameras are used to automatically track multiple fruit bat flies, and a 3D tracking algorithm is developed by solving three linear allocation problems (Hai Shan et al., 2011). In addition, 3D reconstruction is carried out through binocular vision technology to monitor and analyze the 3D behavior of fish and monitor water-quality abnormalities (Jialiang, 2015).

This chapter adopts a simple and efficient method to study the daily behavior of fish. This method will not lead to stressful behavior in fish and more suitable for the environment of daily farming. Double cameras are used to shoot video, and the fish are tracked according to the color features and difference methods of background frame, and the real 3D position is restored, so as to analyze and study the behavioral characteristics of fish.

25.2 EXPERIMENTAL MATERIALS AND METHODS

25.2.1 THE EXPERIMENTAL DEVICE

In this study, the grass goldfish, with its obvious appearance and color, were selected as the experimental subjects. The body length of the grass goldfish was about 19~22 cm, and the body height was about 3.8 cm. The glass fish tank (1, 1, 0.5 m, the water depth 0.45 m) was covered with a layer of black cloth in order to reduce the influence of reflections in the water. The side was occluded or exposed to record the activities of the fish. Two SONY color RGB cameras were installed 2 m from the top of the tank and 1.6 m from the side as video collection devices. The schematic diagram of the device and the actual shooting image are shown in Figure 25.1. The images were then processed in Windows 10 using MATLAB® 2018a.

As the RGB image is greatly affected by light, the change of light intensity caused by the day–night cycle will affect the effect of final shooting. In order to reduce the influence of light, this study was carried out under relatively stable natural light and lamplight. In order to maintain the light intensity to meet the shooting requirements, the LED lights in the laboratory were kept on. In addition, the LED lamp is far away from the fish tank, making the light distribution uniform around the tank. At the same time, we covered the windows but could not avoid the light completely. We just reduced the light change, so as to reduce the influence of light on the effect of shooting and the living habits of the fish.

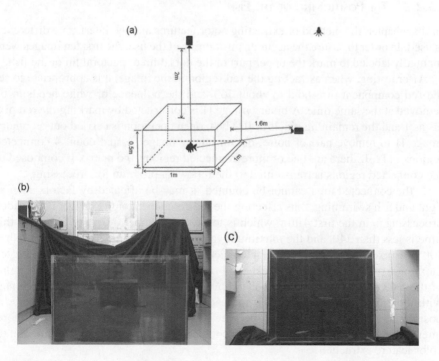

FIGURE 25.1 Image collection system (a) is the schematic diagram of the experimental device, while (b) and (c) are the actual shot renderings of the top and side, respectively.

25.2.2 PREPROCESSING

Choosing a clear and stable dataset is critical. The video format taken by the two cameras is *MTS. In order to ensure the normal processing of MATLAB, we convert the format form *MTS files to ordinary *AVI files using video format conversion software or other methods. After the conversion of video format, in order to facilitate the post-processing of video, it is necessary to align the number of initial frames of the top camera and the side camera. In order to align the initial frames of two video, an artificial light source was flashed once in the laboratory during the shooting of video, which provided a reference for video processing. Before processing these two parts of video, in order to improve efficiency, the frame numbers a1 and b1 of video are determined manually. Then frame numbers a1 and b1 as used to determine the starting frame of the two parts of video and select some regions of a1 and b1 images, respectively. The interframe difference method is used for the number of frames a2 and b2. So a2 and b2 are aligned coordinates. The image frames are started at an interval of 10 s between a2 and b2, and an image is extracted every second. Then the image was cut to extract the area of the fish tank. In order to facilitate the positioning of fish, statistically, the pixel point of the length of the fish body after shearing is 200, and the pixel point on the diagonal of the image after shearing is calculated, with 20% of the pixel point as the subsequent error threshold.

25.2.3 THE POSITIONING OF THE FISH

In this chapter, the method of extracting color features and the interframe difference is used. In order to ensure the accurate positioning of the fish, the first ten images were manually labeled to mark the upper part of the dorsal fin or pectoral fin of the fish.

After testing, when extracting the red region in the image, it is appropriate to set the red component threshold to about 160, and the influence of white needs to be removed at the same time. A binary image I1 is constructed by marking the red pixels as 1 and the remaining pixels as 0. Then, open operation is carried out on binary image I1 to remove part of noise, obtain binary image I2, and count 4 connected regions in I2. If there are one or more connected regions, the matrix L composed of the connected regions is transmitted to the following program for processing.

If the connected area cannot be counted, it may be affected by factors such as light and fish swimming, thus affecting the red component of the image. If the video processing is in the first 4 min, which is to say, the number of image frames at this time is less than 240, and the maximum value of the red component near the fish is 100*100 pixels according to the first ten pictures. We take the average value of the ten maximum red components, and take 90% of the average value as the threshold value of the red component, and repeat the above steps to get the matrix L. If the image with more frames (more than 240 frames) is processed at this time, the long-term positioning of color features will be greatly affected by light changes. Therefore, interframe difference method is used to compensate the influence of illumination on color feature extraction.

The method of background subtraction needs to provide to real-time "background image". The frame difference between the background image and the image without fish detection is carried out to find the region where the fish is located. Assuming that no fish is detected in a frame image, the previous image of the fish is used as the template to look forward to the image where the fish isn't coincident. Replace the area where the fish is in the template to get the "background image". Then, the "background image" and the image without fish detection are differentiated between frames to obtain the region where the fish is, and then the matrix L is obtained.

The statistical matrix is L. If there is only one connected region, k-means clustering algorithm is used to locate the connected region directly, and the position of clustering centroid is obtained. If there are connected regions, the connected region with the largest area is selected and the centroid position of the cluster is obtained by the same method. If the center of mass of the cluster is around the image, that is, the pixel number of the image edge is less than 20% of the image scale, which is mainly affected by the reflection of the glass and the shadow of the fish. Therefore, two connected regions with the largest area are selected and the clustering centers of two connected regions are obtained by k-means clustering algorithm. Then, you remove one center of mass near the edge, and you're left with the last center of mass. This reduces the effect of reflective glass. Since the center of mass may not be the fish, it is necessary to find a point where the minimum Euclidean distance of the fish cluster is at the place where the center of mass connects the fish to ensure that the final position must be on the fish.

In the above positioning, in order to reduce the incorrect anchor point, if the Euclidean distance between a certain anchor point and the previous anchor point of this point is greater than the error threshold, the point is considered as the error point. When the fish swim to the direction of the camera, especially when it is far away, the position of the fish is not obvious in the image due to light, glass, and other factors. At this point, the position of the fish may not be located, so take the previous position as the current position.

25.2.4 REDUCTION OF ACTUAL 3D COORDINATES

By using the above method, the fish in the top and side video were positioned, respectively, and the initial pixel position of the fish was obtained. Due to the camera's perspective distortion and water refraction, there is a certain error between the initial pixel position and the actual position. In order to simplify the program, the errors of the top camera were ignored and only the perspective distortion of the side camera was optimized. The image size of the fish tank captured by the top camera is 1,019*785, while the fish tank is 1 m*1 m. Then a pixel point represents 0.98 or 1.27 mm, respectively, and then the real coordinates are obtained, which is used as the X and Y axes. In images taken by the side camera, the depth of water on the side close to the side camera accounted for 519 pixels, while the side far away from the camera accounted for only 324 pixels. Since the camera is positioned at precisely the same angle as the water, the number of pixels varies evenly from near to far. Therefore, only by combining the position of the top and fitting a line with pixel points representing the actual size, the position of water depth can be restored as the z-axis. The coordinates of X, Y, and Z axes were combined to restore the three-dimensional coordinates of the fish. If the Euclidean distance between the two positions of the fish exceeds the error threshold, the coordinates of the previous position will be taken as the coordinates of this time. For example, exceeding 20% of the vertical diagonal length of water in a fish tank is a false point.

The 3D calibration flow chart of fish is shown in Figure 25.2. The main steps are as follows:

Arithmetic:	Three-dimensional positioning of fish
Input:	Fish are monitored for video
Output:	The three-dimensional position of the fish
Step 1:	Align the start frame of two angle videos and take one frame of images per second. Select the fish tank area, count the approximate number of pixels of fish body length, and select the appropriate error threshold
Step 2:	Manually mark the position of the fish in the first ten frames
Step 3:	Detect the fish area
Step 4:	If the corresponding region is not detected in the image within the first 4 min, the threshold is modified according to the manually marked position and the fish region is detected again
Step 5:	If the corresponding region is not detected in other images, the background image is made by combining the previously positioned image, and the background interframe difference method is used to detect the region where the fish are
Step 6:	Select the suitable region, and k-means clustering algorithm is used to calibrate the center of mass and remove the "shadow" of fish
Step 7:	In the connected area, suitable points were found to be the positioning of fish
Step 8:	Correct the lateral positioning according to the top positioning and restore the pixel coordinates to the real coordinates

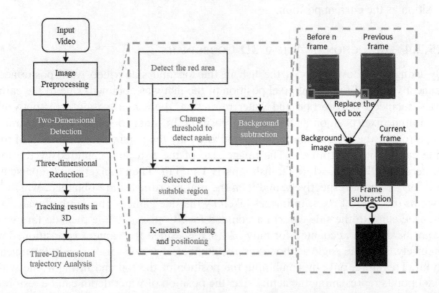

FIGURE 25.2 Flow chart of 3D behavior analysis of fish. (The dashed lines in the gray box flow chart are the measures taken when the dark gray areas cannot be detected.)

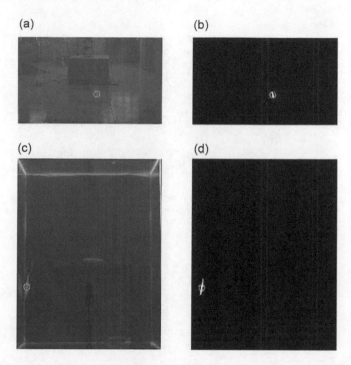

FIGURE 25.3 Top and side positioning effects. (The circle in (a) is the actual location on the side, the area detected on the side in (b), the circle in (c) is the actual location on the top, and the area detected on the top in (d).)

Figures 25.3a and 25.3c show the picture of frame 2,360 on the side and the picture of frame 4,520 above. Figures 25.3a and 25.3c are the positions of fish found by our method. Figure 25.4 shows the three-dimensional movement track of the fish detected by our method. Figure 25.4a shows the 24-h movement track, and Figure 25.4b shows the 10-min movement track.

25.3 THREE-DIMENSIONAL TRAJECTORY ANALYSIS

After the above positioning, if the positioning cannot be completed and the Euclidean distance between the two adjacent registration points exceeds the error threshold, the previous location will be taken as the current location, and these invalid points account for 6.7% of the total. Top and side positioning data and shooting time are shown in Table 25.1 and Figure 25.5. In the lateral positioning, when the fish swims at the shooting angle and is affected by light, the fish captured by the camera is not clear, so there are many invalid points.

Three different experiments were carried out to test the effect of the surrounding environment on fish. For the first time, place the fish tank 12 cm from the white wall. Place a black cloth on the outside of the bottom of the fish tank and a camera on the top for a short time as shown in Figure 25.6a. Because of the influence of illumination and other factors, there are some errors in these locations, but the location of the

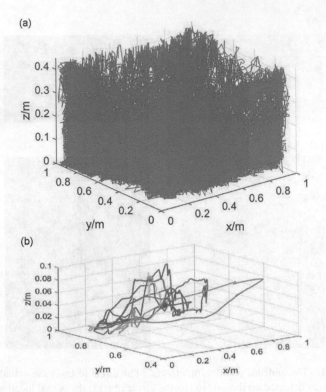

FIGURE 25.4 Diagram of fish tracks. Panel (a) shows the 24-h trajectory of the fish, and panel (b) shows the 10-min trajectory, with the light dots as the starting point and the dark dots as the ending point.

TABLE 25.1
Video Positioning Data Statistics

Number of Experiments	Invalid Points		Total Points		Percentage of Invalid Points	
	Top	Lateral	Top	Lateral	Top	Lateral
1	164	–	8,787	–	1.9	–
2	364	8,407	54,360	54,360	0.7	15.5%
3	140	6,990	48,724	48,724	0.3	14.3%
4	173	11,256	97,458	97,458	0.16	11.5%

wrong locations is still near the fish. In this half hour's track, it can be seen that the fish swims along the side of the wall less. Later, we selected the fish from 15:30 to 17:30 in the afternoon when the light was relatively stable and filmed them for 2 h. And we placed a shield around 25 cm to the left. Take the half hour between 16:10 and 16:40 as an example, the trajectory is shown in Figure 25.6b. We find that fish

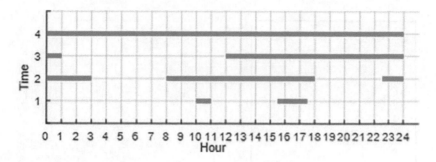

FIGURE 25.5   Video shooting time.

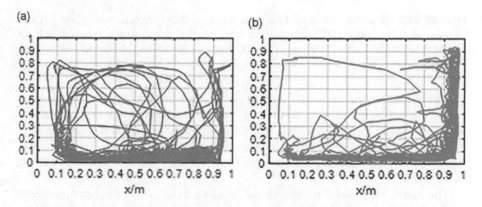

FIGURE 25.6   Schematic diagram of the first experiment track, with the black line as the occlusion. (a) Occlusion one side and (b) occlusion two side.

swim less on one side of the wall and less on the left side. It can be seen that placing the shield near the fish tank will make the fish stress.

For the second experiment, the fish tank was placed in a closed laboratory. A small LED light on the right side of the tank provides less light, and the inside of the tank is shaded by black cloth except on one side away from the window and the bottom. Place a camera on each side of the unshaded top and side, and shoot from 22:40 to the next day 2:55. The activity distribution of fish in the whole period is shown in Figure 25.7a. In the environment with low light intensity, the fish swam on the unshaded side was more, accounting for 49.3%, that is, they used to swim on the side with strong light. Under the same conditions, during the daytime from 8:00 to 18:00, the window of the laboratory was simply shielded, but it was not completely hidden from light. At this time, due to the influence of natural light, the light intensity is stronger than at night, and the swimming of fish is still concentrated on the unshaded side, accounting for 63.6%. This shows that fish have phototaxis.

In the third experiment, similar to the second experiment, we placed a camera on the top and a camera on the side, and only placed black cloth on the bottom of the fish tank. In this experiment, a high-power LED light was used in the above the diagonal of the fish tank near the window, and the light was kept on. The shooting time

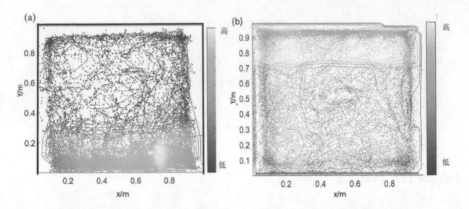

FIGURE 25.7 The distribution of the area of activity of fish in experiments 2 and 3. [Panel (a) shows the distribution of the active area of the fish in the second experiment, and the black line at the coordinate axis is the shelter. Panel (b) shows the active distribution of fish in the third experiment. The dark line represents the direction of the strong light source.]

was from 12:00 to the next day 1:00. The activity distribution of the fish is shown in Figure 25.7b. Compared with the previous two times, the concentration on one side was significantly reduced this time. After statistics, it remained on the side, far away from the strong light source for a long time, accounting for 43.0%. However, only 19.7% of them were near the light source. Thus, fish are more accustomed to moving in mild light.

The fourth experiment monitored the fish 24 h a day, using the same measures as the third. Taking the Euclidean distance between the two positions of the three-dimensional trajectory as the amount of exercise in 1 s, the amount of exercise of the fish was statistically analyzed. As shown in Figure 25.8, the motion point diagram is drawn with 5 min as the unit, and the broken line diagram is drawn with the average amount of exercise per hour. It is found that the amount of exercise of fish is relatively small in the three time periods around 8:00, 14:00, and 21:00, which is similar to the intermittent "rest" behavior of human beings. In addition, compared

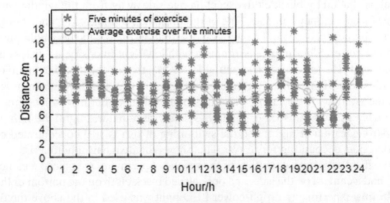

FIGURE 25.8 Twenty-four-hour exercise statistics.

with the exercise in the second experiment to verify the influence of the surrounding environment on the fish, the fish was more active in the light environment. Taking 5 min as the statistical unit, the average amount of exercise of the fish in the second experiment with low light intensity was 6.69 m, while the average amount of exercise of the fish in the third experiment with high light intensity was 9.05 m.

In the previous experiments, we found that even if the experimental environment was changed many times, the fish had the habit of swimming close to the edge, and spent more than 65% of the time within 0.1 m of the surrounding area of the fish tank on average. Through the analysis of the second, third, and fourth experiments mentioned above, it can be seen that fish have a tendency to edge and move around the fish tank, as shown in Table 25.2 and Figure 25.9a. And fish have a tendency to deepen,

TABLE 25.2
Horizontal Distribution Statistics

Distance from the Edge	0–0.1 m	0.1–0.2 m	0.2 m or above	Total
Number of points	127,597	48,881	24,064	200,542
Proportion	63.6%	24.4%	12.0%	100%

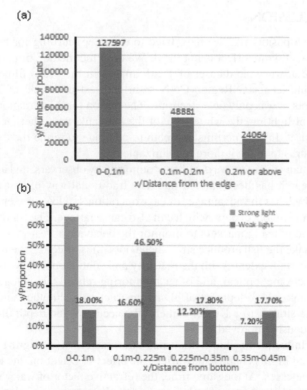

FIGURE 25.9 Level (a) distribution and depth and (b) the bottom of the fish tank is 0 m distribution statistics.

TABLE 25.3
Vertical Distribution Statistics

Distance from Bottom	0–0.1 m		0.1–0.225 m		0.225–0.35 m		0.35–0.45 m		Total	
Intensity of illumination	Strong	Weak	Strong	Weak	Strong	Weak	Strong	Weak	Strong	Weak
Number of points	57,831	11,085	14,999	28,635	11,024	10,962	6,506	10,900	90,360	61,582
Proportion	64.0%	18.0%	16.6%	46.5%	12.2%	17.8%	7.2%	17.7%	100%	

more accustomed to stay in the depth of the tank. Taking the second and fourth experiments mentioned above as examples, fish have a tendency to swim in the lower part of the tank, accounting for 72.7%. The specific statistics are shown in Table 25.3 and Figure 25.9b. However, under strong light, the fish is more used to swimming within the range of 0–0.1 m from the bottom of the tank. Under low light, the fish is more accustomed to swimming within 0.1–0.225 m of the bottom of the tank.

25.4 DISCUSSION

In terms of fish positioning, we have tried to use deep learning for target recognition, but the positioning effect is not ideal. We used the method of manual marking to randomly construct the dataset of 1,231 images in the video filmed in the first experiment, and let faster Region-CNN train and test in a ratio of 7:3. But after several attempts, it was only 66% accurate. The reason for this situation may also be the influence of lighting, the white edge of the fish tank, and other factors when we select the dataset. If time permits, we plan to continue to refine the dataset and try other frameworks to make it more recognizable.

There are many different ways of fish monitoring by observers, and a large amount of related research has been carried out. The traditional way of fish monitoring is often used, which has the advantages of easy operation, but low efficiency, especially in the processing of lots of data, which can also cause fatigue. The traditional method and the method of using markers to monitor the behavior of fish will be invasive to the fish and make the fish produce stress behaviors, thus affecting the analysis of fish behavior to a certain extent. With the maturity and continuous improvement of computers and electronic, optical, and other monitoring-related technologies, combined with the continuous development of 3D reconstruction technology and the application of devices similar to a Kinect camera, the accuracy and depth of fish research can be continuously improved.

Under the influence of illumination and other factors affecting the shooting, the method adopted in this chapter may affect the effect of detection due to poor image quality in some cases. At the same time, the refractive factor of water was not taken into account, and the top camera's perspective distortion was relatively light, so the error of this part was ignored, which finally affected the accurate positioning of the fish to some extent. In addition, due to the influence of various factors, the positioning of the fish can't be accurately positioned to the same position on the fish every time. Especially in the side shooting, the positioning result of fish swimming at the shooting angle is not very good, which has a certain impact on the long-term monitoring and analysis. In the 24-h long analysis of the above-mentioned fish, due to the light change and other factors such as the alternation of day and night, there may be more invalid points compared with other times, which will affect the experimental results to some extent.

Although there are all kinds of shortcomings mentioned above, the method adopted in this chapter can ensure that the fish can be positioned around each time. Although there are relative errors, the errors caused by these factors cannot significantly change the analysis results of fish. Especially when using the same method to study the behavioral differences between multiple fish, these factors have less adverse effects. In the next step, we will optimize the shooting and try to eliminate the influence of water refraction, perspective distortion, and other factors, so as to accurately locate the same part of the fish and restore the 3D trajectory with better effect as far as possible.

25.5   CONCLUSION

This chapter adopts a simple and efficient method to study the living habits of fish. By adopting the above method, the fish in the two-dimensional image can be accurately positioned and the three-dimensional position of the fish can be easily restored. Through this study, we found that fish phototaxis, habitually stay in a temperature and sunlight, and the presence of close proximity to other objects like walls will affect the behavior of fish. In the above fish tank environment, fish is more used to swimming around the tank and stays in the environment less than 0.25 m from the water surface. In addition, in the long-term monitoring, we also found that fish in the 8:00, 14:00, and 21:00 points near the human similar to the indirect "rest" behavior. Using computer systems to automatically obtain the three-dimensional trajectory of fish, rather than labor intensive, can be used as a short or long time to monitor the behavior of fish organisms. In the long run, it provides an easy way to study abnormal or pathological fish behavior. In the next step, we will improve the accurate positioning of fish to study abnormal behaviors of diseased fish and exploratory behaviors of fish in certain environments and try to conduct remote monitoring and timely detect abnormal behavior of fish.

REFERENCES

Aliaksandr P, Petr C, Dalibor Š, Bendik F T, Åsa M O E. Infrared reflection system for indoor 3D tracking of fish. *Aquacultural Engineering*, 2015, 69: 7–17.
Cachat J, Collins C, Kyzar E, et al. 2015. Three-dimensional neurophenotyping of adult zebrafish behavior: Updates, achievements and future directions. figshare. Poster.
Cachat J, Stewart A, Utterback E, Hart P, Gaikwad S, Wong K, Kyzar E, Wu N, Kalueff A V. Three-dimensional neurophenotyping of adult zebrafish behaviour. *PLoS One*, 2011, 6 (3): e17597.

Chan Z, Xunmu Z, Xiaofan Y, Lina W, Shuqin T, Yueju X. Automatic recognition of lactating sow postures from depth images by deep learning detector. *Computers and Electronics in Agriculture*, 2018, 147: 51–63.

Delcourt J, Denoël M, Ylieff M, Poncin P. Video multitracking of fish behaviour: A synthesis and future perspectives. *Fish and Fisheries*, 2013, 14(2): 186–204.

Gang X, Min F, Zhenbo C, Meirong Z, Jiafa M, Luke M. Water quality monitoring using abnormal tail-beat frequency of crucian carp. *Ecotoxicology and Environmental Safety*, 2015, 111: 185–191.

Hai Shan W, Qi Z, Danp Z, Yan Q, Chen Y Q. Automated 3D trajectory measuring of large numbers of moving particles. *Optics Express*, 2011, 19(8): 7646–7663.

Jialiang G. Research and application of three-dimensional behavior monitoring of fish based on computer vision. Yanshan University, 2015.

Leihua L. Abnormal water quality monitoring based on fish movement. Yanshan University, 2017.

Mancera J M, Vargas-Chacoff L, García-López A, et al. High density and food deprivation affect arginine vasotocin, isotocin and melatonin in gilthead sea bream (Sparus auratus). *Comparative Biochemistry and Physiology Part A: Molecular & Integrative Physiology*, 2008, 149(1): 92–97.

Marti-Puig P, Serra-Serra M, Campos-Candela A, et al. Quantitatively scoring behavior from video-recorded, long-lasting fish trajectories. *Environmental Modelling & Software*, 2018, 106: 68–76.

Michael S A, Robert G, Kalueff A V. Developing highER-throughput zebrafish screens for in-vivo CNS drug discovery. *Frontiers in Behavioral Neuroscience*, 2015, 9: 8.

Oppedal F, Dempster T, Stien L H. Environmental drivers of Atlantic salmon behaviour in sea-cages: A review. *Aquaculture*, 2011, 311(1–4): 1–18.

Saberioon M M, Cisar P. Automated multiple fish tracking in three-dimension using a structured light sensor. *Computers and Electronics in Agriculture*, 2016, 121: 215–221.

Shengmao Z, Heng Z, Fenghua T, Zuli W, Zongli Y, Wei F. Research progress of computer vision technology in monitoring fish swimming behavior. *Journal of Dalian Fisheries University*, 2017, 32(4): 493–500.

Shih-Liang T, Wen-Miin T, Chih-Yung S, Tai-Yueh C. Benthic fish behavior characterization with a mechanically scanned imaging sonar. *Aquacultural Engineering*, 2019, 84: 1–11.

Veeraraghavan A, Srinivasan M, Chellappa R, et al. Motion based correspondence for 3D tracking of multiple dim objects. *IEEE International Conference on Acoustics Speech & Signal Processing*. IEEE, 2006.

Viscido S, Parrish J, Grünbaum D. Individual behavior and emergent properties of fish schools: A comparison of observation and theory. *Marine Ecology Progress Series*, 2004, 273: 239–249.

Zhiping X, Xi E C. Zebrafish tracking using convolutional neural networks. *Scientific Reports*, 2017, 7: 42815.

Zion, B. The use of computer vision technologies in aquaculture—A review. *Computers and Electronics in Agriculture*, 2012, 88(C): 125–132.

Index

A

Aflatoxin (AF), 1–2, 29–30, 57–58
  chili pepper, 16–18
  contamination of, 2
  corn/maize, 5–13
  detecting methods of, 37–39
    application-driven key wavelength mining
      method for (*see* Application-driven
      key wavelength mining method)
    flow chart, 5
    fluorescence index, 33–34
    hyperspectral imaging, 3
    illumination compensation and kernel
      segmentation, 32–33
    narrowband spectra, 35–37
    near-infrared spectroscopy, 3–4
    recognition and regression, 34, 35
    sample preparation and image acquisition,
      30–32
  limitation in, 18–19
  pistachio nuts, hazelnuts, brazil nuts, and
    peanuts, 14–16
  spectrum detection, 30
  trends in, 19–20
  wheat, barley, and rice, 13–14
Aflatoxin B1 (AFB1), 2, 7–9, 12, 16–17
AlexNet network, 154, 155, 197, 199–201, 209
  for image-based plant disease detection,
    298, 299
  pest recognition, 285, 286
AlphaGo, 58
Analog simulation method, maize planting
  computer simulation of planting method,
    141–142
  seedling missing spots and missing seedling
    compensation, 142–143
Appearance attribute, of carrots, 167
Appearance characteristic index, of peanut
    seed, 86
Application-driven key wavelength mining
    method, 41–42
  data preprocessing, 44–46
  experiment materials, 42–43
  hyperspectral images by GSM, 49–51
  hyperspectral wave by ASD, 47–48
  key wavelengths selected by weighted voting,
    51–53
  multispectral images by liquid crystal tunable
    filter, 48–49
  recognition methods, 46–47
  sorter design, 53–54
  system integration, 43–44
Artificial neural network (ANN) model, 99,
    242–243
  grade judgment, pears defect extraction, 230
ASD
  hyperspectral wave by, 47–48
  spectrometer, 41, 43
Aspergillus flavus, 1, 2, 6–7, 12, 13, 15
Aspergillus parasiticus, 1, 15
Automatic carrot grading system, 183–184,
    193–194
  control of, 190–191
  defect detection, 188–191
  design of, 184–186
  grading regular carrots by size, 190
  image acquisition, 186–187
  image preprocessing, 187–188
  materials, 184
  performance parameters, 192, 193
  regular carrot grading, 191, 192
  time efficiency, 191, 192

B

Back propagation artificial neural network
    (BP-ANN), 52, 78
  pear grading system, appearance quality, 219
Back propagation (BP) neural network model, 91,
    101, 102
Bananas, by transfer learning, 271–272, 274–275,
    279–280
  application of, 277, 278
  computer *vs.* humans, 277, 278
  convolutional neural network, 274
  experimental setup, 275–277
  image dataset, 272, 273
Barley, 13–14
Bayes classifier, carrot sorting system, 175
Bayes decision function, 168
Blind Signal Separation (BSS), 236
Brazil nuts, 14–16
Bright greenish-yellow fluorescence (BGYF), 2,
    6, 15

C

CaffeNet model, 295–296
Canny operator, edge detection with, 227–231

Carrot appearance quality identification, transfer
        learning, 197–198, 201, 209
    application of, 206–208
    comparison with manual work, 203–206
    convolutional neural network, 199–201
    image characteristics, 199
    performance of model, 201–204
    workflow diagram, 199
Carrot grading system, 167, 169–170
Carrot sorting systems, 168
    using machine vision technique, 167–168, 180
        Bayes classifier, 175
        carrot samples, 168–169
        crack detection, 177–179
        detection, 175–176
        fibrous root detection, 172–174, 176–179
        grading system, 169–170
        image preprocessing and segmentation,
            170–171
        shape detection, 171–172, 178
        surface crack detection, 175
        time efficiency, 178
Chili pepper, 16–18
China
    carrot in, 197
    global carrot yield area of, 197
    maize in, 137, 151
    national standards of, 29–31, 41
    peanut DUS Test Guide, 107
    peanut producing and exporting, 83, 91
Classic pattern-recognition problem, 20
Clustering method
    pedigree clustering of peanut pod, 110
    vegetable seed electrophoresis image
        classification method, 252–254
Computational complexity, carrot sorting, 179
Computer simulation
    automatic carrot grading system based on,
        183–184, 193–194
        control of, 190–191
        defect detection, 191
        defect detection algorithms, 188–190
        design of, 184–186
        grading regular carrots by size, 190
        image acquisition, 186–187
        image preprocessing, 187–188
        materials, 184
        performance parameters, 192, 193
        regular carrot grading, 191, 192
        time efficiency, 191, 192
    in maize planting, 137–139, 148–149
        analog simulation method, 141–143
        field seedling emergence rate, 144–146
        mathematical depiction of problem,
            139–141
        planting methods' yield comparison,
            143–144

    seedling missing spots and distribution
        rule, 146–148
    pear grading system, appearance quality,
        213–214, 223
        algorithm implementation, 217–222
        system development, 214–217
        test performance table, 222
    single-seed precise sowing, of maize using,
        137–139, 148–149
        analog simulation method, 141–143
        field seedling, 145–146
        mathematical depiction of problem,
            139–141
        seedling missing spots and distribution
            rule, 146–148
        two planting methods' yield comparison,
            143–144
Computer vision system, 167–168
Computer vs. humans, transfer learning
    bananas by, 277, 278
    pepper by, 263–264
Conventional detection methods, for fish
        monitoring, 308
Convex polygon method, 167, 171, 172, 178
Convolutional neural network (CNN)
    carrot appearance quality identification,
        199–201
    of deep learning-based aflatoxin detection,
        61–62
    of deep learning method, pixel-level aflatoxin
        detection, 73–74
    development of, 197
    for image-based plant disease detection, 296
    transfer learning
        bananas by, 274
        maize surface and species identification,
            153–155
        pepper by, 259–260
Corn kernels, 6, 7, 9, 11–13
Corn/maize, aflatoxin detection, 5–11
Corp recognition, vegetable seed electrophoresis
        image classification method, 250–252
Correct recognition rate (CRR), 241
Crack detection, carrot sorting system, 177–179
Cross validation method, 63, 64, 77, 123

D

Data preprocessing, application-driven key
        wavelength mining method, 44–46
Deep learning-based aflatoxin detection, 57–59,
        65–66
    CNN of, 61–62
    hyperspectral imaging preprocessing, 60–61
    hyperspectral imaging system and image
        acquisition, 59–60
    peanut sample preparation, 59

using key band images, 62–63
using spectral and images, 63–65
Deep learning (DL) method
for image-based plant disease detection,
295–296
artificial recognition, 301, 302
big datasets validation, 299–300
dataset, 296, 297
data size effect, 302–303
experiment environment, 299, 300
image background, 303
machine recognition and human
recognition, 304
model universal adaptability, 299–302
small datasets validation, 300, 301
symptom variations, 303–304
transfer learning method, 297–298
pixel-level aflatoxin detection
CNN method of, 73–74
models compared for kernels, 77–78
for testing kernels, 75–78
for training kernels, 74–75
Defect detection
automatic carrot grading system, 188–191
pear grading system, appearance quality,
217–219
Difference Radiation Indexes (DRIs), 29, 34, 35
Digital image processing, 118
Distinctness, uniformity, and stability (DUS)
testing
of maize, 99
of peanut, 99
peanut pods, identification and pedigree
analysis, 117–118
analysis and identification model, 121–122
biological basis for seed testing with
appearance, 126–127
candidate features for, 127–128
feature extraction, 119–121
feature selection by Fisher, 122–123
image acquisition, 119
paternity analysis by K-means, 125–126
peanut samples, 118, 119
variety identification by SVM, 123–125
pedigree clustering of peanut pod, 108
DRIs, see Difference Radiation Indexes (DRIs)
Dynamic threshold method, pears defect
extraction, 228

E

Ear rows counting, in maize, 131, 134–135
characteristic indicators of corn varieties, 132
construction of counting model, 132–134
image obtaining, 132
materials, 132
pretreatment, 132, 133

Edge detection operators, pears defect extraction,
227–229
Electric displacement platform, 71
Electrophoretograms, classification recognition
of, 250–251
ENVI4.7, 31

F

Feature extraction, 38
peanut pods
identification and pedigree analysis,
119–121
pedigree clustering of, 109, 110
pear grading system, appearance quality,
218, 219
pears defect extraction, 229–230
Feature optimization methods, 241
Feature recognition, pears defect extraction,
229–230
Fibrous root detection, carrot sorting system,
172–174, 176–179
Field seedling emergence rate
interactions between sensitivity and, 146
on yield influence, 145–146
Fisher algorithm, 117, 121, 241
Fisher-ANN model, 50
Fisher method, 37–39, 62, 117
food detection, using infrared spectroscopy
with, 235, 240, 241, 245
peanut pods, identification and pedigree
analysis, 121–123
Fluorescence index, aflatoxin detection by, 33–34
Fluorescence polarization (FP), 13
Fluorescence spectral imaging, aflatoxin
detection in corn/maize, 9, 11
Food detection, using infrared spectroscopy, 235
with k-ICA and k-SVM, 235–237, 245
algorithm, 238–239
different features selection method, 241–242
different recognition models, 242–243
different samples/features, 243–244
flowchart, 240
materials, 237, 238
Four-feature library, 88
Fourier-transform infrared (FTIR) spectroscopy,
3, 235–237
Fourier transform infrared photoacoustic
spectroscopy (FTIR–PAS), 12
Fourier-transform near-infrared (FT-NIR)
spectrometer, 3, 4, 13, 16
Fracture detection, automatic carrot grading
system, 188–189
Fruit volume detector, 184
FTIR–PAS, see Fourier transform infrared
photoacoustic spectroscopy
(FTIR-PAS)

Fumonisins, 12–13
Fusarium, 19, 20
Fusarium verticillioides, 12–13

G

GoogLeNet model
    for fresh banana, 271, 274, 275, 278, 279
    for fresh pepper, 259–260
    for image-based plant disease detection, 298,
        299–302
    for pest recognition, 284
Google Scholar®, 19, 20
Go program, 58, 70
Grading method, peanut quality grade testing,
    94–95
Grain number, 121, 123
Grating spectrometer module (GSM), 57, 59
    hyperspectral images by, 49–51

H

Hang pepper, 258
Hazelnuts, 14–16
High field seedling emergence rate, 145–146
High-performance liquid chromatography
    (HPLC), 69
Hough transform, 175
Hyperspectral imaging (HSI) system
    aflatoxin detection by, 3
    in corn/maize, 5–11
    deep learning-based aflatoxin detection, 59–60
    by GSM, application-driven key wavelength
        mining method, 49–51
    pixel-level aflatoxin detection, 71–72
Hyperspectral imaging preprocessing, deep
        learning-based aflatoxin detection,
        60–61
Hyperspectral wave, by ASD, 47–48

I

Image acquisition system
    automatic carrot grading system, 186–187
    carrot sorting system, 169, 170
    deep learning-based aflatoxin detection,
        59–60
    peanut cultivars and quality detection, 84–85
    peanut pods, image features and DUS testing,
        119
    pixel-level aflatoxin detection, 71–72
Image analysis method, vegetable seed
        electrophoresis image classification
        method, 255
Image-based plant disease detection, using DL
        method, 295–296
    artificial recognition, 301, 302

big datasets validation, 299–300
dataset, 296, 297
data size effect, 302–303
experiment environment, 299, 300
image background, 303
machine recognition and human recognition,
    304
model universal adaptability, 299–302
small datasets validation, 300, 301
symptom variations, 303–304
transfer learning method, 297–298
Image capture environment, pest recognition, 291
Image-collection system, 92
Image dataset, bananas, 272, 273
ImageNet dataset, 198, 201, 209, 260
ImageNet Large-Scale Visual Recognition
    Challenge (ILSVRC), 285
Image preprocessing
    automatic carrot grading system, 187–188
    pears defect extraction
        background removing and outline
            extraction, 227–228
        defective parts extraction, 229
        spot removal on surface, 228–229
Image processing system
    carrot sorting system, 169
    ear rows counting, in maize, 131, 134–135
        characteristic indicators of corn varieties,
            132
        construction of counting model, 132–134
        image obtaining, 132
        materials, 132
        pretreatment, 132, 133
    pedigree clustering of peanut pod, 107–108,
        114
        clustering algorithm, 110
        experimental materials, 108, 109
        feature extraction, 109, 110
        PCA clustering, 111–112
        PCA method, 110, 113
        statistical characteristics clustering,
            110–111
    quality grade testing of peanut, 91–97
Image processing technology, 118, 151
    fruits and vegetables quality identification, 257
Image recognition, peanut pods based on, 99–104
Image segmentation method, carrot sorting
        system, 170–171
Independent components (ICs), 235, 239, 241
    for SVM and ANN, 244

K

Kernel independent component analysis (k-ICA)
        method, 235
    food detection, using infrared spectroscopy
        with, 235–237, 245

algorithm, 238–239
different features selection method,
     241–242
different recognition models, 242–243
different samples/features, 243–244
flowchart, 240
materials, 237, 238
Kernel support vector machine (k-SVM), 235
food detection, using infrared spectroscopy
     with, 235–237, 245
algorithm, 238–239
different features selection method,
     241–242
different recognition models, 242–243
different samples/features, 243–244
flowchart, 240
materials, 237, 238
K-fold cross validation, 38, 240
k-means clustering algorithm, 117
for fish monitoring, 311
peanut pods, identification and pedigree
     analysis, 122, 125–126
k-nearest neighbor (KNN), 48, 57, 63, 64, 78

L

Laplace operator, edge detection with, 227, 228
Laser-induced fluorescence spectroscopy (FS), 16
Laspeyresiapomonella larva, 291–292
Least square regression (LSR) method, 34
Leaving-one method, 247, 250, 256
Limited data model
carrot appearance quality identification,
     203–205
transfer learning
     accuracy, 162
     performance of, 157, 158, 159
Liquid crystal tunable filter (LCTF), 31, 41
application-driven key wavelength mining
     method, multispectral images by,
     48–49
Logistic regression, 58, 70

M

Machine vision counting, maize, 31–35
Machine vision technique, 183
carrot sorting system using
     Bayes classifier, 175
     carrot samples, 168–169
     crack detection, 179
     crack detection algorithm, 177–178
     detection, 175–176
     fibrous root detection, 172–174, 176–179
     grading system, 169–170
     image preprocessing and segmentation,
          170–171

shape detection, 178
shape detection algorithm, 171–172
surface crack detection, 175
time efficiency, 178
quality grade testing, peanut, 91
seed image process, 84
Maize
distinctness, uniformity and stability testing
     of, 99
ear rows counting, using image processing
     method, 131–135
surface and species identification, transfer
     learning, 151–152
application of, 158, 160, 161
comparison with manual method, 157,
     158, 160
convolutional neural network, 153–155
image characteristics, 153
limited data model accuracy, 162
network model structure, 152
performance of limited data model, 157,
     158, 159
performance of model, 155–157
workflow diagram, 153, 154
using computer simulation, single-seed
     precise sowing of, 137–139, 148–149
analog simulation method, 141–143
field seedling emergence rate, 144–146
mathematical depiction of problem,
     139–141
planting methods' yield comparison,
     143–144
seedling missing spots and distribution
     rule, 146–148
"Man-machine war," 263–264, 277
MATLAB® program, 31, 85–87
MCTS, *see* Monte Carlo tree search (MCTS)
Migration learning, 272, 279
Monte Carlo tree search (MCTS), 58–59, 70
Multiple regression analysis, of yield reduction,
     146
Multispectral imaging system
aflatoxin detection in corn/maize, 11
chili pepper, 16
detect aflatoxin-contaminated hazelnut
     kernels, 15
by liquid crystal tunable filter
     application-driven key wavelength mining
          method, 48–49
Multivariate image analysis (MIA), 13
Mycotoxins, 1–3, 12, 13, 17, 20

N

Narrowband spectra, aflatoxin detection by, 35–37
Near-infrared hyperspectral images
     (NIR HSIs), 16

Near-infrared spectroscopy (NIRS), 1, 2
    aflatoxin detection by, 3–4
        in corn/maize, 12
        in wheat, barley, and rice, 13, 14
Neural network grading method, pears defect
        extraction, 225–226, 233
    ANN grade judgment, 230
    comprehensive grade judgments, 231
    experimental materials, 226–227
    fruit type and defective part extraction, 230
    grade judgment, 231
    image preprocessing, 227–229
    influencing factors of grading, 231–232
    national standard scalarization, 229
    spot removal effect and, 230–231
Nonparametric weighted feature extraction
        (NWFE), 58, 70
Normalized Difference Radiation Index (NDRI),
        29, 33

O

Ochratoxin A contamination, 14
Ornamental fish, behavior trajectory research of,
        307–308, 319
    3D coordinates reduction, 311–313
    experimental device, 309
    positioning of, 310–311, 318
    preprocessing, 310
    three-dimensional trajectory analysis,
        313–318

P

Partial least squares discriminant analysis
        (PLS-DA), 7, 9
Partial least squares (PLS)
    food detection, using infrared spectroscopy
        with, 235, 236, 240, 242, 243
    regression effect of, 34, 35
Pattern recognition test, 247–248
Peanut cultivars and quality detection, 83–84,
        88–89
    appearance characteristic index of seed, 86
    image acquisition and pretreatment, 84–85
    materials for test, 84
    recognition
        on peanut qualities, 88
        on peanut varieties, 88
    recognition model establishment, 86–87
Peanut kernels, quality grade testing of, 91–97
Peanut pods
    based on image recognition, origin
        traceability, 99–104
    image features and DUS testing, 117–118
        analysis and identification model, 121–122

biological basis for seed testing with
        appearance, 126–127
    candidate features for, 127–128
    feature extraction, 119–121
    feature selection by Fisher, 122–123
    image acquisition, 119
    paternity analysis by K-means, 125–126
    peanut samples, 118, 119
    variety identification by SVM, 123–125
    image processing, pedigree clustering of,
        107–108, 114
    clustering algorithm, 110
    experimental materials, 108, 109
    feature extraction, 109, 110
    PCA clustering, 111–112
    PCA method, 110, 113
    statistical characteristics clustering,
        110–111
Peanuts, 14–16
Pear grading system, appearance quality,
        213–214, 223
    algorithm implementation
        defects detection, 217–219
        feature extraction, 218, 219
        grade judgment, 219–222
    system development
        concrete implement, 215
        hardware device, formation of, 214–215
        real object testing, 215, 216
        software system, 216–217
    test performance table, 222
Pears defect extraction, with rich spots and
        neural network grading method,
        225–226, 233
    ANN grade judgment, 230
    comprehensive grade judgments, 231
    experimental materials, 226–227
    fruit type and defective part extraction, 230
    grade judgment, 231
    image preprocessing, 227–229
    influencing factors of grading, 231–232
    national standard scalarization, 229
    spot removal effect and, 230–231
Pepper, by transfer learning, 257–258, 260
    application of, 264, 265
    computer vs. humans, 263–264
    convolutional neural network, 259–260
    image characteristics, 258
    performance of model, 261–263
    three-category recognition accuracy, 263
    workflow diagram, 259
Pest recognition, using transfer learning,
        283–284, 286, 287
    AlexNet, 285, 286
    comparison with human expert, 286,
        288, 289

comparison with traditional methods, 286, 288
image background and segmentation, 291
image numbers and image capture environment, 291
materials, 284, 285
similar outline disturb, 291–293
universal of, 288, 290
Pistachio nuts, 14–16
Pixel-level aflatoxin detection, 69–70, 79–80
CNN of deep learning method, 73–74
deep learning
models comparison for kernels, 77–78
for testing kernels, 75–78
for training kernels, 74–75
hyperspectral imaging preprocessing, 72–73
hyperspectral imaging system and image acquisition, 71–72
peanut sample preparation, 71
Plant diseases, 295
Plant Village project, 296, 297, 299, 300
PLS regression, 11, 236, 242
Plus L reduce R (plusLrR), food detection, 235, 240, 241, 245
Plus-r-remove-q method, 36
Pneumatic control and time-delay techniques, 190–191
Pneumatic grading system, 185, 186, 193
Potato classification model
accuracy and loss of, 206–207
performance of, 207, 208
Preprocessing, ornamental fish, 310
Principal component analysis (PCA), 7–9, 13, 14, 16, 99, 101–103
food detection, using infrared spectroscopy with, 235, 236, 240–242
pedigree clustering of peanut pod, 110, 113
vegetable seed electrophoresis image classification method, 249–251, 254
Principal component analysis–support vector machine (PCA–SVM) method, 58, 70
Princomp function, in MATLAB, 86
Procumbent Speedwell, 283, 284, 288, 290
Protein atlas analysis, 247

Q

Quality grade testing, peanut, 91–92
grading method, 94–9
maintaining integrity of specifications, 93
materials for testing, 92–93
quality recognition model, 94
recognition results of grains' quality analysis, 95–96
specification and grading analysis, 96–97

Quality recognition model, peanut quality grade testing, 94

R

Radial basis function (RBF), 122
Radial basis function–support vector machine (RBF–SVM) model, 29, 34
Radiation Index (RI), 33
Raman spectroscopy, 4, 19
Random Forest, 78
Ratio Radiation Index (RRI), 33
Raytec Vision, 168
RBF–SVM, 78
Recognition methods
application-driven key wavelength mining method, 46–47
peanut cultivars and quality detection
on peanut qualities, 88
on peanut varieties, 88
peanut quality grade testing
accuracy rates of, 96
inaccuracy rates of, 96
results of grains' quality analysis, 95–96
Recognition model, 155, 157, 158
establishment, peanut cultivars and quality detection, 86–87
Recognition time, 79
Reflectance spectra, aflatoxin detection in corn/ maize, 11
Region of interest (ROI), 32
Regular carrot grading, 191, 192
ReLu, 154, 200, 260, 274, 285
Reproducing kernel Hilbert space (RKHS), 239
Rice, 13–14

S

Savitzky–Golay algorithm, 5, 46
Seed identification methods, 151
Seed image process, machine vision based method, 84
Seedling emergence methods, 141–142
Seedling missing spots, 142–143, 146–148
Seed origin traceability, peanut, 99, 103–104
method used to optimize, 101–102
recognition rate of characteristics, 102, 103
test materials, 100, 101
Sequential backward method, 36
Sequential forward method, 36
Shape detection, carrot sorting system, 171–172, 178
SIFT-HMAX, 286
Silicon photomultiplier (SiPM), 53
Single-chip microcomputer (SCM), 183
Single-feature recognition rate, 103

Single-seed precise sowing, of maize, 137–139,
           148–149
    analog simulation method, 141–143
    field seedling emergence rate, 145–146
    mathematical depiction of problem, 139–141
    seedling missing spots and distribution rule,
           146–148
    two planting methods' yield comparison,
           143–144
Sisymbrium Sophia, 283, 284, 288, 290
Skeleton extraction algorithm, carrot sorting, 179
Sobel operator, edge detection with, 227, 228
Sorter process, application-driven key
           wavelength mining method, 53–54
SPA-ANN model, 50
Spectral imaging, 19, 20
Spectral–spatial (SS) combinative deep learning
           method, 57, 59
Spectrum detection, 30, 42
Spodopteraexigua larva, 291–292
Stacked autoencoder (SAE), 70
Support vector machine (SVM), 34, 117,
           236, 242
    peanut pods, identification and pedigree
           analysis, 121–125
    vegetable seed electrophoresis image
           classification method, 250, 251, 254
Surface crack detection
    automatic carrot grading system, 188
    carrot sorting system, 175
SVM regression (SVR), 34

T

Tea plants, pests that affect, 284, 285
Three-dimensional trajectory analysis,
           ornamental fish, 313–318
Threshold-based classification (TBC), 15
Time efficiency
    automatic carrot grading system, 191, 192
    in carrot sorting systems, 178
TOMRA's carrot sorting machine, 168
Traditional machine learning methods, 291
    for image-based plant disease detection, 303
Traditional seeding mode, 138
Transfer learning, 198
    bananas by, 271–272, 274–275, 279–280
        application of, 277, 278
        computer *vs.* humans, 277, 278
        convolutional neural network, 274
        experimental setup, 275–277
        image dataset, 272, 273
    carrot appearance quality identification,
           197–198, 201, 209
        application of, 206–208
        comparison with manual work, 203–206
        convolutional neural network, 199–201

image characteristics, 199
        performance of model, 201–204
        workflow diagram, 199
    for image-based plant disease detection,
           296–298
    maize surface and species identification,
           151–152
        application of, 158, 160, 161
        comparison with manual method, 157,
           158, 160
        convolutional neural network, 153–155
        image characteristics, 153
        limited data model accuracy, 162
        network model structure, 152
        performance of limited data model, 157,
           158, 159
        performance of model, 155–157
        workflow diagram, 153, 154
    pepper by, 257–258, 260
        application of, 264, 265
        computer *vs.* humans, 263–264
        convolutional neural network, 259–260
        image characteristics, 258
        performance of model, 261–263
        three-category recognition accuracy, 263
        workflow diagram, 259
    pest recognition using, 283–284, 286, 287
        AlexNet, 285, 286
        comparison with human expert, 286,
           288, 289
        comparison with traditional methods,
           286, 288
        image background and segmentation, 291
        image numbers and image capture
           environment, 291
        materials, 284, 285
        similar outline disturb, 291–293
        universal of, 288, 290
Transmittance spectra, aflatoxin detection in
           corn/maize, 11

U

Ultrathin-layer isoelectric focusing, 247
United Nations Food and Agriculture
           Organization, 2

V

Vegetable seed electrophoresis image
           classification method, 247–248,
           255–256
    cluster analysis, 252–254
    experimental materials, 248–250
    method of image analysis, 255
    normal electrophoretogram, 255
    optimization method, 248–250

recognition of corp, 250–252
VGG16 network, 299
VGG network, for image-based plant disease detection, 298
Video tracking, for fish monitoring, 308
Visible/near-infrared (Vis/NIR) spectroscopy, 7–9

W

Wheat, 13–14

Y

Yield compensation effect, 142–143